Big Questions in Ecology and Evolution

Big Questions in Ecology and Evolution

THOMAS N. SHERRATT
DAVID M. WILKINSON

OXFORD
UNIVERSITY PRESS

Great Clarendon Street, Oxford, OX2 6DP,
United Kingdom

Oxford University Press is a department of the University of Oxford.
It furthers the University's objective of excellence in research, scholarship,
and education by publishing worldwide. Oxford is a registered trade mark of
Oxford University Press in the UK and in certain other countries

First Edition published in 2009
Reprinted 2009, 2010, 2014

Published in the United States of America by Oxford University Press
198 Madison Avenue, New York, NY 10016, United States of America

British Library Cataloguing in Publication Data
Data available

Library of Congress Cataloging in Publication Data
Data available

ISBN 978–0–19–954861–3

To Kitty Sherratt (1929–2006)
and
To Lionel and Effie Wilkinson

Contents

Preface

A common man marvels at uncommon things; a wise man marvels at the commonplace.

—Confucius.

The motivation for writing this book originally came from our mutual, yet independent, admiration of Paul Colinvaux's 1978 book 'Why big fierce animals are rare' (Princeton University Press), which dares to tackle major questions such as 'why the sea is blue?' and 'why there are so many species?' Colinvaux helped us both realize, more than any teacher did at school, that ecology was a science, and one that asked important questions. He also posed new questions about everyday observations that simply had not occurred to us, such as 'why do birds sing in the morning?' and 'why are tundra plants so close to the ground?' As our opening quotation indicates, it is all too easy for us to accept widespread phenomena like gravity, ageing, and sex without wondering why they occur. However, once the questions are posed, and answers offered, then the reader instantly becomes the detective, weighing up in his or her own mind whether an explanation makes sense and whether there are more plausible alternatives. In this way, one gets closer to the way science is done.

It is now nearly 30 years since Colinvaux's book was published and while it remains fresh and engaging, the scientific community's understanding of the subjects he covered has clearly moved on. To take one small example related to Colinvaux's 'blue sea' chapter, the green photosynthetic picophytoplankton *Prochlorococcus* dominates primary production in the tropical and subtropical oceans and is probably a good candidate for the title of the commonest organism on Earth, yet it was only discovered in 1988, 10 years after the publication of Colinvaux's book.

While the idea for this book was inspired by Colinvaux's approach, this is explicitly *not* an update of Colinvaux's work. There are several reasons for this. Of course, one of the attractions of Colinvaux's book was his style, so the only person who can update Colinvaux is Colinvaux himself. 'Why big fierce animals are rare' was also a product of its time, introducing the science of ecology to a wider audience, just as some universities were starting to offer degrees in the subject. Now that ecology has become firmly established as an academic discipline, we have ended up writing a slightly more technical book than Colinvaux's original. Most importantly, Colinvaux's book was primarily, though not exclusively, ecological, but it struck us that a similar approach could be taken to shed light on many of the main questions in evolutionary biology—why do we age?, why sex?, why cooperate? This has several advantages beyond simply increasing the book's scope—as will be evident throughout our text, ecology and evolution are closely related disciplines. It may sound clichéd, but many ecological questions cannot be fully understood without a consideration of evolution, and almost all evolutionary questions have significant ecological components.

In essence, this book is intended as an introduction to several key ideas in ecology and evolution. However, it introduces readers to these subjects not in the traditional way, but through posing a range of fundamental questions, and discussing the plausibility of solutions that have been offered. These questions lead our entire approach—fundamental ecological and evolutionary concepts are introduced only as and when they are needed to explain the question at hand. Asking big questions and examining solutions have their challenges from an educational perspective—concepts need to be built up. Nevertheless, it is our hope that by introducing the science in this way, our readers will immediately feel involved. Who does not want to know why they age, why so many species engage in sex, why the tropics have so many species, or when humans first started to affect world climate? We hope that our approach also helps to put ecology and evolution firmly on the map: through our book readers can see the immense breadth of the field, its fundamental importance, and learn about some of the exciting breakthroughs that have been made in recent years.

Our book is not intended as a formal textbook, but something designed for background reading, perhaps to support tutorials, which aims to transmit the excitement of the field by discussing major, yet not fully answered, questions. To this end, we decided at the outset to limit our technical language so that readers are more likely to understand the plain meaning of what we are trying to say. Only common species names are used in the text (except for those species without such names), yet backed up with a list of scientific species names, along with definitions of key terms in a glossary. One of the most elegant languages for summarizing the way one sees the world is that of mathematics, but we have invoked very little here because qualitative arguments will serve our purposes. However, we will often describe the results of mathematical reasoning. Similarly, although we discuss aspects of the chemistry of nutrient cycling, we have avoided using chemical equations in our text.

Other considerations have helped shape our philosophy and style. To avoid a 'textbook' look, we have limited the use of graphs but instead have included photographs with the aim of linking theoretical ideas to aspects of natural history that can be seen in the field. Since we are inherently interested in the way ideas are arrived at, we find it natural to describe some of the key players as well as the insights they delivered. However, our accounts are primarily aimed at searching for answers to questions and we could not possibly contemplate the idea of paying homage to every contribution along the way. Thus, although we try to summarize major experiments and insights, we could not consider a comprehensive description of the history of attempts to answer our questions.

Our taxonomic bias has bordered on positive discrimination. Of course, both of us appreciate the attraction of vertebrates but we are also concerned at the continued under-representation of other groups of organisms, such as microorganisms and fungi, which are the prime players in a wide variety of ecological processes. Therefore, whenever several examples could be chosen, we have opted to highlight the most appropriate rather than the most charismatic. We also wrestled for some time with how to approach references. After taking the advice of our students, we decided to cite all of our major sources, but only as indexed notes so that they do not obscure the text. We hope that these references will make the book more useful to advanced students and our professional colleagues. Of course the text can be read without these backup references but we would hope that readers will be sufficiently curious, or disbelieving, to check up on some of the articles we cite.

Finally, a word should be said about our title, and how we came to it. Rather like the movie 'Snakes on a Plane', this was a working title that simply stuck. No doubt some readers may wonder why their pet question has not been addressed. Ecology and evolution are full of 'big' questions: 'why are males often more brightly coloured than females?', 'are complex ecosystems more stable?', 'why are some species common and other species rare?', 'why is nitrogen fixation restricted to so few organisms?', and our selection was primarily motivated by our combined backgrounds and experience. It is not intended as a compendium of all the top questions, or even the 10 most important questions, just 10 fundamental questions that have attracted a lot of interest and can teach us new ways of looking at the world.

Finally, readers should not think for a moment that we are offering the definitive answer to all of the questions we have identified. By their very nature answers to these questions are highly controversial. We have done our best to rule out earlier answers that are now obviously wrong, and to highlight the directions that researchers are currently taking. Nevertheless, it is our fervent hope that the book is read critically, and with alternative explanations in mind.

Acknowledgements

We firmly believe that a book of this nature would lose its style and consistency if it was the product of multiple authors with different specializations. However, one consequence of this decision is that we have had to review a great deal of work outside our own fields of specialization. This has meant many pleasant, but demanding, months buried in the literature, and we wish to thank our Universities—Carleton and Liverpool John Moores-for granting each of us our first-ever sabbatical to devote time to the development of the book. We began writing this book over 2 years ago, but without the support of our home institutions, it would have taken a whole lot longer. Our collaborative research has also benefited from several grants from The Royal Society of London, to fund visits by Dave Wilkinson to Ottawa.

To help us meet the challenge of reviewing a range of subject areas, we have benefited from a wonderful set of colleagues and collaborators who have generously given their time to comment on chapters in various stages of preparation. In particular, we would like to thank Jose Andres, Filipo Aureli, Rod Bain, Chris Beatty, Alison Buchanan, Naomi Cappuccino, Paul Cunningham, David Currie, Silvia Gonzalez, Root Gorelick, Tim Lenton, Paul Martin, Euan Nisbet, Hannah O'Regan, Sally Otto, Paul Rainey, Gilbert Roberts, Nick Royle, Howard Rundle, Graeme Ruxton, Crispin Tickell, Hans Van Gossum, Tyler Volk, Mike Whitfield, and Bill Willmore for their extremely helpful comments and advice. Rod Bain, Alison Buchanan, Root Gorelick, Hannah O'Regan, and Graeme Ruxton deserve special thanks for reviewing multiple chapters. Tim Boland, Rob Laird, Janice Ting, and Richard Webster helped with proofreading. Many other colleagues, too numerous to mention individually, helped in answering questions and providing clarifications. As ever, all inaccuracies and misconceptions remain our own. Redouan Bshary, Adolfo Cordero, Mark Forbes, Silvia Gonzalez, Conrad Hoskin, Richard Law, Edward Mitchell, Stewart Plasitow, Paul Rainey, Sheila Russell, Oliver Sherratt, Rob Smith, Hans Van Gossum, and Richard Webster generously provided photographs or other material.

Ian Sherman from OUP encouraged and advised us throughout the development of the book, while Helen Eaton provided rapid responses when searching for photographs and other materials. Carol Bestley steered the book through to final production, and put up with last-minute changes. We are grateful to them all for their guidance. Finally, we would like to thank one another for support when at times the job seemed insurmountable.

Tom Sherratt
Ottawa, Canada

Dave Wilkinson
Liverpool, UK

1

Why Do We Age?

Figure 1.1 Charles Darwin at the ages of 31, 40, 45 and 71. Image sources: (top left) an 1840 watercolour by G. Richmond; (top right) an 1849 lithograph by T.H. Maguire; (bottom left and right) photographs dated *circa* 1854 and 1880. Copyright Science Photo Library.

Can you tell me why the tortoise lives more long than generations of men; why the elephant goes on and on till he have seen dynasties; and why the parrot never die only of bite of cat or dog or other complaint?

—*Professor Van Helsing in Bram Stoker's Dracula*[1]

In 2004, the amazing 'Flying Phil' Rabinowitz broke the world 100 m sprint record for a centenarian, setting a time of 30.86 s and beating the previous world record time by over 5 s. Despite this impressive statistic, most 20 and 30 year-olds can readily run at these speeds when dashing for a bus, and the overall world record for 100 m currently stands at 9.69 s (set by Usain Bolt at the age of 21). Age-related degeneration in bodily function is familiar to all of us, and is known as 'senescence', or more colloquially, as 'ageing' (Fig. 1.2). Of course, this loss of physiological functioning not only impairs our ability to run: as individuals get older they typically experience an increase in the likelihood that they will die, and also a decrease in fecundity. The incidence rates of cancers and heart attack, for example, are considerably higher in older than in younger

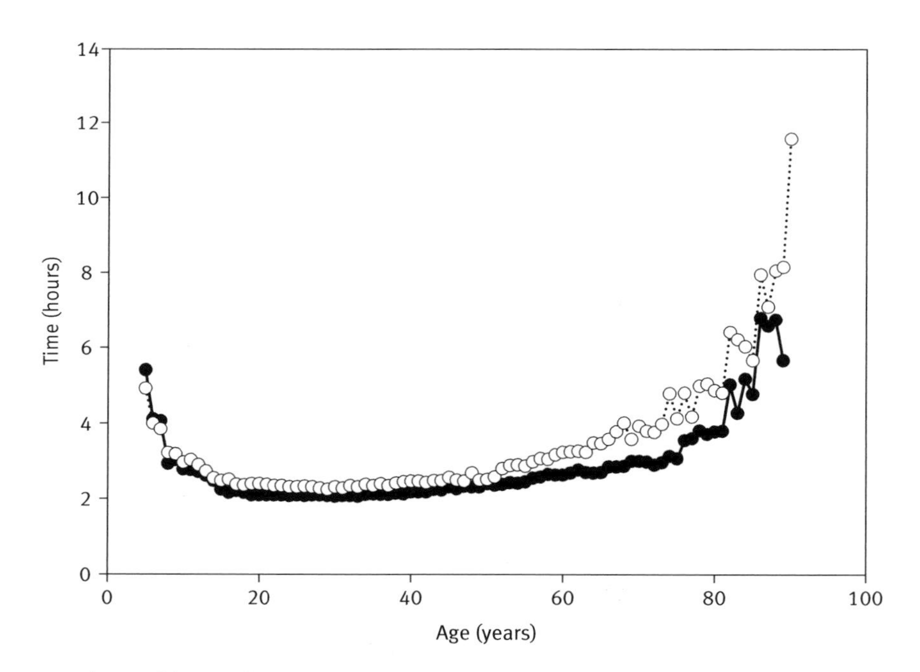

Figure 1.2 The world record times (as of April 2008) for running the marathon, classed according to age and gender (females, open circles; males, closed circles). Despite the fact that these data are not entirely independent (the same athlete can contribute to several data points as they age), and the fact that the available sample size diminishes with age, they show the anticipated trends, with octogenarians taking considerably longer to complete the course than individuals in their 20s. Note that as a stamina event the relationship has a relatively broad minimum range. Data from the *Association of Road Racing Statisticians*, http://www.arrs.net/SA_Mara.htm.

individuals (Fig. 1.3). For these reasons, ageing has been dubbed 'the most potent of all carcinogens',[2] but it has also long been considered as one of the world's worst diseases[3] ('*senectus enim insanabilis morbus est*'[4]—a sickness for which there is no cure).

Live long and prosper?

Organisms die for all sorts of reasons. They may get run over by a bus, they may be eaten by a predator, or they may succumb to a lethal disease. However, even if individuals survive all of these 'extrinsic' challenges, then the odds are that they will begin to experience the signs of senescence. While being eaten by a predator is unfortunate, it is also eminently understandable as a cause of death. Natural selection will tend to act on individuals to reduce the likelihood of this extrinsic mortality (for instance, by promoting higher vigilance or the development of some form of defence) but death from accidents, predators, and parasites cannot be completely avoided. Ageing, however, poses much more of a dilemma for evolutionary biologists. In particular, one might expect that those individuals who managed to slow down the ageing process

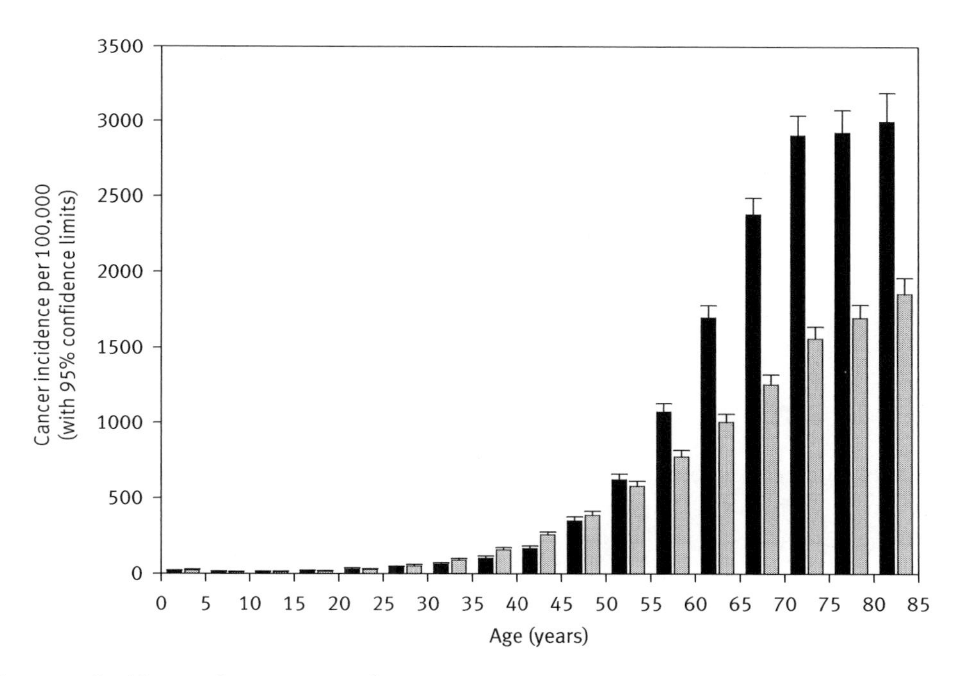

Figure 1.3 Incidence of cancerous malignant tumours per 100,000 subjects in human patients in the Bronx Borough of New York State, sorted by 5-year age classes. Males, black bars; females, grey bars. Note the sharp rise in cancer incidence as individuals age (coupled with what may be an eventual levelling off). Data from *New York State Department of Health*, http://www.health.state.ny.us/statistics/cancer/registry/table6/tb6totalbronx.htm.

would leave more offspring, so that natural selection would favour extreme longevity. As evolutionary biologist, George Williams[5] wrote some 50 years ago, 'It is indeed remarkable that after a seemingly miraculous feat of morphogenesis a complex metazoan should be unable to perform the much simpler task of merely maintaining what is already formed'. In other words, if we can produce vigorous offspring, why do we not continually 'invigorate ourselves' from within?

Organisms are not fridges

Ask a friend or colleague why organisms age, and he or she will probably say something like 'inevitable wear and tear'. Indeed, there are plenty of candidate environmental agents to choose from, ranging from the physical to the chemical. For example, the wings of many damselflies and dragonflies become more tattered as they age, and the mating rates of male damselflies in the field have been found to decline after a few days.[6] Similarly, reactive oxygen species (ROS), otherwise known as free radicals, are generated as by-products of a cell's metabolism and they are widely considered bad news for bodily function—they can damage proteins, lipids, and DNA.[7,8] Paralleling the tatter of damselfly wings, the cells of older organisms, from houseflies to humans, carry an increased concentration of oxidatively damaged compounds (including nucleic acids, lipids, and proteins), particularly in the last third of the organism's maximal lifespan.[9]

Could accumulated damage from life's physical and chemical insults account for the general phenomenon of ageing? At first the idea seems compelling, and indeed it may form part of the correct answer. After all, household appliances such as fridges and dishwashers gradually accumulate annoying faults which lead to their ultimate demise. Yet there are limits to the analogy—wounds can heal, and ROS can be compartmentalized or mopped up with antioxidants.[10] Therefore, in contrast to household appliances, organisms are capable of a degree of maintenance and self-repair—they can potentially *do something* about the damage they accumulate.

Van Helsing's conundrum

The 'wear-and-tear' explanation is also particularly unsatisfactory when it comes to explaining the wide variation in longevity among species. Note at the outset that longevities do not say anything directly about senescence *per se*, but a long life might generally be taken to be indicative of delayed senescence. There are some species in the natural world that live for extremely long periods, which makes one wonder why all species cannot be like that. For example, the oldest bristlecone pines in southeastern California have an estimated age of over 4,700 years (Fig. 1.4) but so far they have shown little age-related reduction in pollen viability or seed germinability.[11] Similarly, in a piece of research that would impress fans of Michael Crichton's *Jurassic Park*, dormant bacterial spores with an estimated age of 25–40 million years have been taken from the guts of extinct bees buried in Dominican amber, and successfully revived and cultured.[12] The dubious record for the longest-lived animal now goes to a quahog clam dredged up in

2007 from the Arctic waters off Iceland (although some might argue that colonial corals are better candidates). At an estimated age of 400–405 years, the specimen—nicknamed 'Ming' (after the Chinese Dynasty)—was a youngster when Isaac Newton was born, yet tragically by the time researchers had recognized its great antiquity, it had already passed away.[13] What do these species have, or lack, that allows them to avoid deterioration over time, and why are not all organisms imbued with these life-maintaining properties?

Jurojin, the Japanese Shinto god of longevity, is frequently portrayed as an old bearded man, carrying a scroll on which is listed the lifespan of all living things. Longevities are highly variable both among and within species, so it certainly helps to have a list. If senescence were only about accumulating damage, then why, as Van Helsing wondered, do some species live an order of magnitude longer than others? Harriet, a Galápagos Giant Tortoise collected from the islands a few years before Charles Darwin visited, died in 2006 at the tender age of 175.[14] Hares, in contrast, do less well in this particular race: the record longevities for hare species are in the region of 5–7 years.[15,16] Similarly, why would the accumulation of damage in a Japanese quail (with a maximum lifespan of 5–8 years) be so different from that of many parrot species (with a maximum lifespan of well over 50 years)?[17]

Figure 1.4 An ancient Great Basin Bristlecone pine (*Pinus longaeva*) in the white mountains of eastern California. Individual trees can live for several thousand years, with the oldest approximately 4,900 years. Photo: TNS.

On observing variation in the longevities of individuals of different species, one might wonder whether all these patterns are simply a direct consequence of the different levels of 'extrinsic mortality'. Thus, unlike hares, tortoises have shells to protect themselves from predators, while ground-nesting quail may be more vulnerable to predators than tree-dwelling parrots. However, it is increasingly clear that the general patterns of species differences in longevity remain the same whether observed under natural conditions or reared in captivity, which suggests that observed lifespan is not always a direct consequence of wear and tear although (as we will see) it may be *shaped by* it.

Remarkably, even different forms of the same species can exhibit significant variation in longevity. Queen honeybees (produced by feeding female workers 'royal jelly' during their larval development) have an average lifespan of about 1 year, while female workers within the same hive typically live for a matter of few weeks.[8] Likewise, late summer migrant adult monarch butterflies in North America not only live over three times as long in the field as the non-migratory forms that emerge in the summer, but also live significantly longer when both types of butterflies are maintained in the laboratory.[18]

A fascinating botanical example of genetically-mediated variation in age of senescence comes from the work of Richard Law and colleagues,[19] who took seeds from populations of the meadow grass *Poa annua* growing in two different conditions: low-density populations (including the disturbed derelict site in Liverpool, UK, shown in Fig 1.5) and high-density populations (including the pasture in Clywd hills in North Wales, shown in Fig. 1.5). Growing these seeds in the same environment, Law and colleagues found that the plants derived from parents in the two habitat types had very different growth forms. Moreover, plants reared from the disturbed low-density populations—which necessarily require a more opportunistic lifestyle—flowered earlier and had much shorter lives than those extracted from the high-density populations (see Fig. 1.5).

Another clear example of a direct genetic influence on lifespan comes from the damselfly *Mnais pruinosa costalis* (Fig. 1.6). This attractive Japanese insect has two forms of co-existing male—an orange-winged territorial fighter which attracts females by securing egg-laying sites and displaying, and a clear-winged non-territorial sneaker which gains access to the female through stealth. The two forms of male have approximately equal reproductive success over their lifetimes, but the clear-winged sneakers live longer.[20] One might be tempted to think that the lower lifespan of the territorial forms arises because of the high energetic costs of fighting and the increased likelihood of it being damaged during territorial contests. However, experiments show that clear-winged males also live longer than orange-winged males when kept individually in the laboratory, in circumstances where there was no opportunity for fighting.[20] Why cannot the orange-winged males simply obtain that set of genetic mutations that allows them to live as long as the clear-winged colleagues? To answer this question, we must take a few detours, by first ruling out some initially attractive but ultimately unsatisfactory theories.

Figure 1.5 The experiments of Law and colleagues[19] involved sowing the seeds of the grass *Poa annua* collected from a range of low-density and high-density sites and rearing them in a common environment. The offspring of plants growing in the low-density sites showed rather different growth forms and, months later, died significantly earlier than the offspring of plants growing in high-density sites. Experiments such as these indicate that rates of senescence are heritable, and potentially subject to selection. Photos courtesy of Richard Law.

Figure 1.6 Two genetically-determined male morphs of the damselfly *Mnais costalis* in Japan. The orange-winged territorial fighter males (above) live a shorter time both in the field and laboratory than the clear-winged non-territorial sneaker males (below), although they have approximately equal lifetime reproductive success. Photos courtesy of Stewart Plaistow.

Life in the fast lane

If one is looking for explanations for the wide variation in species longevities, one might propose (as early researchers did[21,22]) that longevity is simply related to internal metabolism—a 'rate of living'—so that those species with a fast lifestyle simply burn their life candles more quickly. This idea has a long history—physicians such as Galen of Pergamum from the second century AD proposed that individuals age in the same way a lamp runs out of fuel.[23] In more modern terms, one might argue that higher rates of oxidative metabolism could also result in an increase in the rate of production of damaging ROS. The idea is simple and attractive. For example, long-lived tortoises are not exactly the most hyperactive of animals, while dormant bacteria can shut down their metabolism almost entirely. Intriguingly, among mammals the total number of heartbeats appears (very) approximately constant despite high variation in longevity. Elephants (longevity 40–50 years), for instance, beat their hearts at a considerably lower rate (25 heartbeats per minute) than shrews (longevity 1–2 years, and an astonishing 200 heartbeats per minute).[24] Generalizing these observations suggests the question: Do we all only have a fixed number of heartbeats in us?

For all its attractiveness, it turns out that the 'rate of living' theory does not explain a great deal of the variance in longevity. For example, most birds outlive mammals of comparable size[25] despite the fact that both groups maintain stable high body temperatures, while bats have a far longer maximum lifespan than rats (up to 30 years compared to 5 years),[26] despite the fact that they are similar-sized mammals. A comparison of the longevities of 'cold-blooded' (ectothermic) vertebrates, which have low metabolic rates, and 'warm-blooded' (endothermic) vertebrates, which have high metabolic rates, fails similarly to support the 'rate of living' hypothesis.[27] It is also worth noting that in humans at least, those with a sedentary lifestyle do not live longer than those who regularly raise their heart rate through exercise.

Ageing: fact or artefact?

The pure 'wear-and-tear' explanation is essentially a non-evolutionary argument because it puts ageing down to plain old physical and chemical damage, and implicitly assumes that organisms cannot do much about it beyond a bit of tinkering here and there. We have seen already that the wear-and-tear explanation is not entirely adequate, because, for example, even different forms of the same species show parallel variation in longevity under controlled conditions where the environmental damage levels are similar. In addition to plain old 'wear-and-tear', the second type of argument still commonly voiced today (in various contexts) is that senescence is an artificial phenomenon of little relevance to the natural world—humans and pets get old because of the protection we afford them, but natural organisms do not suffer from senescence because they never make it to that age.[28,29]

'Old octopuses become what we call senescent, or senile... and sometimes their actions are very inappropriate', so remarked Jim Cosgrove from the Royal British Columbia Museum, when asked why a mature male octopus recently attacked a small research submarine.[30,31] It turns out that recognizably senile individuals are rarely documented in natural populations, but that is not to say that senescence is not occurring. Indeed, there are now a number of studies that have reported either increases in age-specific mortality or decreases in fecundity with chronological age in species of plants,[32] birds,[33] mammals,[34] and even bacteria.[35,36] There is even evidence that tyrannosaurid dinosaurs showed a rapid decline in survivorship as they got older.[37] Many of these studies have not followed individuals throughout their lives, but a few have. For example, the antler fly is one of life's supreme specialists, breeding exclusively on discarded antlers of deer in North America. From monitoring individually marked adult males on a collection of antlers in the field, it was evident that both their survival rate and their rate of mating declined over consecutive days, and this was despite the fact that adults tend to live on average for less than a week.[38] In an earlier study, individually marked adult female damselflies were also found to exhibit a significant reduction in their rate of egg laying as they age.[39]

Individually marked large herbivorous mammals such as bighorn sheep, ibex, and red deer all show relatively clear evidence of increases in age-dependent mortality in the wild.[40] Spectacular examples of rapid increases in mortality following reproduction, such as that seen in annual plants, many insects, several salmon species,[41,42] and even a marsupial mammal,[43] reinforce the observation that senescence (albeit of a highly acute form) does occur in natural populations. Therefore, while senility is arguably more prevalent in highly cushioned human societies and their pets, examples of age-dependent degeneration occur both in and out of captivity.

Multicellular organisms show signs of ageing, but do single-celled organisms likewise senesce? If an organism reproduces by dividing equally into identical offspring then the distinction between parent and offspring disappears, and such cells would not senesce[44] (by definition, since 'young' would be equivalent to 'old'). In reality, however, it is hard to find any good test cases in which no distinction whatsoever can be made between offspring and parent. Indeed, cellular senescence after multiple bouts of reproduction has now been demonstrated in several unicellular species. These examples include the bacteria *Caulobacter crescentus*[35] and *Escherichia coli*[45] and the yeast *Saccharomyces cerevisiae*.[46] In each of these cases there was some form of decline in the rate of fission (division) over time, and in each of these cases there was some source of asymmetry in that older and more damaged cell components were more likely to accumulate in the originator cell—parents effectively become garbage dumps.[36,47] Therefore, similar to multicellular organisms, single-celled organisms tend to show senescence, indicating that the condition has an extremely long evolutionary history.[47]

Ageing by numbers?

There are several other more sophisticated non-evolutionary (or at least not *directly* evolutionary) explanations of ageing very similar to wear-and-tear, but in these cases the damage is associated with an intrinsic breakdown of the genetic machinery, rather than an accumulation of chemical or physical damage. Paralleling the decline in fission rates in single-celled organisms, it is now recognized that cells within a multicellular body can only go through a limited number of cell divisions before ceasing active division. In the case of human cells, for example, the maximum limit is in the order of 50–60 divisions. This restriction is known (in honour of its discoverer Leonard Hayflick) as the Hayflick limit.[48,49] What causes these Hayflick limits? Accumulation of deleterious mutations within cells may play some roles[50] as well as changes in the quantity and distribution of chemicals ('epigenetic factors') which bind to DNA to influence gene expression, but a widely discussed candidate is the gradual shortening of 'telomeres'. Telomeres are, put simply, disposable buffers located at the ends of chromosomes. They comprise repeating DNA sequences and act as caps, protecting strands of DNA from recombining after replication[51,52] (think of them as similar to the end of a zip fastener). With each cell division, a small amount of DNA is necessarily lost in replication at each chromosome end, resulting in ever-shorter telomeres and altered telomere structure. A key consequence of this shortening may be ineffective buffers and eventual replicative senescence. Intriguingly, cancer cells are different and potentially immortal in that many types can bypass replicative senescence by expressing greater quantities of special enzymes involved in the restoration of telomeres, called telomerases.[53]

Collect together cells with Hayflick limits and you potentially have a cellular explanation for the ageing of the whole organism. Indeed, it has been calculated that the Hayflick limit would allow a developing human foetus to grow and develop through repeated cell divisions just about long enough to complete development before senescence sets in.[49] There is even some evidence that cells from short-lived species reach their Hayflick limits earlier than those of long-lived species.[49] Similarly, several very long-lived animals, such as the American lobster and the rainbow trout, show high levels of telomerase in their cells.[28] Telomeres shorten more slowly in longer-lived birds, and in Leach's storm petrels (a long-lived seabird) they may even lengthen.[54] But are we describing a cause or an effect? Hayflick limits are highly variable both among cell types of the same species and among species. It therefore seems likely that the maximum number of cell divisions is tailored to fit the lifespan of the organism, and not the other way around.[48,49] Of course, we still have to explain why cells cannot be given carte blanche to replicate indefinitely until the organism dies, but as we see from cancers, unlimited growth is not always a good thing.[55] Furthermore, the limits on cell division cannot provide the whole explanation for senescence since many invertebrates, such as adult insects, show little cell division in their bodies[56] yet (as we have seen in the case of antler flies) they still senesce. Finally, telomerase-deficient mice do not tend to show higher rates of ageing[57]; therefore, even if telomeres are primarily responsible for Hayflick limits, then telomeres cannot provide the complete explanation for ageing.

Just as the number of cell divisions may be tailored to fit the lifespan of an individual, another clue to the fact that senescence is shaped by natural selection comes when we consider what parts of a multicellular body tend to deteriorate and when. In vertebrates, circulatory system, nervous system, skin, and muscles all tend to give out more or less simultaneously. Of course this might arise because a single factor links them all (rather like multiple parts of a car malfunctioning when the battery goes[58]), but evidence indicates that the synchrony is much more likely to have arisen because different parts age independently and at similar rates. The relative lack of success of transplantation of old organs into young individuals supports this latter contention (similarly a gearbox from an old car will not be 'born again' when placed in a young car). As Richard Dawkins suggests,[59] from an evolutionary perspective, there is little value in having a long-lived expensive Rolls Royce engine in a short-lived cheap chassis, and the deterioration patterns of animals' bodies largely support this interpretation.

The hows and whys of ageing

Before describing the various explicit evolutionary theories of senescence, we wish to reiterate that we are asking *why* ageing occurs at all, rather than *how* it occurs although we admit at the outset that it is not always easy, or informative, to tease apart these different types of explanation. The 'how' explanations are the staple diet of medically-inclined gerontologists, and they include specific mechanisms such as oxidative stress and changes in protein structure.[10] Indeed, it has been estimated that the number of mechanistic explanations for ageing is somewhere in the hundreds.[60] Collectively, these important insights help to characterize what happens to individuals when they get old. Evolutionary theories do not deny that these processes occur, and indeed they may be central to understanding ageing. However, a satisfactory evolutionary theory should be able to explain *why* more is not done to counteract these processes, and *why* the timing of onset of these processes differs so widely among species. In stark contrast to the number of mechanistic explanations of ageing, there are only a handful of interrelated evolutionary explanations and one of these appears to be a clear front runner, at least for now.

Evolutionary theory 1: ageing and the group

The first evolutionary explanation for ageing was proposed by one of the founders of modern evolutionary biology, Alfred Russel Wallace, and subsequently refined by the German biologist August Weismann. Wallace noted that (circa 1865–1870)[61]: 'for it is evident that when one or more individuals have provided a sufficient number of successors they themselves, as consumers of nourishment in a constantly increasing degree, are an injury to those successors. Natural selection therefore weeds them out, and in many cases favours such races as die almost immediately after they have left successors'. Weismann[62] put forward a similar view in the early 1880s. Building on the idea of accumulated wear-and-tear, these authors proposed that senescence was selected as a way of weeding out the worn-out members of the species, thereby enhancing the survival chances of that species. US President Thomas Jefferson echoed a similar sentiment in a letter to John Adams (Monticello, 1 August 1816): 'There is a ripeness of time for death, regarding others as well as ourselves, when it is reasonable we should drop off, and make room for another growth'. More recently, surgeon and medical historian Sherwin Nuland put the view succinctly when he remarked 'Nature's job is to send us packing so that subsequent generations can flourish'.[63] In Josh Mitteldorf's terms, this is 'aging selected for its own sake'.[64] The phenomenon has even been dubbed *the Samurai law of biology*, following the maxim that 'It is better to die than be wrong', because, according to the logic, programmed death following injury prevents the appearance of 'asocial monsters capable of ruining kin, community and entire population'.[65]

The above explanation is essentially 'group selectionist' in nature (to many evolutionary biologists the term still carries the hallmark of a slur, but we do not intend it that way), in that it proposes that certain traits (ageing in this case) can spread by favouring the group to which the individual belongs, rather than the individual. 'Good of the species' arguments are widely considered too much of a stretch, requiring far too many restrictive assumptions. Indeed, a recent review observed that: 'If you want to have fun at an ageing conference, stand up in the bar and shout, "Ageing is programmed." Then duck as glasses and curses start to fly'.[66] The primary reason for the vehement objections is that it is easy to envisage counterselection on individuals that 'cheat' by living longer than their group mates, thereby leaving more offspring. It is also worth noting that no genes are known to have evolved *specifically* to cause damage and ageing[25] (although adaptive suicide may be a possibility, see later); therefore, rather than being 'programmed', longevity appears only under indirect genetic control. For all the above reasons, arguments based on 'the good of the species' are not particularly convincing.

Evolutionary theory 2: ageing and the family

More plausibly, if the group comprises close relatives that share many of the same genes, then certain traits that enhance the fitness of relatives can spread even if they lower the fitness of the carrier (a phenomenon that comes under the umbrella of 'kin selection', see Chapter 3). There is now a surge of interest to understand this 'adaptive senescence' idea from a kin perspective.[64,67] For example, it has been noted that living longer can generate a larger and more persistent reservoir of disease, from which infection can spread. If individuals are distributed in family-based groups, then this may, in theory, lead to selection for reduced longevity.[68]

Contemporary biologists occasionally offer specific examples of longevity being shaped by this type of kin selection,[69] although in some of these cases death comes altogether too suddenly to qualify as senescence, and the mortality is not necessarily age-dependent. One such example has been reported in a common greenfly, known as the pea aphid. Pea aphids are attacked by a range of enemies, including a parasitic wasp. The next generation of wasps emerge from their hosts relatively soon after parasitism (about 2 weeks), and they may potentially infect the young of the host's colony mates (which, thanks to parthenogenesis—see Chapter 2—are genetically similar). Since parasitism dooms any aphid to death before reproduction, then parasitized aphids are effectively 'dead hosts walking'. Under these conditions, one might reasonably expect that aphids should commit suicide (taking the parasite with them), thereby protecting their kin from further parasitism.[70] McAllistair and Roitberg claimed evidence for this 'adaptive suicide hypothesis'[70] when they found that parasitized pea aphids from some locations were more likely to drop off the plant than unparasitized hosts when approached by a ladybird predator (although one might wonder why they did not simply offer themselves up to the predator, satiating it in the process). The common gut bacterium *E. coli* provides a fascinating (and somewhat more convincing) microbial version of the same general phenomenon—when individual cells are attacked by bacteriophages, they then stop producing a short-lived antidote to a long-lived toxin that they simultaneously produce. This not only causes their own death and that of the phage, but also prevents their clone mates from being infected.[71,72]

Note that the above examples refer to a form of conditional suicide, not ageing *per se*. There are other examples in which a relatively early death of an individual may favour relatives, but in these particular cases predators rather than parasites are thought to play a mediating role. For example, certain species of camouflaged moths are known to have shorter post-reproductive lives compared to related warningly-coloured distasteful moths.[73] In an early application of kin selection logic, it was proposed that cryptic moths die soon after finishing reproduction to reduce the chance of close relatives being subsequently spotted by predators that have cued into their disguise.[73,74] Likewise, it has been proposed that distasteful prey species tend to live longer to provide predators with more opportunity to work out that this type of prey (and its similar-looking relatives) are distasteful.[73,74] It is possible that these arguments are going a little too far. Of course, such an outcome could simply arise as a by-product of the defence itself without the need to invoke kin selection. Indeed, there is now mounting evidence that chemically-protected (venomous or distasteful) species tend to have a longer maximum lifespan than non-protected species,[75] and it is highly unlikely that kin selection explains all of this variation.

Despite these reservations, there may frequently be reproductive benefits in staying around long enough to provide parental care to offspring, and even grandparental care to grandchildren, even if you have finished reproduction entirely. After all, in evolutionary terms, the rate of survival of offspring is just as important as the number of offspring that are produced. To take a concrete example, in a recent detailed study of births, marriages, and deaths in populations of Canadian and Finnish women in the eighteenth and nineteenth centuries, it was found that women with a prolonged post-reproductive lifespan had more grandchildren, primarily because being around to help with grandchild care allowed the grandparents' own offspring to reproduce earlier and more frequently.[76] Somewhat surprisingly, evolutionary theories of how 'intergenerational transfers' might affect senescence are only now being developed[77,78] yet they may help explain why certain species (including humans) have such long post-reproductive lives.

In summary, there are no experimental data that senescence is actively programmed—no such genes have been found. The occasional fascinating examples of adaptive suicide do not tend to involve age-dependent senescence *per se*. Even if organisms were ever found to deteriorate 'on purpose', it is unlikely that such a trait could ever have evolved for the good of the species (although dying soon after reproduction to protect kin remains a possibility). The precise *timing of senescence* may well be malleable however, and kin selection may play some role in influencing when ageing kicks in, which can sometimes be well after an individual's own reproduction is complete. How can senescence be subject to selection yet not be directly genetically programmed? It gets easier to understand when we shift our perspective: genes are not the root cause of ageing, but they can help us defend against ageing so long as they are selected to do so.

Evolutionary theory 3: ageing and the individual

When considering the early ideas of Wallace, one also thinks of Charles Darwin, who had a remarkable track record of being right. Darwin's notebooks contain the unanswered question 'Why is life short?'[79] (thereby incidentally echoing the sentiments of his grandfather Erasmus in *The Temple of Nature: 'How short the span of life'*), indicating that he at least wondered about the subject of longevity. Indeed, in later editions of *The Origin of Species*, Darwin included a chapter on 'Miscellaneous Objections' where he reacted caustically to the suggestion that living long should be so advantageous that longevity should always increase over evolutionary time.[80] He first observed that seeds or ova may sometimes be the only way that an organism can survive a harsh winter (thereby postulating one reason for the life cycle of annual plants), but then made a more sweeping statement noting that 'longevity is generally related to... the amount of expenditure in reproduction and in general activity. And these conditions have, it is probable, been largely determined through natural selection'.

It was to take half a century before more explicit evolutionary solutions to help explain longevity were provided, and several of these solutions embraced the idea of a trade-off as Darwin had implied. There are currently three well-known, yet highly interrelated, individual-based theories, two of them explicitly genetic, and one of them more concerned with overall process of allocation of resources to reproduction and maintenance. We now consider these theories, as well as one or two other theories that are currently gaining interest.

Ageing as a consequence of benign neglect

The first individual-based evolutionary theory is known as the 'mutation accumulation theory', and is generally credited to Nobel laureate for medicine Peter Medawar in the early 1950s.[81,82] Medawar himself stood on another intellectual giant's shoulders, those of J.B.S. Haldane, in arguing that the strength of natural selection in removing individuals with deleterious mutations which act late in life after reproduction would be relatively weak.[83] One such example is Huntington's disease, a rare debilitating neurological disorder that is inherited genetically but only produces severe symptoms as people enter their 40s and 50s, long after many have reproduced. It stands to reason that the strength of natural selection in removing these late-acting genes will be correspondingly weaker. There may appear to be an element of circularity creeping in here, but there is not. Even without ageing, individuals are at continual risk of death (and infertility) from a variety of agents including accidents, predation, and disease. What this means is that even in the absence of ageing, the probability of an individual living to a given point of time declines as the time interval increases. Clearly, there is no guarantee that the occasional parent that happens to survive extrinsic challenges for a long period of time will also produce offspring that are lucky enough to survive for a similar length of time. With less raw material around for natural selection to work on, the relative intensity of natural selection maintaining survival and fertility in a given age class will get progressively weaker as individuals age beyond their point of first reproduction. Without strong counter-selection to do something about it, populations are not as effectively purged of mutations with late-acting deleterious effects, so they can begin to express their effects in occasional old individuals, causing death in old age.

Medawar's theory of mutation accumulation does not refer to the accumulation of mutations within an individual during its lifetime, only that mutations that can cause harmful effects in late life are not as effectively purged from a population over the course of many generations. In effect, the mutation accumulation theory suggests that senescence is a form of 'benign neglect'—natural selection just does not care very much about oldies. Mutations with late-acting deleterious effects can be thought of as undefused time bombs, present in an organism's genetic code but not exerting a harmful influence until late in life. If there are such genes, then natural selection largely ignores them.

In fact, the whole mutation accumulation process may have the potential to be self-reinforcing because once senescence sets in through an accumulation of mutations with late-acting effects, then natural selection might care even less about oldies. Here the fundamental evolutionary cause of ageing is 'extrinsic mortality'—it is the same basic reason why many individuals in natural populations die, but in the case of senescence it does not act directly. The theory has been placed on a formal mathematical footing by researchers such as Bill Hamilton[84] and Brian Charlesworth[85], and it feels as though the theory captures some important elements of truth. One has to wonder why the deleterious effect of certain genes might be expressed only late in life, but this may be due at least in part to the build up of damage and metabolic products to a 'tipping point', beyond which they are seriously deleterious. More serious challenges come when one considers examples of 'acute senescence' exhibited by a diverse range of organisms such as salmon[41] and annual plants, which die almost immediately after they reproduce. In these cases, it is hard to envisage late-acting deleterious mutations so suddenly catching up with the organism. As we will see, some direct trade-offs between reproduction and longevity may also be involved.

Ageing as the price one has to pay

The second major theory was anticipated by Medawar, but expounded most forcefully by the eminent evolutionary biologist George Williams.[5] In 1957, Williams proposed what would now be considered a 'life-history' solution to the problem, namely that ageing is a consequence of the actions of genes that favour early survival and reproduction over late survival and reproduction. This is a 'live now, pay later' phenomenon, in which early reproductive success is actively purchased at the cost of future reproductive success.

Pleiotropy (Greek: 'many changes') is a widely recognized genetic phenomenon and occurs when a single gene has more than one effect on its carrier. However, Williams went one step further and proposed a temporal form of pleiotropy in which the same gene can have one effect when expressed in a young organism, but another effect later in life. A hypothetical example of 'antagonistic pleiotropy' proposed by Williams was of a mutation that promotes calcium deposition—such a mutation might be beneficial for young vertebrates because it accelerates bone growth in early life, but the same process might eventually be bad news for older organisms because it leads to hardening of the arteries. Borrowing from Haldane and Medawar's argument, it is easy to see that a reproductive advantage expressed early in life would spread even if it came at a cost of a similar-sized disadvantage late in life, because of the premium placed on youth. This is not simply a consequence of extrinsic mortality, but also has something to do with the rush to reproduce ('turnover'). For example, any mutant form that died after producing two offspring in 1 year (leading to 2^n of its type after n years) would spread more rapidly than forms which left three offspring after 2 years (leading to only $3^{n/2}$ of its type after n years). Of course, not all genes work in this pleiotropic way—some may be positively beneficial at whatever age they are expressed, and natural selection may be able to downplay the effects of others by evolving mechanisms to effectively switch gene expression off at ages when they become deleterious, but all it takes is a *proportion* of genes with unavoidable side effects and those that live long enough will be paying for the consequences of a well-spent youth.

Unlike the mutation accumulation theory, the antagonistic pleiotropy theory can be thought of as a theory in which an optimal selective balance is achieved[44]—senescence is a by-product of adaptation—rather than simply a case of ignoring old individuals entirely. With the mutation accumulation theory, extrinsic mortality alone reduces selection to prolong the reproductive life of individuals, while in the antagonistic pleiotropy theory there is also an intrinsic counterbalance, with deleterious effects the price one has to pay for early success. Athletes using performance-enhancing drugs may capture just this sort of trade-off, frequently paying the cost of impaired health later in life, for improved performance now.

Ageing as a consequence of a balancing act

There is one more influential theory currently circulating, the 'disposable soma theory' (DST) of Tom Kirkwood.[86] This idea was put forward in the late 1970s, and it is effectively a reformulation of the above life-history theory in terms of strategic investments, concentrating on the role of repair. Before describing this theory, it is important to draw a distinction, as August Weismann[62] recognized in the nineteenth century, between germ cells (cells containing genetic material that may be passed to offspring) and somatic cells (Greek—'body' cells, not directly involved in reproduction). Almost by definition, changes to the soma are not heritable—chop off your arm, and your offspring will still have arms. Instead, somatic cells (in vertebrates at least) may be seen simply as a vehicle—a 'disposable' means to an end—to get the genetic material contained in some fortunate germ cells into the next generation. In contrast, germ cells achieve a form of immortality simply by being passed on, rather like a baton in a relay race, from generation to generation.

Unlike the 'antagonistic pleiotropy' theory which emphasizes genetic effects, the DST focuses simply on patterns of resource allocation towards propagation of the germ line and general maintenance of the somatic cells. The theory works on the assumption that somatic maintenance (such as DNA repair and the use of antioxidants to mop up ROS) is a metabolically costly activity involving both physical infrastructure and running costs, and that resources invested in general maintenance and repair are not available for development and reproduction. The theory proposes that it is advantageous for organisms to allocate most of their resources to development and reproduction (i.e., propagation of the germ line), and only sufficient investment in somatic maintenance to keep the organism in reasonable condition for the expected duration of its life. Of course, we can recast DST in terms of antagonistic pleiotropy and end up in much the same place: under Kirkwood's theory, a mutation that increases the allocation of resources to reproduction has the antagonistic pleiotropic effect of decreasing allocation in somatic maintenance and repair.

The great value of the DST is that it brings focus back on the fundamental processes of damage and repair, recognizing that much damage can be extrinsic as well as intrinsic, and asking why bodies have not evolved to repair all of the damage they experience. Indeed, one might wonder why an organism does not work out a way to simply 'balance its portfolio' so that it invests sufficiently in maintenance to keep itself going, occasionally reproducing, indefinitely into the future. Thanks to extrinsic mortality and turnover (all else being equal, it is better to produce offspring earlier than later), strategic investments that bring about early rewards are preferred over late rewards. In economic terms, there is future discounting (the future benefits are tempered by the probability of living long enough to realize them), so life is not about balancing a portfolio—it is about maximizing profit before the trader's market abruptly closes.

Other evolutionary theories

While we have described the 'big three' explanations there are several additional theories, although some are not yet explicitly evolutionary. One of the most promising is based on 'reliability theory', an approach borrowed from engineering to understand how systems with irreplaceable redundant components can exhibit increased failure rates as time goes on.[87,88]

For example, if certain key genes are liable to get damaged during the natural course of an organism's life, then organisms may simply evolve multiple copies of the same gene to serve as backups. Eventually, however, there will be a limit to selection on the number of redundant genes because the vast majority of individuals are likely to have died for other reasons before they are ever needed. In the same way, houses may have electricity and a backup generator in case the electricity fails, but few home owners would consider having a backup for the backup because such contingencies so rarely arise. Ultimately, those few that happen to live long enough (or are protected in captivity) will eventually experience damage for which there has been little or no selection to do anything about. In contrast to the mutation accumulation theory, the deleterious effects do not arise directly from some deleterious late-acting genes, but from genes which get damaged and thereby fail to work (hence ageing arises as a consequence of damage to beneficial genes, rather than functioning genes that are actively deleterious). The theory is an attractive one, and may help explain the late-life plateau in mortality rates seen in many species[87] (see also Fig. 1.3 for some evidence of a levelling effect in human cancer rates).

Are ageing rates evolutionarily maleable?

Before evaluating the relative merits of the above evolutionary theories, we can first ask a more fundamental question—whether evolutionary theories in general explain our observations of ageing better than non-evolutionary theories alone. First, we note that the 'wear-and-tear' non-evolutionary theory suggests that ageing is predominantly extrinsically driven, and so is not something that can be shaped by changing the nature of selection, whereas the evolutionary theories each assume it is much more open to modification by natural selection.

Many studies, such as those comparing the longevities of human monozygotic (identical) and dizygotic (non-identical) twins, suggest that roughly one-quarter of the variability in lifespan (although not necessarily senescence) is explained by genes.[89] So, if you want to live long, choose your parents well.[90] Moreover, there is now ample evidence, particularly drawn from intensive laboratory studies on yeasts, fruit flies, and nematode worms, that longevity is under a degree of genetic control. For example, a mutant form of a gene that encodes components of an insulin (or insulin-like growth factor) signalling pathway in the nematode worm *Caenorhabditis elegans* can extend their lifespan by three times or more compared to the wild type.[25]

Therefore, longevity is mediated by intrinsic genetic factors, not just extrinsic ones. It is also clear from laboratory experiments that longevity is manifestly something that one can alter through altering selection pressures. For example, in experiments where fruit flies were selected for late first reproduction (simply by discarding all offspring produced by young individuals), their average lifespan was dramatically increased.[91,92] A similar effect has been observed when selecting directly from families of fruit flies with longer mean longevities.[93] The high flexibility in longevity means that we can rule out 'wear-and-tear' from extrinsic sources as providing the *entire* explanation.

Ageing and extrinsic mortality

There is another important way to discriminate between the evolutionary and non-evolutionary explanations, but the predictions from evolutionary theory are not as clear-cut as they might at first seem. All of the individual-based evolutionary theories argue that the intensity of selection to keep an organism in good health is mediated by the probability of being killed by extrinsic factors, because there is no fitness advantage in keeping an organism going for any longer than it would naturally live. Hence, if these evolutionary explanations were broadly correct, then one potential prediction might be that ageing would be slower (and longevity longer), the lower the level of extrinsic hazard.[5] In contrast, the 'wear-and-tear' non-evolutionary explanation makes no such prediction, unless one wants to argue that mortality factors such as predators actually increase the level of physiological damage.

As one might expect, it is challenging to separate out the effects of extrinsic hazards from senescence on longevity in wild organisms because extrinsic hazards themselves reduce longevity, but by choosing the appropriate statistical model one can hope to tease apart their contributions. One can also bring the same species into captivity and see how they fare. A number of comparative studies have now been conducted which lend support for a negative association between longevity and the degree of extrinsic hazard, so that, for example, there is evidence that bird and mammal species with low baseline mortality rates also tend to exhibit reduced rates of senescence. More direct evidence comes from selection experiments with rapidly reproducing species such as fruit flies. In some ingenious experiments lasting over 4 years, Stearns and colleagues managed (after a few adjustments) to select for lower rates of intrinsic adult mortality in laboratory populations of fruit fly by decreasing extrinsically imposed adult mortality rates.[94] However, it is important to note that these basic predictions have not always been confirmed, and there may be some good reasons for this.[95,96] For example, if extrinsic mortality is increased but this increase only affects younger organisms, then *longer* lifespan should evolve because younger animals make on average a smaller contribution to reproductive success.

There are other complications involved in relating extrinsic mortality to ageing, and puzzling experimental results which cry out for an explanation. Reznick and colleagues recently compared the life histories of Trinidadian guppies (small tropical fish) derived from populations that had co-evolved with predators (high-predation environments), with guppies derived from upper reaches of streams where fewer predators occur (low-predation environments).[97] As might be expected, the high-predation guppies matured earlier than the low-predation guppies under laboratory conditions free from predation. However, contrary to expectation, the guppies from high-predation environments had *lower* mortality rates throughout their lives, and hence *longer* average lifespan. What is going on? We can explain the situation if we make slightly different assumptions about the way extrinsic mortality acts. For example, if predators preferentially attack the most senescent individuals (so that extrinsic mortality weeds out the old), then there may be selection for overall slower rates of senescence, at least early in life, despite higher predator mortality.[96] In short, one of the classic lines of evidence traditionally held up to support the evolutionary view of senescence—that an increase in extrinsic mortality leads to the evolution of increased intrinsic mortality—is only at best a coarse prediction. The relationship can be much more subtle, as recent empirical and theoretical work has begun to highlight.

More patterns

Once we adopt an evolutionary life-history perspective to senescence, then certain results from among-species comparisons in longevity become easier to understand. For example, it is widely appreciated that longevity tends to increase with increasing body size in both birds and mammals.[24] Thus, elephants tend to live longer than dogs, who in turn live longer than rats.[48] Although the reasons for this association are not entirely clear (many factors could contribute to influencing body size, and teasing apart cause and effect from comparative studies is notoriously difficult), a plausible explanation is that both longevity and large size are simultaneously selected as a consequence of low extrinsic mortality. Large size is predicted to evolve under conditions of low extrinsic mortality, because individuals that grow and delay reproduction are more likely to gather a return on their investment, especially if large size means reduced predation. Flight may also be important in reducing predation, which may help explain the relatively slow ageing in birds and bats compared to similar-sized non-flying mammals.

Which evolutionary explanation?

The current best answer to 'why do we age' is therefore a simple one. Extrinsic mortality—including accidents, bad weather, famine, predators, and parasites—eventually kills the majority of individuals directly. Those that happen to survive to old age experience senescence because selection largely overlooks this age bracket and/or because selection has favoured individuals that seek early benefits even at the expense of late costs.

One might now begin to ask whether, of the big three at least, the mutation accumulation or the trade-off theories (including antagonistic pleiotropy and the more specific DST) better explain the facts. First and foremost, it is important to stress that the three theories are not mutually exclusive (all three could be correct) and they share a number of assumptions and predictions, which makes distinguishing them particularly challenging.

If you want the bottom line, the jury is still out. Current opinion generally favours the trade-off argument, in part because trade-offs between survivorship and reproduction, and early vs late reproduction, have been widely reported in both laboratory and natural populations. For example, captive salmon that are castrated live much longer than those that are not.[48] Similarly, giving male fruit flies more access to females reduces their longevity. The possibility of a trade-off is not a new idea: even Aristotle thought that each act of copulation had a life-shortening effect.[98] Moreover, many of these reproductive trade-offs have often been found to be under a form of genetic control, providing candidate examples of pleiotropic genes. For instance, in an intensively studied natural population of swans, the age of first reproduction and age of last reproduction were positively correlated, so that early reproducers would tend to cease reproduction earlier than late reproducers.[99]

Molecular genetics is a time-consuming process and specific examples of late-acting genes with deleterious effects remain relatively thin on the ground.[100,101] Nevertheless, advances are being made and as technology improves it should become much quicker to amass relevant data. Promislow recently argued that if the antagonistic pleiotropy theory was correct then those genes with different (i.e., pleiotropic) effects should be more likely to be involved in senescence.[102] By analysing detailed information available on the yeast genome and the proteins made by yeast genes, he provided support for the hypothesis by showing that the estimated average degree of pleiotropy exhibited by proteins associated with senescence was greater than proteins with no known association. However, we need many more concrete examples before we can begin to generalize.

Ageing is not all about genes

In accepting an evolutionary argument for ageing, there may be a temptation to believe that ageing is *all* in our genes. Thus, if it was not for those darn genes with late-acting deleterious effects that are not purged from the population, or genes which give us a right royal hangover after the excesses of the reproductive party, then perhaps we would live indefinitely. We have come full circle from wear-and-tear, but we wish to stress that without damage, the gene-based view would also be far too one-sided. Certain damselflies have lower rates of reproduction later in life because their wings tatter, while certain bacteria may decline in their rate of reproduction because of cellular damage—we may ask why damselfly wings are not made of sturdier stuff, or why bacteria cannot sort out their problems, but at the heart of ageing comes the damage.

Consider, for example, the observation that various strains of fruit flies selected for extended lifespan also exhibit an increased resistance to oxidative stress, through the enhanced activity of antioxidant enzymes.[53] Do we age due to ROS, or a lack of further selection to do anything about it? In a way, both assertions may be right—gerontologists and evolutionary biologists have simply been tackling the problem from different perspectives. The former mechanism provides one reason why things go wrong in cells, and the latter helps explain some of the variability in species responses.

Repairing the repair mechanism

Some damage may ultimately be irreparable, whatever resources are thrown at it, and sometimes the repair mechanisms themselves can break down. Indeed, even if there was a mechanism to repair the repair mechanism, then this too may break down. It may be that, thanks to Medawar's selective shadow, there is very limited selection for 'higher-order' or backup repair mechanisms, so that some of the signs of ageing may occur because the repair mechanisms themselves have broken down, not because there is less investment in them. More work is needed to understand what happens when genetic repair mechanisms for fixing damaged soma themselves go wrong, and to elucidate their potential role in the ageing process. In these instances more than any other one can see how 'starting from scratch' is more appropriate from a natural selection perspective than keeping the original afloat. When the somatic boat is sinking, it may be a better option to put one's energy into releasing the germ-line lifeboats (which have been far better protected and capable of selective screening) than attempting to plug the hole.

Extending life

What does all of this mean for the prospect of life extension? Williams[5] was in no mood to pull his punches when he explored the implications of his 'antagonistic pleiotropy' theory, assuming that such pleiotropic genes would be common: 'This conclusion banishes the "fountain of youth" to the limbo of scientific impossibilities where other human aspirations, like the perpetual motion machine...have already been placed by other theoretical considerations'. The basic argument is that there would be just too many things to fix to counter the effects of ageing.

Many others in the field are considerably more optimistic, although there is always the possibility that such views are tainted by the need to keep research grants coming in. Researchers' recent elucidation of the entire genomes of classical laboratory animals, such as yeasts, fruit flies, and nematodes, have led to a surge of interest in ageing from a perspective of protein chemistry and molecular genetics. For example, scientists have now investigated the process of ageing in the fruit fly by simultaneously measuring the activity of a large number of genes and counting the proportion of genes that show changes in expression with age.[103] It turns out that about 6%–7% of over 13,000 gene products assayed showed significant changes in expression levels.[104] Even if this underestimates the number of genes involved in human senescence then, as Michael Rose recently argued,[105] we may soon have the technologies to develop therapies to deal with this 'many-headed-monster', no matter how many heads it has.

Research on ageing has also provided a few other solutions to life extension which are not quite so technological. Calorific restriction (while avoiding malnutrition) has also been found to extend the lifespan of a wide range of animals including rats, fruit flies, and nematodes.[106] The mechanism by which caloric restriction extends lifespan is unclear, and it is even possible[107] (at least in fruit flies) that it extends longevity by reducing death rate rather than postponing senescence *per se*. One hypothesis is that dietary restriction slows metabolism, thereby slowing the production of toxic products such as ROS, but it may in part be linked back to reproduction if poorly fed individuals are not reproductively active. Of course, even if the same phenomenon were firmly established to hold for humans, it is questionable whether many people would want to take this course of action. We should also stop to think about the fundamental ethical, social, and ecological implications if we humans could find a way to dramatically increase our longevities. To explore these issues would take a whole new chapter.

Challenges

Despite the fact that examples of ageing are everywhere, and many of us humans will die of an age-related illness, there have been remarkably few evolutionary theories for senescence proposed, and many university courses in biology do not even cover them at all. Perhaps one reason why the subject is not currently more popular amongst evolutionary biologists is the general perception that the subject is pretty well sown up. After all, the topic was considered by some of the brightest biologists of the past century. However, the field continues to be highly controversial (as we have seen, researchers cannot even agree whether life extension will ever be possible), with a wide range of different perspectives ranging from the medical to the evolutionary.

One question is whether the model organisms commonly used to investigate ageing in the laboratory, such as short-lived fruit flies and nematode worms, are widely representative and there is a concern over side effects of intensive laboratory culturing. For example, fruit flies in laboratory cultures are typically maintained (for convenience—even fruit fly scientists have lives) on a 2-week generation time, and subsequent extensions of longevity in selection experiments may arise in part from relieving the population from the earlier selection regime.[108] Similar arguments apply to mammalian cell culture in which some immortalized cell lines show rapid 'genetic drift', forcing cell biologists to revive old frozen stocks of cells to obtain responses similar to those found prior to the 'drift'. In the future, it will be important to conduct research on ageing with a more diverse set of species, especially longer-lived organisms,[109] if we are to achieve a broad understanding of the phenomenon of ageing.

There are several intriguing phenomena appearing on the horizon for future researchers to get to grips with. For example, in the Bob Dylan song 'My Back Pages' are the lyrics: 'Ah, but I was so much older then, I'm younger than that now'. It turns out that the phenomenon of 'negative senescence' (defined formally as a *decline* in mortality with age after reproductive maturity, coupled with an *increase* in fecundity) is not just blowing in the wind, but is a very real possibility which is only now being taken seriously.[110] The very possibility that ageing can effectively be sent into reverse is a fascinating prospect. Continued growth after reproductive maturity may well explain several cases of this negative senescence. For example, mortality decreases in several coral species as they grow and age; while fertility is thought to increase by 10-fold or more once certain snails grow past sexual maturity.[110]

New directions will also need to be taken. While much research has been done on sources of damage, far less has been done to investigate the mechanisms that bring about repair. As noted earlier, some structures may be irreparable no matter how much energy and resources you throw at the problem, while sometimes the repair mechanisms themselves may break down, and it is important to explore the implications of this fact—there must be limits on the extent of selection to repair faulty repair mechanisms. A recent review of ageing research from an evolutionary perspective[111] noted that 'The field is clearly indebted to Medawar and Williams, but we should not be too much in awe of them. The time has come to stand on the shoulders of these giants, and reach farther than they might have imagined possible'. This is a fundamental field with clear applications, and it continues to need imaginative minds to help it develop.

2

Why Sex?

Figure 2.1 A male and female of the damselfly *Nesobasis heteroneura* in a copulation wheel. The female is taking sperm from the male (marked H71), who had earlier loaded it into his accessory genitalia. Photo courtesy of Hans Van Gossum.

> *—Why all this silly rigmarole of sex? Why this gavotte of chromosomes? Why all these useless males, this striving and wasteful bloodshed, these grotesque horns, colours,... and why, in the end, novels like Cancer Ward, about love?*
>
> *—W.D. Hamilton, 1975*[1]

There is considerable confusion about the meaning of sex. It can be taken to mean gender (male or female), a form of recreation ('hot sex'), or a form of procreation ('sexual reproduction'). To most biologists, sex is none of the above. Instead, one definition of sex would simply be a process that combines genetic material from more than one individual.[2] By this definition, sex is not in itself reproduction. For one thing, reproduction is not a necessary consequence of sex (many bacteria can simply share DNA through hooking up via conjugation), and sex is not always needed for reproduction (dandelions, greenfly, and starfish, to name a few, can all produce viable offspring without it). In fact, the specific act of combining genetic material can be thought of as the precise opposite of reproduction since it typically involves the coming together of the genetic material of two cells ('gametes') to create one ('zygote'), rather than the splitting of one cell into two. All that said, in eukaryotic species (like us, with chromosomes housed discretely within a nuclear membrane) sex is a *precursor* to reproduction. Indeed, in some species including humans, other mammals, and many insect species, sex is an essential step in the production of offspring.

To understand what sex is, we must first cover some basic genetics. Let us begin by thinking about the process of combining two gametes to produce a zygote—a fertilized egg. Naturally, if you simply combine all of the genetic material present in the nucleus of one of your typical cells with that derived from some lucky mate, then any resultant zygote would contain double the number of chromosomes. For example, humans are diploid, and have 23 pairs of chromosomes (one chromosome from each pair derived from each parent). Therefore, if you simply combine chromosomes from the diploid cells of two potential human parents, then the resulting zygote would have 46 pairs of chromosomes, and if two such individuals mated then their offspring would have 92 pairs of chromosomes. Clearly such accumulation of genetic material would quickly get out of hand, and cells would rapidly become obese with chromosomes. Most diploid species solve this problem by producing haploid (one set of chromosomes) gametes that eventually fuse to form diploid zygotes. The process of producing haploid gametes from diploid cells (or more generally, halving the number of sets of chromosomes) is known as 'meiosis' (Fig. 2.2). We now describe this process in more detail.

Dance of the chromosomes

Each sexually reproduced diploid offspring has pairs of chromosomes which do pretty much the same job (they are 'homologous'), one derived from each parent. When it is this offspring's time to reproduce, then, as we have seen, it must produce haploid gametes through meiosis (some organisms, such as mosses, spend much of their life in

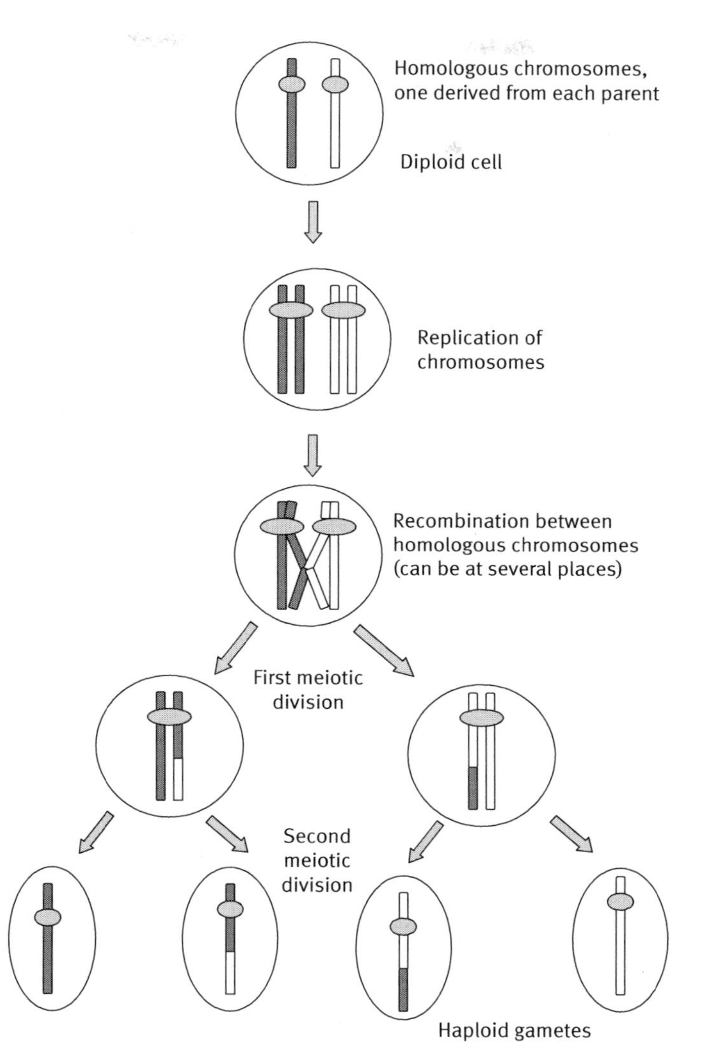

Figure 2.2 The essentials of meiosis. Note that for simplicity here we depict just one pair of homologous chromosomes and one crossover point between adjacent chromosomes. Note also that generally the sex chromosomes that influence gender do not recombine, especially if they are different in size.

the haploid state but the haploid form still has to be generated from the diploid form at some point in the sexual life cycle). Meiosis is rather strange because, despite the fact that it is all about halving chromosome numbers, its first stage is a *replication* of chromosomes (Fig. 2.2), seemingly making the problem worse rather than better. There is good reason for this, and indeed many researchers consider events at this stage crucial to understanding why sex evolves at all.

In the process of generating haploid gametes, the homologous chromosomes derived from the individual's parents—kept separate in its genome until now—get a chance to recombine with one another. Thus, following replication, the homologous chromosomes are allowed to physically touch and 'crossover' at one or more points, generating recombined chromosomes which contain genetic material from both parents, now together on the same chromosomes. Two separate divisions then take place in which chromosomes randomly segregate (remember there can be a number of different homologous chromosomes, not just one pair), halving the chromosome number in the cells until we are left with haploid gametes. These haploid gametes all have one full set of chromosomes, but the individual chromosomes will have come from physically recombining the genetic material from the individual's parents.

So, while meiosis appears primarily as a mechanism for halving the number of chromosomes, the process also involves recombination, creating new combinations of genes on the same chromosome that were not exhibited by either parent. This genetic recombination arises *before* the haploid gametes from different individuals get to fuse. The fusion of two gametes (the act of sex) is sometimes called 'outcrossing' and generates diploid offspring with homologous chromosomes. Overall, therefore, the genes in any sexually reproduced offspring will have come from two pairs of grandparents. Many of the genes donated by each set of grandparents will have become mixed with one another on chromosomes through recombination in the parent's meiosis that generated the sperm or egg, and a similar set of recombined chromosomes from the alternative pair of grandparents will sit along side them in the diploid state thanks to outcrossing.

The problem of sex

So, why sex? First, we need to reformulate our question in rather more explicit terms. We could ask 'how did sex originate?', or alternatively, 'why is sex currently maintained?'. We also have to be clear whether we are considering prokaryotes (chromosomes not held in a nuclear envelope and do not exhibit meiosis) or eukaryotes, because the underlying mechanisms of sex (and possibly the adaptive value) may be altogether different. In this chapter, we will concentrate on the question of why sex is maintained in so many eukaryotic species when (as we will see) asexuality appears to provide so many advantages. Our choice of emphasis is primarily because this particular question has proved more amenable to testing, and consequently it is the part of the puzzle where the greatest in-roads have been made. However, this is not to say that the origin of sex is unimportant or uninteresting—indeed it has almost certainly been unfairly neglected compared to work on the maintenance of sex.

So, we wish to understand why sexual reproduction, once evolved in eukaryotes, is not replaced by something far simpler and potentially less costly. The reason for putting the question in this way is that there are plenty of alternative solutions to sexual reproduction, which on the face of it look a whole lot simpler. Some organisms are parthenogenetic (Greek: 'virgin birth') in that they can produce eggs that successfully develop without fertilization by a male (such as some species of water flea, greenfly, lizard, and turkey[3]), while other organisms can reproduce without sex from a single unspecialized cell or group of cells, such as the bulb or tillers of plants, or budding in starfish. Compared to something such as budding, sexual reproduction appears to be a mightily complicated thing to orchestrate; therefore, you might wonder why any organism would go to all the effort of recombining DNA and outcrossing. Moreover, why should an organism that has got along pretty well in life go ahead and spoil a good thing by producing offspring that differ from itself: 'if it ain't broke, why fix it'?

Let us look deeper into these issues. There are also all sorts of hurdles to contend with in sexual reproduction, which you simply do not have to put up with if you reproduce asexually. For one thing, an asexual organism can reproduce by itself, but a sexual species has to find a mate. Finding a mate is frequently costly in terms of time and energy, as exemplified by many flowering plants that have to bribe pollinators with nectar simply to come visit them (likewise, think of humans bribing potential mates with the help of flowering plants). Even when there is a mate in sight, then (especially if you are a male) you may have to 'strut your stuff' and compete with a range of suitors. Some individuals may remain unmated, and even if you get to mate, you may have sexually transmitted diseases to contend with too.

Sex can be a big drain on resources, but there is also a more subtle and potentially even more serious disadvantage of sex when compared to asexuality. Put bluntly, males give every impression of being a complete waste of genetic resources. In sexual reproduction, the unit of reproduction is the pair (in many species, a male and female) whereas in asexual reproduction it is the individual. Let us assume for the time being that females set the limit on fecundity, so that each female can only produce a set number of offspring whether she reproduces sexually or asexually. This is justifiable if one assumes that the females have to create large gametes with cytoplasm and nutrients, while the male (if one is needed) provides little more than its genes. If we overlook any potential benefits of male care, then any asexual individual would be expected to produce about twice as many offspring *per parent* as a sexual one, because the males (assuming they constitute about half the offspring) effectively stand around and do nothing. This is the so-called twofold cost of sex.[4,5] Put in a slightly different (but more direct) manner, sex reduces the efficiency with which genes are transmitted from generation to generation because it involves the production of males. If females alone determine fecundity, then a gene that allows asexuality should, all else being equal, spread like wildfire in a population.

Problem, what problem?

Given all of these anticipated problems with sex, one might expect sex to be a rare phenomenon, but it is not. It is not just the birds and the bees—almost everywhere you look in eukaryotes from basidiomycota fungi to bamboo to beavers, they are all doing it (Fig. 2.3). Almost all of the known 42,300 vertebrate species (with the exception of a bucketful of fish, amphibians, and reptiles) are sexual.[6] Many other species known for their ability to reproduce without sex, such as parthenogenetic water fleas and greenfly, reproduce sexually now and again. There are occasional species, such as certain species of dandelion or lizard, that seem to reproduce without standard sex, but they typically have close relatives that engage in sex, suggesting that sexuality is a very ancient state and that modern asexual species are derived from sexual species. This suspicion is supported by the fact that many parthenogenetic species (the 'automixic parthenogens', as opposed to 'apomictic parthenogens'—see Glossary) actually employ meiosis—indeed, as an aberration of normal sexual reproduction, some authors frown at the use of the term 'asexual reproduction' for them.[7]

Intriguingly, some species are sexual in some parts of their distributional range, yet asexual in other parts. For example, recent work in the Azores Islands has led to the discovery of a female-only population of the damselfly *Ischnura hastata* that reproduces from unfertilized eggs[8] (Fig. 2.4). This species is sexual in North America and other parts of its distributional range (including the Galápagos Islands). Why this particular population is parthenogenetic is currently unclear, but, based on similar cases, it is possible that it is driven by a maternally inherited intracellular parasite that forces the species to produce only females, thereby increasing its own fitness.[9]

The argument that modern cases of species that reproduce without sex have derived this mode of reproduction from sexual species is also supported by the fact that the genes that facilitate meiosis, hence sexual reproduction, appear to have evolved very early in the evolutionary history of eukaryotes.[10,11] The occasional examples that buck

Figure 2.3 A pair of mandarin ducks (female on the left, male on the right). Why is there such a thing as a male? Photo courtesy of Richard Webster.

the trend, which include the several hundred species of bdelloid rotifers (the term 'species' is especially vague for asexuals, see Chapter 4), have been given celebrity status ('an evolutionary scandal'[12]), because they represent a whole group of anciently asexual species.[13] Despite these occasional exceptions, the fact remains that the taxonomic distribution of obligate 'asexually' reproducing species is extremely spotty, indicating that while this type of reproduction may occasionally be successful in the short term, sexual reproduction tends to beat it hands down.

Overcoming the twofold cost

Let us not lose track. Bdelloid rotifers may be unusual in having no sexual relatives but it is not the bdelloid rotifers that are a scandal, it is the sexuals we have to explain.[14] Similar to Winston Churchill's description of Cold War Russia, sex is widely considered something of a 'riddle, wrapped in a mystery, inside an enigma'. It is the 'queen of problems in evolutionary biology'[15] and 'the outstanding puzzle in evolutionary biology'.[16] Remember, this is not some mysterious cosmological dark matter evolutionary biologists are trying to explain, or some problem with string theory in the ninth dimension—look out the window and you will see plenty of examples of it. Indeed think of yourself: you are here because your mother and father combined their gametes.

As we have just argued, one of the biggest obstacles to overcome in explaining the persistence of sex is the 'twofold cost of sex'. Perhaps if we could challenge the

Figure 2.4 Reproduction without fertilization in a damselfly. On the Azores, *Ischnura hastata* is parthenogenetic, laying viable diploid unfertilized eggs. Elsewhere in the world, this damselfly species is sexual. Photo courtesy of Adolfo Cordero.

assumptions underlying this old chestnut, then all we would have to do is to find some minor advantage to sex and the problem would go away.

We have noted already that if the male partner contributes to the rearing of offspring such that a couple can rear twice as many offspring (or more) as an asexual loner, then we can cancel out the twofold cost of sex. However, parental care occurs in only a small proportion of taxa with sexual reproduction—so this cannot represent a universal explanation for the presence of males. Besides, DNA fingerprinting studies frequently show that males can be duped into rearing offspring that are not their own—under these conditions a mutation doing away with the need for sex could potentially spread among females, with males helping to rear offspring that are descended from the female alone.

Female gametes are generally much larger and costlier to produce than male gametes (indeed, by definition, females are the gender creating the larger gametes); therefore, males contribute proportionately fewer resources to a newly formed zygote—a condition known as anisogamy. Under high anisogamy, one might expect that the number of zygotes an asexually reproducing female could produce would be similar to the number of zygotes a sexually reproducing female could produce. However, turning the argument on its head, the twofold cost of sex would be rendered largely irrelevant if the gametes lack any form of size difference, such that they are effectively identical ('isogamous'). In effect, in isogamous species the gametes contributing to a zygote are more or less equally provisioned; therefore, one might expect an asexually reproducing isogamous female to be able to produce about half of the offspring of a sexually reproducing female which can combine its investment with the father.

In isogamous taxa (including ciliate protozoa, unicellular algae, and many fungi), males and females do not (by definition) exist, but even here there are often specific mating types (such as + and –). Can isogamy render the twofold cost obsolete? In theory, yes. Unfortunately, however, this excuse does not apply to most eukaryotes in which female gametes represent a substantially greater investment than male gametes. Moreover, somewhat ironically, if one looks at the distribution of sex across different species, the frequency of sex appears lower in isogamous species than in anisogamous species, despite the anticipated higher costs of sex in anisogamous species.

Note also that so far in describing the twofold cost we have been assuming a 50/50 sex ratio, but that cost disappears if sexual females happen to produce almost entirely females (the gender that limits fecundity). In most instances, a highly biased sex ratio is unlikely to happen (for the simple reason each male in this fantasy situation would tend to sire more offspring than females on average, leading to individual selection to reduce the bias). Alternatively, or in addition, it is possible that any asexual form that arose in a population of sexual forms may suffer from male mating attempts that harm asexuals more than sexuals.[17] Such a phenomenon might prevent the return to asexuality, but it is unlikely to be general.

Perhaps females do not have to be the fecundity-limiting gender? It is likely that they are not limiting in some circumstances, but there is ample evidence that they are in most. In particular, when one compares the maximum reproductive output of males and females when they are allowed many sexual partners, the female output is far less than the males. To take a particularly eye-watering example, the maximum number of offspring produced by a human female in her lifetime is reputed to be a staggering 69 (the eighteenth century Russian peasant Feodor Vassilyev who gave birth to 16 pairs of twins, seven sets of triplets, and four sets of quadruplets).[18] In contrast, the maximum number of offspring (allegedly) produced by a human male is 888 (this credit goes to the charmingly titled 'Ismael the Bloodthirsty', emperor of Morocco from 1672–1727).[18]

Finally, perhaps there is always some subtle benefit to sexually reproduced offspring—such as an improved ability to survive cold winters in the sexually produced egg stage (as seen in greenfly), or an ability to disperse widely as a zygote (compare vegetatively reproduced plants with sexually produced seeds)—which is not directly associated with genetic mixing itself, but nevertheless provides the crucial advantage to sex. We cannot overlook the possibility that sex brings certain non-genetical benefits, but it seems an extremely roundabout way to arrange overwintering or dispersal *per se*, and besides, there are asexual ways of achieving just the same solutions (e.g. some water flea species can overwinter using parthenogenetically reproduced eggs, while asexual testate amoebae can 'encyst' and sit out conditions as extreme as the Antarctic winter).

In a related way, it has been postulated that sex and recombination provide opportunity for the repair of double-stranded DNA damage, using the sequence of nucleotides on the homologous chromosome derived from the other parent as a template.[19,20] However, while it may arguably provide an explanation for the origins of sex in prokaryotes,[21] it does not offer an obvious explanation for the crossover events that generate recombination. Alternatively, it has been proposed that meiosis plays an important role in resetting developmental programmes to multicellular animals with tissue differentiation, but these ideas still need to be developed and explored.

Therefore, in most cases we may be forced to admit that asexuality will beat sexuality when it comes to sheer quantity, and it seems a far stretch to argue that sex is maintained primarily to allow overwintering, more effective dispersal, or DNA repair (although such properties may provide an added bonus). If it is a numbers game, then asexuality is the best bet.

Evidence for the twofold cost

For those who doubt whether the twofold cost could ever be incurred, let us look at some experiments. In one experiment, the growth rate of a population of sexual mud snails (more on this charismatic mud snail, *Potamopyrgus antipodarum*, later) was significantly lower than the growth rate of a population of mud snails of the same species that happen to be able to reproduce asexually,[22] indicating that traits facilitating asexuality have a high capacity to spread. In another study conducted on newly emerged adult psychid ('bag worm') moths, researchers could not find any difference in the total number of viable eggs that were produced by sexual females and by asexual females of a closely related species,[23] which again raises the issue of why an individual should waste its resources on males. If you cannot beat the asexuals on *quantity*, then *quality* may well hold the key. Thus, sex may well have more about enhancing the viability of offspring (at least some of them) rather than enhancing their sheer number.

Theories for sex

By far the most commonly accepted view among evolutionary biologists today is that sexual reproduction allows the production of a diverse array of genotypes, enabling the creation and spread of advantageous combinations of traits, coupled with the efficient removal of harmful combinations of traits. We will now look at some 'quality-based' explanations that have been proposed to account for the maintenance of sex, and explore their implications. Our approach will be roughly broken down to genetic explanations and ecological explanations, although these two approaches can frequently be seen as two sides of the same coin, especially when considering the benefits of sex as a means of bringing together advantageous mutations. In this way, our ecological explanations tend to focus on the selection pressures that bring about an emphasis of quality over quantity, while genetic explanations are aimed at elucidating the means by which sex delivers these desirable outcomes.

Sex in history

One of the first researchers to recognize the evolutionary conundrum that sex posed (and to suggest a solution) was the nineteenth century biologist August Weismann whom we met in Chapter 1. In his 1889 classic,[24] Weismann reasoned that sex could not solely have arisen to produce offspring, because there are so many other ways to reproduce. Instead, he proposed that the advantage of sex was that it provides variation for natural selection to act on. Writing in 1904, he noted[25]:

> [T]he communication of fresh ids [genes] to the germplasm implies an augmentation of the variational tendencies, and thus an increase of the power of adaptation. Under certain circumstances this may be of direct advantage to the individual which results from the amphimixis [sex], but in most cases the advantage will be only an indirect one, which may not necessarily be apparent in the lifetime of this one individual, but may become so in the course of generations and with the aid of selection. For amphimixis must bring together favourable as well as unfavourable variations, and the advantage it has for the species lies simply in the fact that the latter are weeded out in the struggle for existence.

Thus, Weismann saw sex as a means of increasing variation in offspring—getting some really good gene combinations, and allowing genes that do not combine well to be weeded out—which might be good both for the individual parent and the population. The geneticist Hermann Muller, with post-Mendelian knowledge of the mechanisms underlying meiosis and heredity, subsequently concurred with the '*genius of Weismann*' writing, 'the major value of recombination is the production, among many misfits, of some combinations that are of permanent value to the species'.[26] Evolutionary geneticist, statistician, and all-round genius Ronald A. Fisher had come to a similar conclusion at about the same time,[27] suggesting that sex may even liberate an advantageous mutation if it arose in a genome with several deleterious mutations (an idea memorably described as 'a ruby in the rubbish'[28]).

This is not the place to discuss population genetics at length, but to understand the basic logic, suppose that a gene of type A is *deleterious*, that is, harmful in some way (the form of a gene is known as an 'allele'). If an asexual diploid parent had this form of the gene on one of its chromosomes, then (assuming no meiosis) it would pass this deleterious gene to all of its offspring. Bad news kids. In fact, bad news grandkids and great-grandkids too. In contrast, sexual reproduction can produce gametes through meiosis that lack the gene due to segregation; therefore, not all offspring will inherit it. Now, let us instead consider two *advantageous* genetic forms, A and B, arising in separate genes. In an asexual species A and B can only jointly occur in the same individual by separate chance mutations in its asexual ancestors (A followed by B or *vice versa*), which may take rather a lot of generations. However, in a sexual lineage mutations to A and B can occur in entirely separate individuals, and these two solutions subsequently combined through sex (even if they arise on the same chromosome in different individuals then recombination can eventually get them together). As one might expect, getting together two advantageous genes will be even easier in a sexual population if 'fit' females (with a gene of form A say) tend to choose 'fit' males (with a gene of form B say). In short, sex may speed up evolution by rapidly finding good genetic combinations (and weeding out poor ones).

Graham Bell, who professed a distaste for naming theories after people as though they were diseases, called this general 'adaptability' explanation (as far as it applies to *long-term* advantage of combining beneficial mutations over numerous generations) after the English folk song and comic opera 'The Vicar of Bray',[15] and the name has stuck. The fictional vicar changed church allegiances whenever a new monarch of different religious persuasion came to the throne ('Unto my Dying Day, Sir, That whatsoever King may reign, I will be the Vicar of Bray, Sir!'); in a similar manner, recombination and sex may allow new varieties to be thrown up that are better suited to the prevailing circumstances. Up until the mid-1960s, the Vicar of Bray ideas went unchallenged, but at that point evolutionary biologists, notably George Williams, began to express serious doubts.

Species vs individual selection

Before discussing these doubts, let us go back to basics. When talking about a benefit, we must ask 'to whom?' and 'how does the benefit come about?'. Copulation is pleasurable to humans and probably to many other organisms, but nobody has yet offered a good evolutionary argument that sex is maintained solely because it directly pleasures the parents. We must, therefore, look for a benefit to offspring, and possibly beyond. If sex somehow benefits some of the offspring in their struggle for existence (such that, for example, one of the sexually produced progeny now has the ideal combination of A *and* B forms of genes so that it can outcompete its contemporary asexuals with only A *or* B forms), then we can begin to explore when and where such outcomes will happen. We develop this idea much more fully below. If, however, the argument is that sex has evolved for the longer-term success of the species, then the argument is *much* harder to justify. Natural selection has no foresight and works only on the material that it currently has available, so unless sex provides some form of quality advantage to offspring *now*, then asexuals will tend to spread.

Despite the above reservation, there is a chance that sexuality might still hang on (and even prevail) in the longer term because of its effects on species persistence, even if it is disadvantageous in the short term. For example, one might imagine a scenario in which there is a dynamic flux, with any mutant asexual forms rapidly taking over a sexual species when they occur, but with sexual species adapting better to environmental changes in the longer term (and hence persisting as a species for longer). One specific mathematical model of this scenario suggests that sexual species will remain common in a system if no more than one mutation conferring asexuality in any given sexual species arises in the time it takes 100 sexual species to go extinct, but these conditions seem rather restrictive.[29,30] 'Genetic imprinting', which we will not go into here, may do the job of preventing successful mutations for asexuality arising in mammals[31] (to quote Kondrashov[32]—'mammals are probably the only group in which virgin birth is impossible without a miracle'), but such genetic constraints are not sufficiently general to hold back the tide of asexuality.

Occasionally, the above type of 'group selection' argument is given sympathetic treatment in textbooks because it seems to explain certain patterns. In particular, the Vicar of Bray has the advantage in explaining why asexual species are relatively recent and, therefore, patchy in distribution—they keep turning up only to be snuffed out by the greater adaptability of sexual forms. Some botanists have argued that this sort of process can explain the evolution of asexual blackberries and dandelions.[33] However, individual selection can explain all of the above too, but as a secondary phenomenon. Thus, the immediate selection for sexual reproduction via benefits to certain offspring will also help bring about a more persistent sexual species in evolutionary time, even if the primary advantage is gained through immediate rather than long-term benefits. After saying all that, there may be an inherent bias when we point to asexuals turning up only sporadically on the tree of life. If sexual reproduction plays an important role in speciation (see Chapter 4), then is it any surprise that we see a largely sexual tree?

Sex as a way of combining advantageous alleles

Let us return and focus on the immediate benefits to a proportion of offspring of getting new combinations of advantageous genes rapidly together on the same genome.

Mathematical models suggest that the process could work to help explain the maintenance of sex, but it tends to depend, among other things, on having relatively small populations and relatively rare beneficial mutations. To see why we need these conditions, note that for sex to be of particular help in combining beneficial mutations, we need conditions where several beneficial mutations to genes are unlikely to rapidly accumulate in the same asexual lineage. Such an outcome will happen when the population is small and/or when beneficial mutations are rare (for the same set of reasons).

That said, sex can sometimes be particularly helpful compared to asexuality in combining advantageous alleles even in larger populations. For example, if there are a number of different genes that can have beneficial mutations and the population is large, then models show that asexual populations will have a really hard job of accumulating beneficial mutations when different clones with different beneficial mutations are continually battling it out (a phenomenon known as 'clonal interference'). In contrast, sex releases this 'speed limit' by allowing the cobbling together of each of these individual advantageous mutations.[34]

Before going any further, let us pause for a moment. One might reasonably point out that if sex were all about combining good parts of solutions, then parents would eventually run out of ways to improve offspring any further. One might even propose that it would pay an individual to become asexual once it has reached the 'pinnacle' of success. It turns out that the nature of what constitutes a 'beneficial mutation' may change in time and space, in part because of interactions between the effects of different genes, and in part because of interactions within and between species—ecology for short. In fact, thinking about ecology and environment has generated some of the most highly regarded explanations for the evolution of sex, and we will return to them shortly.

Ratchets and hatchets: purging deleterious alleles

Let us keep with genetics for the time being and reverse the above logic. We now ask whether sex is good at purging deleterious mutations (as Weismann had also implied), as opposed to generating beneficial combinations of new ones. Of course, sex does both, but it helps in delivering our story to keep the two phenomena as separate as we can for the time being, and historically this is the way answers to the question have developed.

Two main 'deleterious mutation' theories have been proposed. The first one was proposed by Hermann Muller who argued that in small populations the genomes of asexual lineages will inevitably 'ratchet up' very slightly deleterious mutations over multiple generations, through the chance loss of the individuals with the very least mutations.[35] The overall effect is now known as 'Muller's Ratchet', and is somewhat akin to a gradual reduction in quality as the same lecture notes are photocopied from a selection of the earlier year's photocopies. As noted earlier, sexual reproduction allows a degree of mutation clearance (such that parents with deleterious mutation A do not have to pass it to all their offspring) and rescues any beneficial genotypes that might have been lost by recombination, thereby preventing this slide into deleterious oblivion. The same processes can therefore prevent the slightly deleterious genes being ratcheted up.

Muller's Ratchet will be the fastest when deleterious mutation rates are high, selection against deleterious mutations is weak (otherwise they would be rapidly purged by selection), and population sizes are small (collectively ensuring that there are not many individuals with the least number of mutations, so they can by chance be lost). It is easy to envisage a long-term species benefit to sex in such circumstances (so, as above, one might invoke a type of group selection to explain sexual reproduction in this way), but the fact remains that the intensity of individual selection represented by the Ratchet is generally considered too low to compensate for the twofold cost. Besides, a quick bit of sex every few generations is, in theory, enough to halt the Ratchet, so the theory does not explain obligate sexuality.

There is however a deleterious mutation theory which may well help explain the maintenance of sex through individual selection: the 'mutational deterministic hypothesis'. The theory has been postulated by several researchers but developed most fully by Alexey Kondrashov (and has since, in deference to the ratchet, been dubbed 'Kondrashov's Hatchet'[32]). The good news is that the Kondrashov's Hatchet does not depend on the chance loss of the fittest individuals (hence the term 'deterministic') and it therefore applies even to very large populations. While most versions of the Ratchet assume that individual deleterious mutations all act independently, the Hatchet only works if one can assume that each individual harmful mutation has an even greater effect on reducing fitness than simply the sum of their individual effects ('synergistic epistasis').

We know that through its effect of combining genotypes, sexual reproduction will (under many, but not all, conditions) increase the variability in the number of harmful mutations, generating some offspring with very few, and others with a high number of deleterious mutations. The whole process has been likened to combining motor cars with defects (e.g. one vehicle may have a working engine but a faulty gear box, while the other has a reasonable gear box but its engine has gone). Through sex you can turn two wrecks (i.e. parents) into a runner plus a wreck (i.e. offspring). Now, thanks to synergistic epistasis, any individual with a very high number of deleterious mutations will be rapidly eliminated from the population. The theory works in the short term because sexually produced offspring will tend to include some individuals with even fewer deleterious mutations than either of their parents (not the case for any competing asexually produced offspring) and because over the slightly longer term, asexual lineages will accumulate more deleterious mutations than sexual lineages. If one or both genders choose mates on the basis of their quality, then sexual selection may act even further to filter out individuals over the longer term with a high deleterious 'mutational load'.

There is now little doubt that sexual reproduction reduces the frequency of deleterious mutations in a population compared with asexual populations, but does it explain the maintenance of sex? Some have argued that natural selection would do a good job of quality control, purging the less fit individuals from the population, without the need to invoke additional mechanisms. There is also the problem that Kondrashov's Hatchet, as typically formulated, compares the success of asexual and sexual lineages rather than asking directly whether a rare asexual mutant can invade a population of sexuals. To compensate for the immediate twofold cost of sex, the Hatchet model not only requires synergistic epistasis, but also requires that the rate of deleterious mutations per genome is high. Indeed, mathematical models suggest that the deleterious mutation rate per (diploid) genome per generation would have to exceed 1 for a sexual population to resist being taken over by a rare asexual form.[36] Slightly more realistic models push this even higher.[37] The jury is still out on whether synergistic epistasis among deleterious mutations is common, but the indications are that the rate of creation of deleterious mutations per genome does not exceed 1 in smaller animals with short generation times, indicating that the Hatchet cannot provide a general explanation for sex.[38] As one might expect, it is notoriously difficult to get a handle on deleterious mutation rate and we are not quite there yet, in terms of reaching firm conclusions. Indeed more recent estimates suggest a value of 1.2 for fruitflies.[39] The great value of the Hatchet theory is that it dares to stick its neck out by making a clear and testable prediction, and it may still play an important role in explaining the evolution of sex even if one has to invoke additional mechanisms to overcome the twofold cost.

So, let us summarize where we are so far. Sexual reproduction is ubiquitous in nature, but it is not obvious why that should be so, especially given the much vaunted 'twofold cost' of sex. Put simply, why do not asexual species take over? Sex cannot be maintained because it benefits parents themselves directly, and it is unlikely to be maintained because it provides a role in overwintering or DNA repair. The notion that sexual reproduction is maintained because it allows populations to combine advantageous traits from different individuals is an attractive one, but this type of solution would make much more sense if it could explain why a sexual female in a population of sexuals would on average do better than a rare asexual mutant, rather than explaining why sexuality is good for the long-term success of the population. There is also the issue of why new advantageous traits are continually possible, which we will return to in a moment. Genetic theories that stress the other side of the equation, the purging of deleterious mutations, are more about maintaining stasis (reducing the accumulation of harmful mutations) than climbing 'adaptive peaks'. One theory, Kondrashov's Hatchet theory, appears to make most sense from a perspective of immediately overcoming the twofold cost of sex. However, while deleterious mutations will almost certainly be found at higher rates in asexual variants, it is unclear whether the underlying conditions required for the theory to overcome the twofold cost are generally upheld, making it the *prima face* reason for sex.

Sex meets ecology

Let us now return to the potential role of sex combining advantageous traits, this time from a complementary ecological perspective. In short, it may pay to create genotypic variety when the local environment that your offspring will face is different from the environment that you are currently facing. The advantage of sex in these circumstances has been likened to buying tickets for a lottery.[16] Whereas asexuality generates identical offspring with the same number, sexuality generates offspring with different numbers on their tickets, increasing the chances that one (or a few) will win the prize. The general idea is an attractive one and it appears to fit the bill. For example, sex is more prevalent in long-lived than short-lived organisms. There may be a whole variety of reasons for this, but it is worth noting that the longer you live the more likely the environment will have changed for your offspring compared to when you were young.

Many mature oaks in Britain today were seedlings in the last 'little Ice Age' over 400 years ago, so perhaps sex has evolved in these species and many others as a means of coping with variation in non-living factors, such as temperature and rainfall. If the general argument were correct, then one would expect sex to predominate in environments that vary considerably in space and time (such as fresh water compared to the sea or higher altitudes compared to lower altitudes). In fact, the association between sexuality and environmental uncertainty is quite the opposite from that predicted through physical and chemical variation in the landscape: sex is more common in taxonomic groups that appear to reside in stable uniform environments than in fluctuating heterogeneous environments.[15] Hence, environmental variability, more specifically variability in physical and chemical conditions, does not seem to explain the maintenance of sex.

The capriciousness of others

All is not lost however. It is important to note at the outset that we can anticipate that not all types of environmental change will have the same effect in promoting sex. For example, if the environment randomly changes in time (or space) between a few fixed extremes, then the conditions in the next generation (or next patch) is anybody's guess and the average success of offspring may be similar, whether or not they differ from the parent—why change something that works, especially if there is a chance that the environment will be broadly the same? As one might expect, the greatest advantage of sex will come when it is highly unlikely that the currently most successful genetic combination will provide high success in the next generation, that is, the environment is positively '*capricious*'.[15,40] In other words, for sexual reproduction to have the greatest chance of resisting take over by mutant asexuals, conditions elsewhere or at another period of time should not simply be unpredictable, they should be in some way 'opposite' from the current conditions.

In this light, ecological interactions such as competition and parasitism become much more promising as agents maintaining sex because the form of variability they require may be sufficiently capricious to promote sexual reproduction. This is simply because the nature of the interaction evolves too, with competitors and parasites being continually selected to find ways around the current common defences. Indeed, the observation that asexuality is more common in unstable habitats suggests that asexual reproduction tends to be selected for in species that do not have to face interacting with others: here the quantity of offspring is the key requirement to establishment. In more crowded habitats, however, it becomes more a case of persistence: the quality of one's offspring may have greater influence in determining a parent's success than the quantity.

One attractive idea is that sex is maintained in more crowded systems because it provides a competitive advantage to offspring—'In a saturated economy it pays to diversify'.[41] Graham Bell called this competition theory the 'Tangled Bank'[15] hypothesis, after the famous concluding passage used by Darwin in *Origin of Species,* who invoked the term to conjure up a complex habitat generated by a few simple laws. The argument goes that sexual reproduction leads to new genotypes, some of which will stand a better chance of surviving than the existing genotypes because these offspring will be in less direct competition with others, exploiting a slightly different niche. The dynamic is not hard to envisage—for instance, mixtures of wheat and barley typically produce higher yields than a monoculture (although this outcome may be mediated by additional factors, such as making life harder for specialist herbivores and diseases). In addition, the competition mechanism is even more powerful if one assumes, not unreasonably, that dispersal is frequently local and that by being different an offspring will face less intense competition from its parents or siblings. Indeed, researchers have shown experimentally that the intensity of competition is lower among sexually produced siblings than asexually generated sibs in sweet vernal grass.[42] Mathematical models have shown that the Tangled Bank theory can work so long as there is a 'genotype–environment' trade-off (such that no individual species type has a competitive advantage on all types of resource that are available) among other more technical restrictions.[15,43,44] If the tangled bank theory were valid then sex should be favoured when competition is most intense, such as in high-density populations, and in relatively stable-crowded environments. In support, sexual forms of greenfly and water flea tend to arise at high densities—indeed you can induce greenfly to generate sexual offspring by brushing them, so simulating the touch of other greenfly in a crowded habitat.

Running with the Red Queen

Unfortunately, other related theories, including the 'Red Queen hypothesis' which is probably the most popular ecological hypothesis to date, can also help explain the above relationship. The term 'Red Queen' derives from the fictional character in Lewis Carroll's *Through the Looking Glass,* who must run constantly just to keep in the same place. The metaphor was originally invoked by palaeontologist Leigh Van Valen[45] to describe a system in which there is substantial evolutionary change but no real progress. Similar to one of those nightmares in which you are running towards a door that never seems to get closer, hosts may be continually finding new ways to avoid being too heavily parasitized, but the parasites themselves are continually finding new ways to unpick these defences: the end result, despite plenty of change, is no net gain. This stasis is particularly likely for a host species because parasites often have short generation times, and are therefore likely to be able to evolve rapidly to meet any barrier before hosts can try out a different approach. Of course, it is not just hosts that face the Red Queen. Employees periodically get offered pay rises but the value of this increase almost inevitably gets eaten away by inflation, so while there is more money in peoples' pockets from one year to the next, they rarely become substantially richer. Any form of antagonistic coevolution, including competition, can in theory set up a Red Queen effect but it is with hosts and parasites that the name has become most closely associated.

So, as the saying goes, sex may be good for your health. The idea that sex can provide a selective advantage through reducing the parasite burden of at least some offspring has been around since the late 1970s when Hamilton, Levin, Jaenike, Bremermann, and others developed the first formal models.[46] Imagine a sexual population in which a mutation arises that makes one of the individuals asexual. All else being equal, asexuality is likely to spread because the asexual individual produces more offspring (the two-fold cost of sex). However, such genetic uniformity in the host represents a parasite's dream, and over time the asexual clone may well find itself subject to disproportionate parasitism. This is because parasites themselves will be selected in part on their efficiency in exploiting the common and unvarying host types. With the clonal form being subject to more intense parasitism than the sexual form, individuals engaging in sexual reproduction will once again be at a selective advantage until the clone is significantly reduced in numbers. In essence, these coevolutionary arms races tend to produce a form of time-delayed negative frequency-dependent selection in which it pays to be different from the rest. Defences in one generation may be of little benefit in the next generation. Similar to clothing fashion, once everybody is wearing it, the clothes are no longer fashionable. Such a dynamic (whether it arises in hosts attempting to avoid being parasitized or people trying to be fashionable) has all the hallmarks of capriciousness. Thus, many people would not be seen dead wearing what our parents found cool and trendy in the last generation.

The Red Queen arguments depend on several conditions, including the assumption that a host's susceptibility to parasites can be influenced in part by its genes. Experimental evidence to support this contention is not hard to find. For example, Carius and co-workers[47] investigated the effects of the sterilizing bacterial parasite *Pasteuria ramose* on the water flea *Daphnia magna*, an inhabitant of temporary ponds. Water fleas tend to reproduce parthenogenetically (via apomixis, and therefore not involving meiosis) for a number of generations and then engage in sex (they are 'cyclical parthenogens'). In a series of laboratory experiments it was found that different genetic clones of *Daphnia* differed in their susceptibility to the bacteria, but no single clone of *Daphnia* was superior to all of the available isolates of the bacteria; likewise, there was no bacterial isolate that was superior in infectivity to all of the available clones.

Red Queen arguments also depend crucially on genotypes having an advantage when they are rare. There are several good examples of this, but one of the best comes from the small freshwater mud snail, *P. antipodarum* introduced earlier in the chapter.

The species probably originated in New Zealand, and has now spread throughout Australasia, much of Europe and North America—indeed the entire US population of the mud snail before 1995 was thought to have derived from a single asexual clone introduced by accident. In New Zealand, there are individual populations of this snail in which only obligate sexual (diploid) forms occur, only obligate asexual forms occur (the asexuals are triploid and probably arose as a hybrid of two sexual species—see Chapter 4), and populations where both forms co-occur. Like most species, the snail is attacked by a bunch of parasites, most notably trematode worms from the genus *Microphallus*, which (as the name implies) are parasitic castrators. In one study of an obligately asexual population of these snails on South Island New Zealand, Lively and Dybdahl[48] found that common clonal genotypes were more susceptible to infection by the parasite than rare clonal genotypes, supporting the idea of a frequency-dependent advantage. Moreover, they found that trematodes that co-occur with the same snail population were more infectious to the snail hosts than trematodes that were shipped in from outside,[48] indicating a degree of coevolution (and that the parasite had the upper hand).

Many other studies also point to a relationship between genetic diversity and parasitism. For example, Baer and Schmid-Hempel[49] artificially inseminated the founding queens of a species of bumblebee with a cocktail of sperm of either low diversity (sperm from four male drones from the same colony) or high genetic diversity (sperm from four drones, each from unrelated colonies). The resulting colonies founded by these queens were then allowed to develop, exposed to natural parasites (mites and flies) under field conditions. As might be expected, the high-diversity colonies had fewer parasites and showed greater reproductive success, on average, than low diversity. Although the authors could not infer the precise mechanisms by which parasitism is reduced, it seems likely that the diversity in the workers reduced the rate that incoming parasites could establish, or at least spread.

The ingredients are here, but when it gets down to mathematical modelling the Red Queen solution to sex, some reservations start to appear. Part of the problem is that the Red Queen hypothesis rests on an advantage to rarity, not in sex itself.[50] Therefore, if any new asexual form can hang on for long enough before a new asexual mutant arises then asexuals can benefit from their own relative rarity, as well as their capacity to reproduce. Perhaps the additional benefit of sex in purging deleterious mutations has some role to play in redressing the balance here. Alternatively or in addition, if a host is more likely to encounter a parasite transmitted by the host's mother than expected by chance, then the fitness of offspring will be higher if they are genetically dissimilar to the mother—sex will help to 'distance yourself' from your parents.[51] Another problem with the Red Queen is that sex can break apart resistant host genotypes before they have started to become seriously exploited by parasites, in the process creating less fit genotypes. The general feeling today is that host–parasite coevolution is not in itself sufficient to prevent the replacement of a sexual population by an asexual lineage,[52] unless there is rather extreme (but not unheard of) selection to avoid parasitism (such as when parasites kill or castrate infected hosts, or when individuals with high parasite burdens simply do not get mated).

No shortage of theories

How do we begin to discriminate between these rival explanations, many of which are formally correct in that they could, if the conditions were right, explain sex? Looking at patterns in the real world, and conducting experiments, is clearly the way to go. One approach is to examine species which contain both sexual and asexual forms (either at different times, or different places) and look at the factors that are associated with sexuality. In fact these 'partly sexual' species are a lot more common than the species comprising entirely obligate asexual clones. The first trend is that there are more asexual forms of a species found as one moves towards the poles,[15] although there are plenty of exceptions (many greenfly produce frost-resistant sexual eggs in the autumn, but only at high latitudes). Why is there a general tendency for more asexuality towards the poles? One of many factors that changes with latitude is species diversity (see Chapter 6), so it is possible that sexual reproduction may provide a particular selective advantage when organisms must interact with others.

Another interesting pattern is that asexual forms of a species tend to have a wider geographical distribution than the sexual forms.[7] We have already seen how a few clones of the mud snail have taken over North America. Weeds such as the dandelion (Fig. 2.5), which is sexual in the Mediterranean yet asexual in northern Europe and throughout

Figure 2.5 A field of dandelions in the United Kingdom. Despite their flowering heads, implying cross-fertilization through pollination, do not be deceived: dandelions typically produce viable seeds without sex, a trait that may allow them to colonize and rapidly spread into new areas. Photo: DMW.

North America, are also well known for their wide physiological and ecological tolerances. Gardeners in North America and northern Europe do not be misled by the bright yellow flowers—they produce seed identical to the parent. It may well be that the asexual clones comprise 'general purpose' genotypes suited to a range of conditions, and only begin to suffer when faced with more specialist competition or parasites.[7] The third pattern is that in species that can switch from one form of reproduction to another at different times of year, there seems a general trend for the timing of asexuality to coincide with periods favouring population growth—low densities and suitable growing conditions.

In a similar manner, and probably for similar reasons, asexual reproduction seems more common in more disturbed habitats. Take the New Zealand freshwater snail mentioned earlier. Lively surveyed a variety of populations in streams and on the fringes of lakes, using the proportion of males as an index of the extent of sexuality in each population.[53] Overall, he found that there was a significantly higher proportion of males (hence sexuality) in lakes than in streams. Since lakes are arguably more stable environments than streams, then this observation in itself again raises doubts as to whether sex has evolved to deal with temporal fluctuations in the non-living environment. More importantly, Lively found a positive correlation between the proportion of males in populations and the percentage of individuals infected by parasites (principally *Microphallus*). No, this trend was not due to males being more susceptible to parasites, and it appears that sex is not important for the parasites' transmission.[53] The positive correlation overall was taken as support for the Red Queen hypothesis, since the theory seems to predict that sex should be most common in populations where there has been a history of parasitism. Indeed the correlation held even when statistically accounting for the habitats (stream or lake) the snails were derived from.

Psychid ('bag worm') moths are attacked by a variety of wasp parasitoids, which (rather like the life form in the Sci-Fi film 'Alien') lay eggs in the caterpillars and eat the host from the inside out. In a study looking at the distribution of sexual and related asexual psychid moth species in Scandinavia, Kumpulainen and co-workers[23] found, as Lively had done, that there were proportionately more sexually reproducing moths (at least as judged from the survivors) in those populations where there were more parasitoids. Coevolution is a curious thing, however, and the prediction can go both ways. One might argue, for example, that if sex was so good at enabling a species to avoid parasites, then it should be populations with more asexuals that should have the most parasites, not the sexuals. The nature of the correlation will depend in part on who is winning the evolutionary race at that time. Even more importantly, it leaves open the question of why parasites are relatively rare in some sites, when certain sites frequently contain invariable clones that are there for the taking. Perhaps more direct evidence for asexuals being more susceptible to parasites comes when we compare the parasite burden of coexisting asexual and sexual forms, and here again there seems to be some reasonable evidence that asexuals have the higher burdens, at least in certain species of fish[54] and lizard.[55]

A range of experimental work has also been conducted to test whether, and why, sex can be maintained. In one set of experiments, Goddard and colleagues[56] artificially created different strains of yeast (isogamous, so there is no twofold cost) that differed in their ability to engage in facultative sexual reproduction (by craftily deleting genes necessary for normal meiosis and recombination from one set of strains). In a relatively benign environment good for growth, the obligately asexual and the facultatively sexual strains grew equally well. In contrast, in a harsher set of conditions (warmer temperatures, and certain changes to the growth media), the relative rate of growth of the sexual strain was significantly higher, supporting Weismann's view[24] that sex can be maintained because it supplies more variation to act on. Of course, we must bear in mind that in an anisogamous species, the asexuals may outgrow the sexuals in a benign environment. We must also question the nature of the environmental challenge and wonder whether it would be sufficiently capricious to continually maintain sex. That said, this type of experiment is a useful affirmation that sex can indeed provide a short-term advantage if the circumstances are right.

In another approach, Burt and Bell[57] attempted to distinguish between the Red Queen and Tangled Bank theories by examining the average number of crossover events during meiosis in a variety of species. They reasoned that if the Red Queen model was correct, then hosts with long generation times should face greater selection to recombine their genes because the parasite (with far shorter generation times) will be well ahead of the game and new variants are even more desperately needed. In contrast, if the Tangled Bank theory was correct, then the larger the number of offspring (a likely correlate of the intensity of competition among siblings) the more important it will be that they are different from one another, so recombination should correlate with fecundity. Overall, they found that long-lived mammals had more crossover events per chromosome pair than short-lived mammals (all of which are obligately sexual) even when one controls for factors such as body size. In contrast, the mean number of crossovers per chromosome pair was independent of fecundity, leaving the authors to conclude that the Red Queen hypothesis better fitted their observations. We have to admit that it is not entirely clear why the optimal level of recombination should vary in the ways described, although clearly there is a cost to too much recombination in that it scrambles good gene combinations. On-going theoretical work elucidating the optimal level of recombination assuming 'modifier genes'[6,58] may well shed further light.

The bottom line

What are we to conclude from all this? The bottom line is that our understanding of the evolution of sex is still a bit of a mess, but it is getting a whole lot better. Plenty of explanations have been proposed and many of them work under some sets of conditions on paper (or in a computer simulation), but there is still no broad consensus. Ecological explanations in which asexuality is favoured for rapid growth, while sexuality is selected to deal with complex and ever-changing selection pressures of a biological environment, have perhaps found most favour. However, one must bear in mind that comparative work that supports these ideas tends to concentrate on ecological associations; therefore, it is not surprising that the role of sex in purging deleterious mutations gets de-emphasized.

Recognizing that many of the individual theories had their limitations, but that in concert they could explain many of the features of sex, West and colleagues have advocated a more pluralistic approach.[59] Indeed they have even argued that some theories, such as the Red Queen and Kondrashov's Hatchet, work together in an entirely complementary manner in that they can explain sex much more robustly when together than alone. Kondrashov himself reacted as the majority of scientists would[60]:

> I do not like this possibility because such a beautiful phenomenon as sex deserves a nice, simple explanation and messy interactions of very different processes would spoil the story. Of course, this does not mean that such interactions are not, nevertheless, essential.

Yet in a general sense, one has to ask why sex should be only about collecting advantageous mutations, or only about purging disadvantageous mutations, when it can simultaneously do both. Rather like sex itself, the pluralist approach can cobble together parts of potential solutions. More and more theories are now being proposed which borrow, for example, a bit of Ratchet and a bit of Tangled Bank. It would be nice to clear them all away and find a really simple single elegant solution to sex, but life may not be like that.

The future

What of the future? Part of the excitement of this research area is that finding a solution to sex, amongst the most general and puzzling of all biological phenomena, is still somewhat of a 'Holy Grail'. The downside is that there is more to gain from suggesting a new solution (especially if it is radical) than testing an established theory, so new ideas tend to get pushed out at a greater rate than people can test them. It seems to us that more experiments are badly needed, including more direct tests of evolutionary outcomes when asexual strains are introduced to sexually reproducing populations, although these will necessarily be limited to rapidly reproducing species such as microorganisms. We also need more empirical genetic work to take on the challenge of quantifying (and elucidating) the underlying genetic mechanisms, such as the rates of deleterious mutations. More work to catalogue the distribution of sexual and asexual forms from a perspective of parasites and competition would also be welcome. Put slightly differently, it would be great if Curt Lively's elegant study of New Zealand snails were just one of a large number of studies we could call on when thinking about the Red Queen.

There are issues for theoreticians to help clear up too. One question centres on the relative importance of recombination during meiosis in providing an advantage to sexual reproduction. In Bell's analogy,[15] sex is a qualitative phenomenon which acts as a switch, while recombination acts as a rheostat, a quantitative fine-scale adjuster. Yet many evolutionary biologists tend to treat sex and recombination as synonymous, and where a distinction is made, they go out of their way to emphasize the importance of recombination. In perhaps an extreme example, Muller argued[26] that 'of its two major features... only recombination itself has any evolutionary value'. Others see recombination as somewhat less important. To quote Cavalier-Smith,[10] 'Although recombination has some advantages [in meiosis] they are probably small compared with those of ploidy reduction'. We have little doubt that recombination is an important force in generating variability, but it is currently unclear (to us at least) how important it is in the evolution of sex, compared to the act of outcrossing.[61] Of course, sex involves both the coming together of chromosomes from different parents and recombination, but theoretical models which tease apart the relative contribution may well shed light on the major selective forces.

So sex is not as easy to understand as one might first think. But thank heavens it has evolved. Without sex, life would be very different, and most likely pretty boring. It is not simply that humans would have lost something pleasurable to engage in, it is that we would have lost much of life's colour and variety. Flowering plants would no longer need to attract pollinators, birds of paradise would no longer dazzle us with their brilliance, and, as Hamilton argued, we might no longer have love.

3

Why Cooperate?

Figure 3.1 A living bridge created by ants. © Hung Meng Tan/iStockphoto.

All untaught animals are only solicitous of pleasing themselves, and naturally follow the bent of their own inclinations, without considering the good or harm that from their being pleased will accrue to others. This is the reason, that in the wild state of nature those creatures are fittest to live peaceably together in great numbers...

—*Bernard Mandeville* (1732)[1]

An altruistic act is one in which an individual incurs a cost that results in a benefit to others.[2,3] Giving money or time to those less fortunate than ourselves is one example, as is giving up one's seat on a bus. At first, one might consider such behaviour hopelessly naive in a world in which natural selection seemingly rewards selfishness in the competitive struggle for existence. As the saying goes, 'nice guys finish last'. Yet examples of apparent altruism are commonplace. Meerkats will spend hours in the baking sun keeping lookout for predators that might attack their colony mates.[4] Vampire bats will regurgitate blood to feed their starving roost fellows,[5] while baboons will take the time and effort to groom other baboons.[6] Some individuals, such as honeybee workers, forego their own reproduction to help their queen and will even die in her defence. The common gut bacterium *Escherichia coli* commits suicide when it is infected by a bacteriophage, thereby protecting its clones from being infected.[7,8] If helping incurs a cost, then surely an individual that accepts a cooperative act yet gives nothing in return would do better than cooperators? What, then, allows these cases of apparent altruism to persist? In his last presidential address to the Royal Society of London in November 2005, Robert M. May argued, 'The most important unanswered question in evolutionary biology, and more generally in the social sciences, is how cooperative behaviour evolved and can be maintained'.[9,10]

In this chapter, we document a number of examples of cooperation in the natural world and ask how it is maintained despite the obvious evolutionary pressure to 'cheat'. We will see that, while it is tempting to see societies as some form of higher organism, to fully understand cooperation, it helps to take a more reductionist view of the world, frequently a gene-centred perspective. Indeed, thinking about altruism has led to one of the greatest triumphs of the 'selfish gene' approach, namely the theory of kin selection. Ultimately, as the quote from Mandeville indicates, we will see that cooperation frequently arises simply out of pure self-interest—it just so happens that individuals (or, more precisely, genes) in the business of helping themselves sometimes help others.

A word on semantics. The field of cooperation is full of misunderstandings, so we want to define some terms before we begin. Following West and colleagues,[3] we treat cooperation as any behaviour that provides a benefit to another individual (the recipient), and that is selected for because of its beneficial effect on the recipient. This is our 'umbrella term' for helping others. In contrast, altruism is a subset of cooperative behaviour in which the act of helping is costly to the actor and beneficial to the recipient. It is this form of costly cooperation that is particularly challenging to explain and therefore forms the basis of most of the examples we consider. The reverse side of the coin is a behaviour that is costly to both the actor and the recipient: researchers call such interactions 'spite', although it has been far less intensively studied and we will not be discussing it further.

Acts of cooperation can be observed in a wide range of biological contexts ranging from the programmed death of individual cells within a multicellular organism, to the export of phosphates from mycorrhizal fungi to plants. Owing to the enormity of the field, we will restrict ourselves here to understanding acts of cooperation between individuals of the same species ('intraspecific cooperation'), although many of the explanations also apply to these other contexts. Several interrelated routes to intraspecific cooperation are currently recognized,[2,11–13] and we will briefly review them in turn.

Keep it in the family

First and foremost, cooperative behaviour can sometimes spread and be maintained if it favours the survival and reproduction of close relatives. Few people might ever stop to wonder why parents go to such lengths to look after their children and even their grandchildren—after all, it seems such a natural thing to do. However, take examples of altruism beyond this level of relatedness and somehow it seems to stretch our credulity. Charles Darwin was among the first to suggest that cooperation might be linked to family relatedness when he wondered how sterile castes of insects[14] could arise as a consequence of natural selection: 'This difficulty, though appearing insuperable, is lessened, or, as I believe, disappears, when it is remembered that selection may be applied to the family, as well as to the individual...'.

To see how altruism might spread if it favours close relatives, consider this. The probability that two (diploid) full brothers share the same form of a gene by common descent is 1/2 and the same probability for cousins is 1/8. With this in mind, the geneticist and polymath J.B.S. Haldane is alleged to have quipped (apparently over a beer in the—now demolished—Orange Tree Pub on Euston Road in London[15]) that he would lay down his life to save at least two of his brothers, or eight first cousins. To see why this makes sense on an evolutionary level, let us assume that there was a form of gene coding for such heroic life-saving behaviour, and repeated circumstances under which relatives might need to be saved. Under these conditions, a life-saving form of a gene would tend to spread in a population because on average more altruists would be saved in the population than are lost, while non-altruists stand by watching relatives, typically sharing the same non-altruistic trait, perish. The end result is that, all else being equal, the proportion of the population exhibiting the life-saving form of the gene would go up. While J.B.S. Haldane,[16] R.A. Fisher,[17] and other notable scientists toyed with the idea, it was Bill Hamilton who recognized its importance and who single-handedly developed a formal quantitative theory, in which he introduced the logic of 'inclusive fitness'.[18–20] This general body of theory is now known under the heading 'kin selection' (a term which we use here), although strictly speaking this refers to the narrower subset of conditions in which individuals assist other individuals that share the same copy of a gene through their close genealogical relatedness.

Kin selection works when there are multiple families or groups of individuals that preferentially interact with one another. Although natural selection may favour non-cooperative individuals over cooperative individuals within these families (after all, a brother who lacks the life-saving form of the gene might leave more offspring than a brother prepared to sacrifice himself), the proportion of cooperators may nevertheless increase if families with cooperators do better overall than families without cooperators. Kin selection *is* natural selection: while a form of a gene can spread because it enhances the carrier's own survival and reproduction (the textbook case which we all know and love), we now recognize that it can also spread in a population because it assists relatives who will tend to share copies, even if this occasionally comes at a cost to the bearer's own reproductive success.

Perhaps the easiest way to understand the logic of kin selection is through Hamilton's rule,[20,21] which states that a form of a gene that causes an individual to perform an altruistic act will tend to spread so long as $r\,b - c > 0$, where b (broadly speaking) is the fitness benefit to the recipient from the altruistic act, c is the fitness cost to the altruist, and r is a measure of relatedness. The formula is effectively a shorthand for a full population genetics model, and, strictly speaking, it is only correct if we define the terms in very particular ways. For example, r formally measures how genetically similar two individuals are when compared to two random ones in the population with which the altruist will compete for entry into the next generation (it can be negative).[22] Through its simplicity, Hamilton's rule serves to highlight the composite minimum conditions for kin-based cooperation, which are both ecological (mediated through b and c) and genetical (mediated through r). The rule also renders the implications of kin selection much more transparent: all else being equal, altruism is more likely to evolve among related individuals. However, it is important to note at the outset that high relatedness does not in itself guarantee cooperation. In particular, if all competition to enter into the next generation takes place within the family, then it is hard to envisage circumstances in which altruists would do better than non-altruists (brothers that risk their lives to save brothers, would, under these circumstances be evolutionary losers).[23]

Before proceeding, let us take the time to clear up a few potential misunderstandings, many of which were admirably dealt with by the evolutionary biologist Richard Dawkins over a quarter of a century ago.[24] Controlling for differences in genome size and the like, we share something like 98% of our DNA with chimpanzees[25] (and more than half of our genomes with bananas[26]), so we should be selected to cooperate with them, right? Wrong (well, not on the basis of kin selection anyway). The reason is that in kin selection, it is *differences* in relatedness above the average that matter, not the size of average itself (hence the caution with defining r). Similarly, one might think that kin selection dictates that an individual should share its resources much as one would divide a cake, so that siblings would get a slice that is four times the size of their cousin's, while more distant relatives would each get very thin slices. Wrong again. Kin selection has nothing to do with proportional representation—for example, evolutionarily speaking, if the costs of cooperation (c) are consistently higher than the benefits (b) to those helped, then one should favour one's own reproductive success and not be altruistic to anybody. Finally, one might also question whether there is such a thing as a form of a gene that prompts an individual to perform a given altruistic act. There are now numerous examples from the field of behavioural genetics[27] which show that all sorts of behaviours (including anti-predator behaviour, cleaning behaviour, dietary preference, and movement behaviour) are heritable and under a degree of genetic control. Moreover, the whole kin selection argument works just as well if a given genetic form simply makes it more likely that an individual would perform the act—'all-or-nothing' genetic 'hard wiring' is not necessary.[28]

So, a gene for cooperation can spread simply if it helps copies of itself get into the next generation; however it manages to do so. It is that simple. A basic prediction of kin selection theory is that altruism, when it arises, should typically be directed towards close kin, or at very least those who are likely to share the same altruism trait. This second condition is inserted because in theory, a gene that simultaneously codes for a distinguishing trait (such as a green beard, as memorably suggested by Richard Dawkins[29]) *and* a predisposition to help others, may be able to spread in a population even if those helped were not relatives.[29,30] That said, there have been relatively few good examples of 'green beard' cooperation in nature[30] (perhaps because 'cheats' that have the distinguishing trait yet lack altruism can undermine cooperation[31]). However, the basic prediction of kin-based cooperation appears broadly supported: when cases of cooperation are reported in the literature, they very frequently involve the helping of close relatives.[21]

An impressive act of kin-based cooperation can be seen in the social amoeba ('slime mould') *Dictyostelium discoideum*, in which solitary cells start to aggregate under harsh conditions to produce a 'slug'.[32] Some cells eventually produce the spores that can colonize new areas, but only at the expense of a minority of other cells that collect together to hold the spore-producing cells aloft. The cells involved in producing the colony are frequently (but not always[33]) genetically identical, so this extreme altruism can be readily understood in terms of kin selection—here helping others reproduce is tantamount (genetically) to helping yourself.

Another example of cooperation comes from alarm calls. It appears that the alarm sounds of Belding's ground squirrels, made on the sighting of terrestrial predators (such as weasels, badgers, and coyotes), reduce the caller's own survival, but help to alert relatives.[34] Why do the squirrels engage in these heroics? Well, kin selection readily provides an answer; indeed it is possible to at least approximately confirm the validity of Hamilton's rule in these circumstances. Moreover, the fact that such calls are made relatively more frequently by females (who disperse less and are therefore more likely to have relatives around them) is further indication of a role of kin selection.[34] Interestingly, the alarm calls made in the vicinity of birds of prey have a rather different effect—they tend to increase the caller's own chances of escaping (presumably by notifying the predator that they have been spotted).[35] Therefore, while calling may continue to help kin, here it also arises as a consequence of self-preservation, emphasizing that you do not always have to hurt yourself in order to help a relative. Of course, in these cases one would be hard-pressed to view alarm calling as cooperation at all, and it turns out that alarm calling does not appear to be influenced by relatedness of group members.[35]

There are other ways to test the validity of kin selection arguments. For example, one can also turn the logic on its head and ask whether physical attacks towards members of the same species are more generally directed towards non-relatives over relatives. Clearly, if you preferentially eat relatives over non-relatives then any such behaviour, if heritable, would rapidly be selected against because by eating relatives one is also likely to be removing copies of the same genes. The short answer is that if examples of kin discrimination are found, then almost invariably the interests of kin are favoured. For example, when given a choice, cannibalistic spadefoot toads[36] preferentially ingest unrelated individuals over their siblings.

When one thinks of cases of cooperation in the natural world, the colonies maintained by wasps, ants, and bees (collectively the Hymenoptera) often come first to mind (see Fig. 3.1). Here, sterile masses of individuals work by gathering food, cleaning, and repelling predators, all in the service of a small reproductive minority. Why such servitude? We have already pointed out that it is a question that also concerned Darwin. In a series of seminal papers,[18,19] Hamilton proposed that their unusual genetics ('haplodiploidy') might help to explain the prevalence of cooperation in these groups, an insight which simply bowled over the renowned biologist E.O. Wilson[37] when he first read it in 1965. Wilson's initial reaction was negative[37]: 'Impossible, I thought; this can't be right. Too simple' yet by the early afternoon, he gave up: 'I was a convert, and put myself in Hamilton's hands. I had undergone what historians of science call a paradigm shift'. To see the potential role of genetics in maintaining this extraordinary degree of sociality ('eusociality'), bear in mind that while female ants, wasps, and bees have two sets of chromosomes (one from the father and one from the mother), males develop from unfertilized eggs and therefore have only one set (hence 'haplodiploid'). One implication of this is that females are more related to their full sisters (relatedness 0.75) than they would be to their own offspring (0.5). It is easy (with hindsight) to begin to envisage how such asymmetries, created by the unusual genetics, can help to explain the prevalence of cooperation in this group. As Dawkins memorably put it, 'this might well predispose a female to farm her own mother as an efficient sister-making machine'.[29]

Unfortunately, while haplodiploidy may *help* to explain the evolution and maintenance of cooperation in the highly social ('eusocial') insects, it cannot provide the complete solution.[38,39] One problem is that eusociality is also seen in other animal species such as termites (Fig. 3.2), thrips, and naked mole rats which have standard diploid genetics. Likewise, there are plenty of haplodiploid species (including Hymenoptera, but also many species of beetles) that are not social. Indeed, once one factors in the 0.25 relatedness of sisters to brothers, the overall relatedness among siblings in haplodiploids is not especially high. Multiple queens and multiple matings with different males further act to reduce the average relatedness of offspring within a colony. While

Figure 3.2 A termite mound in Namibia. Termites (order Isoptera) are highly social ('eusocial') yet diploid species, with the vast majority of individuals forming sterile castes. The mounds formed by many species (notably in tropical savannahs) are often complex and are believed to be built in such a way to facilitate thermoregulation. Photo: TNS.

the manipulation of sex ratio and the ability of females to preferentially favour their sisters may each play a role in enhancing relatedness between the donor and recipient of cooperation, more emphasis today is placed on identifying other factors that may help tip the balance. For example, recent work suggests that while high relatedness may help in the initial establishment of eusocial colonies (and continues to help maintain cooperation), in honeybee colonies the 'harmony' of the colony may be maintained by social sanctions that reduce the number of workers that act selfishly by attempting to lay their own eggs.[40] Coercion through punishment can hardly be considered an act of cooperation, but even if relatedness does not provide the whole explanation, then it may help level the playing field considerably, making cooperative behaviours more likely.

For sexual species, sometimes it may be in an individual's direct interest to help its mate, because by helping one another, couples can collectively ensure that their shared offspring have a better chance in life. One recent example of this phenomenon comes from a study of a colony of nesting seabirds, called guillemots (see Fig. 3.3), on the Isle of May in Scotland.[41] These birds are sexually monogamous and tend to lay one egg each year on cliff edges, which are looked after by both parents. Unfortunately, the offspring

Figure 3.3 A guillemot preening ('allopreening') its mate. Allopreening is common among birds and mammals, and one might wonder why individuals engage in it when there seems to be an obvious temptation to sit back and let others groom you, without offering anything in return. Photo courtesy of Sheila Russell.

do not always survive to fledge, and they are killed for a variety of reasons—including aggression from neighbours. As if life was not hard enough, the adults can also get infested with ticks. It can be challenging for a bird to preen itself, especially around its head and neck, and the birds help one another by removing parasites from their mates and even from their colony neighbours (these latter actions tend to get reciprocated[41]). Interestingly, those sexual partners who tend to preen one another a lot also tend to have higher long-term breeding success. Of course, correlation does not demonstrate cause, but it seems likely that these simple acts of cooperation between sexual partners in removing parasites can be translated into mutual fitness benefits.

Another well-known example of cooperation may also involve both kin selection and other more direct forms of return. Providing 'parental' support to young that are not your own is reasonably common in birds, mammals, and some fish.[42] For example, in populations of birds such as white-fronted bee-eaters, Florida scrub jays, and long-tailed tits, and mammals such as meerkats and brown hyenas, there are non-breeders that help raise young produced by dominant breeders. Why don't non-breeders go it alone, rather than help look after another's offspring? Kin selection may play an important role—indeed, groups in cooperatively breeding species are typically made up of extended families, so that subordinates will often be helping their relatives. Moreover, studies have shown that helpers sometimes provide their closer kin with preferential care.[43] However, there is also an increasing recognition that direct fitness benefits may also be important in maintaining cooperation in this type of system, perhaps even more important than kinship itself.[42] In some cases, helpers may be forced into helping behaviour to avoid punishment, but by helping they may also sometimes increase their chance of inheriting the breeding territory of the breeding pair (they are 'paying the rent'). Likewise, the increased survival chances from grouping together may sometimes outweigh the costs of helping. In meerkats, for example, the foraging success and survival of all group members increases with the size of the group.[42] It is these and other non-kin routes to cooperation that we now need to consider.

You scratch my back

Kin selection may help to explain many cases in which individuals incur costs that benefit others, but as we have already seen, some of these examples involve additional phenomena that further maintain cooperation. At an extreme, how do we explain examples of cooperation among non-relatives? Included in these examples are vampire bats that regurgitate blood to feed their hungry roost mates,[44] and impala that groom unrelated individuals within the herd.[45] Perhaps by helping others, the donor might subsequently be helped by the receiver when its own need arises? Evolutionary biologist Robert Trivers presented just such an explanation, referring to the phenomenon as 'reciprocal altruism'.[46]

One of the challenges, of course, is understanding how a form of reciprocal altruism can ever be stable. Imagine a situation in which a partner helps you. What is to prevent you from 'doing the dirty' by accepting all their cooperative acts, yet never reciprocating when their need arises? To understand how cooperation might be maintained under these circumstances, mathematical modellers have spent huge amounts of time and effort (some may say too much) elucidating the types of strategy that would do well in a simple game, known as the two-person iterated Prisoner's Dilemma.[47,48] In each round ('iteration'), two players decide simultaneously whether to 'cooperate' (C) or 'defect' (D), just as two prisoners accused of a joint crime may decide to cooperate with each other by staying quiet, or defect on their criminal partnership by talking to the police in exchange for a lighter sentence. According to the defining inequalities of the Prisoner's Dilemma, mutual cooperation ('CC') pays more to both players than mutual defection ('DD'), but defecting while your partner cooperates ('DC') pays the defector most of all (reflecting a 'temptation' to defect) and a sole cooperator least of all ('CD', the 'suckers pay-off'). In a one-off game, the most rewarding strategy is always to defect because whether your partner cooperates or defects, your best option is to defect. Of course, the irony here is that if you both do the 'logical' thing, then you would both gain less than what you would have if you had both cooperated. Many scientists consider the Prisoner's Dilemma the key metaphor for understanding cooperation, primarily because it captures the temptation to defect, but also because it can reflect some of the damaging effects of pure self-interest.

In an iterated two-player game (played repeatedly, where the number of rounds is not known in advance by the players), things get a lot more interesting because the set of potential strategies is enormous—especially if long memories of previous interactions are allowed. To begin to explore the type of behaviour that would do well when players have a certain probability of meeting again, world-renowned political theorist Robert Axelrod staged computer tournaments in which people were invited to submit potential strategies (such as 'Unforgiving'—which cooperates until it receives a defection, and thereafter always defects and 'Random'—which cooperates with 50% probability).[47,49] Each strategy played against all others in the contest and against itself (a 'round-robin' tournament). It turns out that relatively cooperative strategies tend to do well overall in these types of tournament, so that partners do not necessarily have to lock into mutual rounds of defection. However, these cooperative strategies were not 'naive' in that they will not allow themselves to continue to be suckered by defectors. In particular, a strategy called 'tit-for-tat' (TFT—cooperate on first move, thereafter follow the partner's previous move) was highly successful. Although TFT did not tend to win individual matches, its points difference meant that that it performed well on aggregate, fairing relatively well against both cooperative and non-cooperative strategies. This strategy is thought particularly effective because it is nice (in that its starting move is to cooperate), retaliatory (in that it follows a defection from the partner with defection), and forgiving (in that it subsequently matches any cooperative act with cooperation). In a moment, we will discuss whether there is any evidence of organisms using TFT-like strategies in the natural world. However, we note at the outset that researchers are not implicitly assuming that players 'do the math'—instead the assumption is that successful cooperative rules can spread in a population by being inherited (if the behaviour has a genetic component), learned, or simply copied through imitation.

Before looking for evidence of reciprocative strategies in nature, we should say that theorists have since spotted some weaknesses with TFT.[50] For example, if mistakes are occasionally made (either in execution or in interpretation), two tit-for-tatters can get stuck into indefinite rounds of defection, rather like a washing machine stuck in a cycle. Similarly, if TFT ever becomes established then it can gradually be taken over (through simple chance drift) by more unconditional cooperative strategies that cooperate similar to TFT, but exhibit a disastrous naiveté if defectors ever re-enter the system. It turns out that a Win–Stay, Lose–Shift strategy [WSLS—keep to your strategy if your previous exchange was high paying, namely DC (continue to defect) or CC (continue to cooperate), but otherwise change it (i.e. cooperate if DD and defect if CD)] exhibits fewer deficiencies—it can correct occasional mistakes, yet it is also streetwise enough to sucker naive cooperators if it discovers them by chance.[51]

A word of caution. Researchers have had a field day investigating the success of new strategies. It is all too easy to get sucked into this abstract world in which, for example, 'tat-for-two-tits' competes with 'Grim Reaper' on a two-dimensional lattice. Field biologists have long been scratching their heads, asking themselves what it all means for cooperation in the natural world. In recent years the situation has improved. For example, theorists have increasingly recognized that cooperative decisions are rarely made simultaneously,[52] and even neatly alternating games remain something of an abstraction. Similarly, researchers have begun to consider games in which individuals choose how much to invest, rather than only allowing the discrete options of C or D[53,54] and they become much more interested in partner choice (you do not have to continue to play with a partner who continually defects—you can walk away).[55–57] Perhaps even more importantly, they have begun to observe that not all cooperation among non-relatives neatly falls into the Prisoner's Dilemma paradigm,[58,59] and we will deal with some of these alternatives later.

One other challenge is how cooperation gets started. Clearly a lone cooperator in a sordid pit of defectors does not stand much chance. Indeed, if the population is big enough, even two or more cooperators may never get a chance to meet. One solution is to assume that cooperators tend to associate—either by recognizing one another through their helpful deeds, or by simple spatial proximity. Axelrod showed that a small cluster of players who used TFT could establish themselves in a population of defectors, and even take over the population.[47] Many other spatially extended games have subsequently reported similar effects.[60,61] In essence, although not always noted, this spatial explanation gets us back to kin selection as a primary force, at least in terms of getting cooperation started. After all, if cooperative players interact with their nearest neighbours and those nearest neighbours are likely to contain copies of the same altruism genes, then the r relatedness coefficient (as far as altruism is concerned) becomes very high.

A good example of spatial interactions facilitating the initial evolution of cooperation may be seen in the formation of particular types of microbial mat. Populations of the bacterium *Pseudomonas fluorescens* rapidly diversify when maintained in unshaken broths,[62] and a particular form—known as the 'wrinkly spreader' (produced by single mutations with large effects[63])—tends to build up at the liquid–air interface, creating a surface scum[64] (see Fig. 3.4). The ability to live at the boundary layer allows the wrinkly spreader bacteria to avoid the oxygen-deprived conditions deeper in the water column, but it comes at some cost. To form and maintain the mat, the wrinkly spreaders have to make a cellulose polymer (glue), a metabolic cost that is not borne by other forms of the bacterium. Despite this cost, the mat can initially develop by kin selection (individuals helping to bind to the surface are initially genetically identical), but it undergoes periodic collapses when freeriders (who avoid the cost of making the glue) start to invade the mat.[64] Details of the underlying dynamic are still being worked out,[65] but the phenomenon may be of more than simply academic interest. For example, although the underlying evolutionary dynamics are likely to be radically different in different systems, biofilms can form on surgical implants such as heart valves or catheters and thereby affect human health.

Let us get back to reciprocal altruism. Can direct reciprocation explain examples of cooperation that we see around us? There do appear to be some good examples of reciprocation maintaining altruism, but not many (you can often tell when a neat idea is short of well-documented examples when different texts cite the same cases). The classical example is reciprocal blood sharing in vampire bats, in which females regurgitate blood meals to roost mates who have failed to obtain food in their recent past.[5]

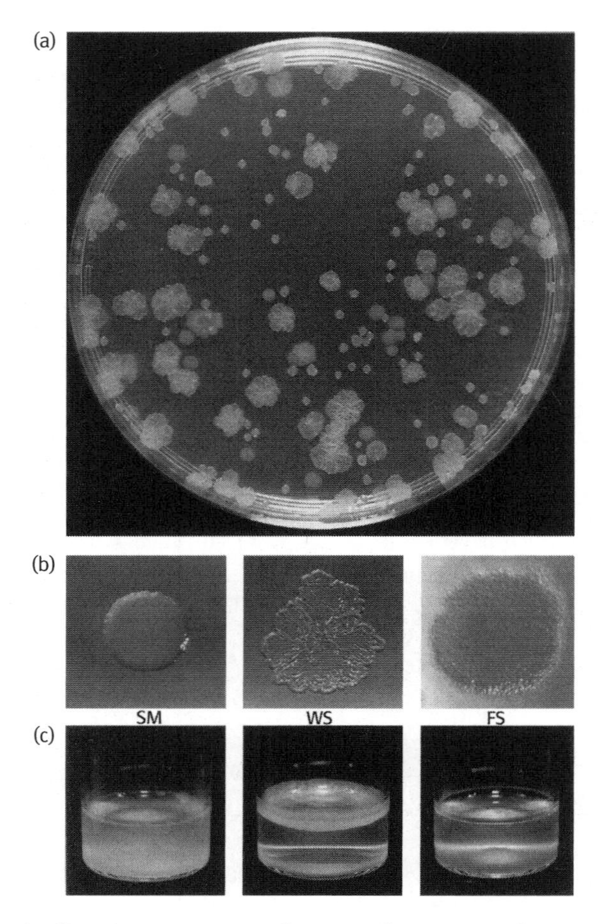

Figure 3.4 Phenotypic diversity among *Pseudomonas fluorescens* colonies. Populations in this experiment were each founded from a single 'smooth' morph (SM) cell and incubated without shaking, thereby generating a spatially heterogeneous environment. (a) After 1 week, the populations show substantial phenotypic diversity which can be seen on plating on to a flat dish. (b) Most of these variants can be assigned to one of three morph classes: SM, wrinkly spreader (WS), and fuzzy spreader (FS). (c) To gain access to the oxygen-rich surface, wrinkly spreaders pay a metabolic cost to stick to one another and the side of the beaker, in much the same way that ants cooperate to form living bridges (see Fig. 3.1). Thanks to Paul Rainey for the photo, and for the ant analogy.

Here the cost of giving for a well-fed bat (in terms of time to starvation) is considerably outweighed by the benefits of receiving for a hungry bat (individuals will starve if they are denied food for more than 60 h), so that (borrowing from Hamilton's rule) 'b' is typically much greater than 'c'. Yet while nest mates are often related, there appears to be more structure to the interaction. In particular, experimentally starved bats who received blood, subsequently gave blood to the former donors more often than one would expect by chance. Likewise, in laboratory experiments cotton-top tamarin monkeys gave more food to a trained conspecific who regularly offered them food in the past compared with an individual who never gave them food.[66]

Studies of grooming have also thrown up some clear cases of reciprocal altruism. For example, on the African savannah, impala (often non-relatives) approach one another and begin grooming.[45] Similar to the vampire bat story, the benefit, in this case removing parasites, may be high, but the costs in terms of time, fluids, and energy may be relatively low. Here, individuals hand out grooming in bouts ('parcels'[67] of 6–12 licks), and the number of bouts received and delivered are remarkably well matched. Of course, here defection can be taken to mean simply walking away or doing nothing and, while the relationship is based on reciprocation, it seems very likely that parcelling up the cooperative acts in this way helps reduce the temptation to defect.[67] Business deals often show a similar structure to avoid exploitation—half paid in advance and the other half paid when the job is complete.

Male red-winged blackbirds in North America also appear to cooperate, sometimes coming to the aid of neighbouring males in defending their nests and territories from potential predators such as American crows.[68] One possibility is that the helpers are in fact the true fathers of some of the offspring on the neighbouring territory and are selected to help out of sheer self-interest, that is, simple parental care. It is also possible that the males in neighbouring territories are related. Alternatively, or in addition, the helper may benefit directly by removing any potential predator from the neighbourhood ('not in my backyard') and any benefit to the neighbour is incidental. Such a dynamic is called a 'by-product mutualism'—assisting others comes simply as a direct consequence of helping yourself.

In each of these cases, the failure of a neighbour to help you should not necessarily elicit any form of retaliation. In contrast, one would expect some form of retaliation (such as a refusal to help them) if the birds were using TFT-like cooperative strategies—I will help mob your predators, if you help mob mine. Olendorf and colleagues recently put these contrasting explanations to the test[68] and ruled out any kin-based explanations on the basis of genetic analyses. However, they also looked for evidence of reciprocity by examining patterns of nest defence against a stuffed crow and simulating cheating by making it appear that a neighbour was not helping with the defence (a 'defection'). As anticipated, male blackbirds tended to decrease their defence against a potential nest predator after their neighbour appeared to defect in the earlier trial, suggesting that reciprocation was having an important role in maintaining cooperation.

While reciprocity seems clear cut in the earlier example, it does not necessarily imply that the blackbirds are caught in a Prisoner's Dilemma (with pay-offs ordered such that $DC > CC > DD > CD$). For example, one could conceivably argue that while seeing off a predator is always in one's own direct self-interest, and individuals would naturally prefer others do the job for them, doing the job yourself is better than nobody doing it. Under these conditions, the game becomes an iterated 'snowdrift' game (otherwise known as 'chicken'[48]) with pay-offs ordered so that DC > CC > CD > DD. Why not always cooperate in these instances? Despite the incentive to remove a common threat, one might nevertheless be prepared to occasionally pay a retaliatory cost by doing nothing, reminding others that the removal service does not come for free.

The possibility that some cooperative relationships are not the outcome of a Prisoner's Dilemma will be dealt with in a later section ('Escaping the Dilemma'). Furthermore, we note in passing that while many relationships are appropriately represented as the outcome of repeated pairwise interactions between players (similar to those discussed earlier), some cooperative relationships (such as fighting and hunting coalitions) are more appropriately seen as games with more than two players ('*n*' player games). Perhaps the best-known multiplayer game in all of ecology is the 'tragedy of the commons'[69] which is essentially the outcome of an *n*-player version of the Prisoner's Dilemma. Here is the rub: if everybody else is cooperating by not overgrazing a piece of common land (or overfishing a stretch of ocean, say), then it may pay you to introduce an additional goat (or boat) because the benefit goes directly to you, yet the cost (in terms of overexploitation) is shared among the community. Similar to the two-player game, while 'defect' is the pay-off-maximizing strategy in a one-off game, it has the tragic effect of ruining the system for everybody since everybody chooses likewise to defect. Many modern-day cases of overexploitation may be understood using just this metaphor, and while more cooperative solutions exist, this generally requires the introduction of other factors. For example, a form of self-restraint can sometimes spread over evolutionary time through kin selection (parasites may do this within hosts), but not when the members of a community are unrelated. Likewise, the medieval commons that Hardin based the term on were not so 'tragic' since they had various rules and regulations governing their use[70]—rules enforced by the threat of punishments. We will return to the role of punishment maintaining cooperation later in this chapter.

Reputation matters

By its very nature, direct reciprocity requires repeated dealings among the same sets of individuals; therefore, it cannot apply to cases of helping strangers we might never see again. Indeed, as previously noted, the only logical thing to do in a simple one-off Prisoner's Dilemma game is to 'Defect'. However, what if others were looking on? Perhaps by helping others one might gain sufficient reputation as a 'nice' individual that strangers would be willing to help you when your own need arose. So instead of 'You scratch my back, and I'll scratch yours' one could consider another, seemingly even more vulnerable, guiding principle 'You scratch my back and I'll scratch someone else's'.[71] In other words, even if helping others has a short-term cost, it may pay in the long run to help others because it builds up a reputation that is rewarded by third parties. This is called the 'indirect reciprocity'[72] route to cooperation and it may seem a bit strange at first. Bear in mind, however, that what matters is that acts of cooperation are returned, not who returns them. Here cooperation is simply a way of buying your entrance into the cooperation ball.

Examples of the importance of maintaining an untarnished reputation are widespread in human societies. For example, 'The Winslow Boy' by Terrance Rattigan[73] tells the story of young Ronnie Winslow, who was accused of stealing a postal order from his Naval College and, subsequently, expelled. On the basis of a true story, the play (and film) tells of how his family launch a long and costly campaign to clear the boy's name, believing in his innocence, and fearing the worst for their family's honour and the boy's future, if this blot on his reputation is not removed. Likewise, many large businesses have entire public relations departments dedicated to managing and promoting their reputations. In fact, there are now Internet-based companies that allow you to manage your personal online global reputation. As one might expect, there are plenty of good examples of how having a good reputation is good for opening up business with new clients. For example, eBay in part relies on reputation to maintain honest transactions when it provides scores of partner satisfaction. Being a good person or good company to deal with does not in itself explain cooperation, but it begins to suggest a role for reputation in partner choice.

Recent experimental evidence for the importance of reputation in facilitating cooperation has come from an analysis of the contributions to an 'honesty box' for drinks in a university departmental coffee room.[74] Bateson and colleagues looked at contributions to the box when images (always posted above the recommended price list) of a pair of eyes were alternated on a weekly basis with images of flowers. The amount of milk consumed turned out to be the best indicator of total consumption, but remarkably almost three times more money was paid per litre in weeks when there were eyes portrayed, compared to when there were flowers portrayed.[74] Of course this experiment was only conducted in one location, but the effect size is impressive and it seems to indicate that individuals do not want to be observed cheating the system.

Growing from earlier arguments by Richard Alexander,[75] mathematical modellers have demonstrated the theoretical plausibility of cooperation via indirect reciprocity by showing that behavioural rules can evolve in which individuals are more prepared to help strangers if these strangers have a reputation for cooperating.[72] Yet precisely how reputation is built and lost affects whether and how cooperation will spread (e.g. some models use an 'image score',[76] which simply goes up or down whether the observed potential donor cooperates or not, while others use 'good standing'[77]—which does not go down when individuals fail to help others with poor standing)—we will not go into details here. Of course, since reputable individuals tend to provide assistance to similar reputable individuals, then kin (or 'green beard') selection may also play a key role here.

Can indirect reciprocity explain cooperation in humans? Martin Nowak and Karl Sigmund report a story of an elderly academic who made it a point to attend the funerals of colleagues, his reason being that 'otherwise, they won't come to mine'.[78] Clearly, if it all works out to plan, then these apparently altruistic acts will be returned not by recipients, but rather by third party observers. Humans frequently help others who may never have an opportunity to reciprocate[79]—acts of giving to charity and donating blood come to mind. Sometimes in giving in these ways one receives badges or bracelets, and it is possible that these are interpreted as signals of generosity which bring rewards of their own. As the evolutionary biologist Manfred Milinski has noted, there is a German phrase that captures the underlying motivation 'Tue Gutes und rede darüber'—'Do good and talk about it'.[80] Nevertheless, we should not get carried away—the reasons why people give to others are varied and complex, and increasing one's reputation is surely only part of the motivation.

Staged laboratory games support the view that indirect reciprocity can play a role in facilitating cooperation among humans.[81] For example, in a recent experiment,[82] human subjects were repeatedly given the opportunity to give money to others (at a slightly lower cost to themselves so that $c < b$), having been informed that they would never knowingly meet the same person with reversed roles (all donations were anonymous). Yet the history of giving and not giving were displayed for participants to see at each interaction. It would be a pretty strange (and boring) outcome if the participants simply kept their money (a control experiment with no public information might potentially have produced this result according to a one-off PD), and indeed volunteers did tend to give to others. More importantly, the authors found that donations were significantly more frequent to receivers who had been generous to others in earlier interactions.

There are far fewer examples of indirect reciprocity in non-humans, and even here they include examples of cooperation between species rather than within species. One recent example comes from work on cleaner fish mutualisms.[83] The cleaner fish *Labroides dimidiatus* remove skin parasites from their fish clients, but there is an apparent temptation for them to take a little more at the expense of the client by feeding on the client's mucus. Clients are faced with the challenge of getting cleaners to feed against their preferences if they are to come away unscathed. Field observations indicate that client fish almost always invite a potential cleaner to draw closer and inspect them if they have had the opportunity to see that the cleaner's previous interaction ended without conflict. In contrast, the clients invite particular cleaners far less frequently if they observed that the last interaction of the cleaner ended with conflict, such as a chase away. In some follow-up experiments, Bshary and Grutter[84] found that clients allowed to observe cleaner fish in staged encounters, spent more time next to 'cooperative' cleaners than next to cleaners with an unknown cooperative level (see Fig. 3.5). Therefore, a good reputation is good for the cleaner's business and it may be an important way in which clients avoid 'defections'.

The study of information gathering by animals ('eavesdropping') for strategic purposes is still in its infancy.[85] Clearly, we need more examples of indirect reciprocity in natural systems before biologists as a whole can take it really seriously. Perhaps it places too many cognitive demands on individuals to be a widespread route to cooperation in non-human societies?[72] However, the whole area of indirect reciprocity is a lot less intuitive than direct reciprocity and for that reason it is hugely intriguing for theoretical

Figure 3.5 Clients and their cleaners at Ras Mohammed National Park, Red Sea, Egypt. The picture shows a pair of cleaners *Labroides dimidiatus* inspecting a much larger long-nose parrotfish, *Hipposcarus harid*, with another long-nose (who can potentially observe the interaction) nearby. Recent work has shown that clients preferentially associate with cleaners who appear to have done a good cleaning job on others. Photo courtesy of Redouan Bshary.

biologists. Benjamin Franklin once quipped: 'It takes many good deeds to build a good reputation, and only one bad one to spoil it'. Who knows, maybe theoreticians will one day be able to put some even more exact numbers on this idea.

Cooperate or else...

There is a sense of 'fair play' in human societies. For example, people regularly challenge others when they 'queue jump' (typically but not always when they have jumped ahead of the challenger, indicating that self-interest is part of the motivation). In the well-known 'ultimatum' game in economics, two players A and B have to agree on how a monetary reward has to be shared.[48] Player A (the proposer) has one chance to suggest how the money is to be shared (e.g. 60% to player A, 40% to player B), but player B (the responder) can accept or reject the proposed division. If the bid is rejected, then both receive nothing, but if the bid is accepted then the proposal is implemented. What would you propose if you were given 100 dollars (or whatever your local currency)? The logical optimum is to offer the responder an almost negligible amount (1 cent say, because 1 cent is better than nothing). Perhaps you would be rather more generous and keep 90% yourself, providing a 10% 'sweetener' to keep the respondent happy? Even this may be insufficient, because a common result in this type of game is that responders tend to reject proposals if the offer is anywhere less than about 25% (even when the sum is quite considerable).[86] This suggests (assuming they have understood the game correctly) that individuals are prepared to incur costs (or at least forego benefits), simply to punish others that appear to behave unfairly.

Other primates may also exhibit what we might think of as a sense of justice. For example, capuchin monkeys will voluntarily share treats with other monkeys that helped to secure them.[87] In an intriguing paper entitled 'Monkeys reject unequal pay',[88] Brosnan and de Waal investigated what happens when capuchin monkeys previously trained to exchange a pebble for a piece of cucumber started to see others being rewarded with a more favoured food (a grape)—the monkeys tended to go on strike, refusing to exchange a pebble for cucumber even though the alternative was no reward at all.

Perhaps punishment can play a role in maintaining fair play and hence cooperation? We have to be careful here, because some may not see it as cooperation at all, bordering more on enforced slavery than on acts of 'kindness'. Nevertheless, when we see apparent examples of altruism we need to ask whether the threat of punishment is helping to maintain it.

Punishment may be a more common factor maintaining cooperation than we think. Recent work on captive pigtailed macaques indicated that a small subset of individuals engage in occasional policing activities, that is, they intervene (seemingly impartially) in conflicts between others.[89] By temporarily removing these key individuals, the extent of policing was lessened and as a consequence the group became an altogether less hospitable place to hang out in, with grooming and play reduced and spatial distances increased.[89] Therefore, it appears that policing not only prevents conflicts, but also helps to bind societies together.

University undergraduates are suitable subjects to work with when exploring theories of cooperation because they are readily available to researchers, and they are usually up for anything from smelling sweaty T-shirts[90] to playing computer games.[91] In a series of staged repeated games among human volunteers, Fehr and Gächter showed that students are prepared to take on costs in order to punish those who had earlier shirked their opportunity to contribute to a public good.[92] According to Fehr and Gächter, this altruistic punishment is simply a consequence of a 'negative' emotional reaction to the sight of somebody freeriding,[92] although there may be positive pleasure in seeing 'justice' done. As one might expect, those that were punished for not contributing learned their lesson and cooperated more in subsequent rounds. Moreover, games that prevent altruistic punishment altogether see a marked reduction in the mean amount of cooperation over time. The behavioural tendency to punish non-cooperators and thereby maintain cooperation has been called 'strong reciprocity'.[93,94]

Strong reciprocity may explain many examples of human-based cooperation, but it is difficult to understand how altruistic punishment might evolve as a consequence of natural selection. After all, if altruistic punishment is costly, then an individual who freerides and lets others do the policing would tend to leave more offspring. A recent study staging repeated games among university students, found just this—costly punishment puts the individual who uses it at a disadvantage, so 'winners don't punish'.[95]

The temptation to sit back and let others punish defectors has been termed a 'second-order defection'[10] or 'twofold tragedy'.[96] Therefore, if strong reciprocity can explain cooperation, perhaps it has only replaced the problem with another one further down the line—why should you be the one to punish?

Kin selection may provide one potential solution to this question, but note that kinship can reduce the underlying incentives to defect in the first place.[96] For example, in many eusocial Hymenoptera, worker-laid eggs are killed by the queen and other workers, thereby reducing the incentive of workers to cheat. In a comparative analysis, Wenseleers and Ratnieks[40] found that fewer workers reproduce when the effectiveness of policing worker-laid eggs was higher, indicating that these sanctions were an effective deterrent. However, higher relatedness among colony workers lead to less policing, not more, a result which is consistent with the view that less policing is needed when workers are highly related. As Hammerstein argues, self-restraint based on kin selection can achieve for free what expensive policing could bring about.[96]

Perhaps apparently altruistic punishment has evolved for reasons other than promoting cooperation, such as coercing individuals into submission and establishing dominance hierarchies.[95] In some situations certain punishers may have more to gain than others, at least partly getting around the issue of who should do the policing. For example, social 'queues', in which subordinates wait for their turn to inherit dominant breeding status, are seen in a wide variety of organisms ranging from wasps to fish. Sometimes, as in the case of a number of fish species, the hierarchy is maintained by individuals exercising control over their growth rates and never being large enough to pose a physical threat to individuals higher than themselves in the pecking order.[97] Why don't fish try jumping the queue, just as humans occasionally do? The answer to both parts of the question may be similar in both fish and humans: subordinates who transgress by growing larger have a high chance of being forcibly evicted by dominants.

Another intriguing example of the interrelationship between different ways of maintaining cooperation comes when we consider wars fought by humans. As we all know, wars are violent conflicts between two or more groups of (generally) genetically unrelated individuals. While ant colonies can engage in battles over territories, it appears that no other species besides humans has succeeded in creating large-scale cooperation among genetically unrelated individuals.[98] Why (except in a small minority of cases) do individuals not attempt to avoid the costs of fighting, while enjoying the benefits when his/her countrymen come back victorious? Punishment may play a key role—individuals can be tried for desertion if they attempt to avoid fighting, but there may be more subtle solutions. Panchanathan and Boyd[99] recently analysed one solution to this general problem by considering the importance of an individual's reputation for contributing to the collective good—we are back to indirect reciprocity again. Simply refusing to help those who have been observed not to contribute (with no loss to one's own reputation) can help maintain cooperation through indirect reciprocity. So, while active punishment may help keep individuals cooperating, denial of rewards may play a similar role.

Escaping the dilemma

All adaptive explanations of altruism involve taking the 'altruism out of altruism', either by showing how the actions can selectively benefit individuals carrying the same traits, or by showing how the nature of the interaction is such that it is in the interests of the altruist to cooperate. However, this commonality should not be taken to mean that all cooperation can be related back to the Prisoner's Dilemma.

The Axelrod and Hamilton paper on cooperation is rightly regarded as a scientific classic[49] but as David Stephens put it in 1995, 'We've been trying to shoehorn every example of cooperative behavior into this Prisoner's Dilemma since 1981'.[100] As we noted earlier, there is a distinct possibility that many cooperative interactions are not based on this game at all: some may not even involve a temptation to defect. Take for example lions. In a series of experiments using acoustic playbacks conducted by Heinsohn and Packer,[101] female lions showed differences in the extent to which they participated in warding off potential intruders. Some individuals consistently led the approach, while others consistently lagged behind. However, there was no punishment for these laggards (save occasional apparent 'dirty looks'), indicating that 'loafers' were tolerated and that the costly act of defending a territory was not maintained by reciprocity. Perhaps the laggards contribute to the group in other ways, or the leaders (so long as there are not too many of them) stand to gain more than the loafers (rather like a producer and scrounger game[102,103]). Whatever the reason, it is clear that not all acts of cooperation (defined as costs that result in benefits to others) have their roots in the Prisoner's Dilemma.

Another fascinating example of cooperation which is almost certainly not represented by a Prisoner's Dilemma comes from recent work on a species of fiddler crab on the northern coastlines of Australia, where males aggressively defend their burrows from wandering males (intruders). Backwell and Jennions[104] found that male fiddler crabs may sometimes leave their own territories to help neighbours defend their territories against these floating intruders. Why be a good Samaritan? It turns out that reciprocity cannot explain it because the ally that came to the neighbour's assistance was always bigger than the neighbour itself.[104] Here it may directly benefit a resident to help its neighbour to defend a territory so it can avoid having to renegotiate the boundaries with a new and potentially stronger individual.[105] In this way, there is no temptation to cheat—large allies are helping themselves, and it is only incidental that helping the neighbour keep its territory is part and parcel of maintaining *status quo*.

Partner choice may provide another way in which the Dilemma is avoided. In the classical Prisoner's Dilemma game, players have an option of cooperating or defecting. However, if players are able to terminate a relationship at no cost and can look for another partner, then cooperation can readily spread.[106] The reason is simple: cooperators getting on well can stick together for longer and 'defectors' can simply go hang. The mechanism is similar to that of reputation building and indirect reciprocity, but here an individual learns about its partner directly by interacting with them repeatedly. In essence, a competitive biological 'marketplace' will almost inevitably help extract greater degrees of cooperation from potential partners[55–57,107] and moves the game to one in which defection is no longer a profitable option.

What if apparently cooperative behaviour was in some way an unintended by-product of some other activity? As noted in our introduction, we do not treat such examples as *bona fide* cases of cooperation and neither do several modern commentators[3] because the behaviour has not arisen in order to benefit the recipient. A vulture flying to some corpse may incidentally attract lions, or a bird hunting for fish may incidentally flush prey to another predator. Connor[108] (with Leimar[109]) called this a 'by-product benefit' and defined it as occurring when 'an individual benefits as a consequence of traits in another, which traits have *not* evolved for the purpose of influencing the individual'. While by-product benefits may be common in natural systems (one man's garbage is another man's gold, an idea crucial to understanding nutrient cycling), theoreticians have not spent much time on understanding the dynamic because it is frequently seen as 'a solution without a problem'. Nevertheless, this does not make such behaviours unimportant or uninteresting. Indeed, our interpretation of cooperation gets tested further when we observe that some individuals may actually pay a cost to acquire or enhance the by-product benefits produced by another (a phenomenon known as 'pseudoreciprocity'[109]). Many lycaenid caterpillars, for example, produce sugary secretions that are consumed by ants and in turn the ants protect these individuals from predation. The sugar may be viewed as an investment, yet the protection may (arguably) be part of general territorial ant defence. Producing the sugar fits the definition of cooperation (it is made to benefit the ants), and the dynamic even includes elements of reciprocity (caterpillars clearly ramp up the delivery of droplets if they perceive imminent threat from predators[110]). However, the relationship is rather asymmetric, in that the ants may be doing little more than protecting their assets.

Therefore, the bottom line is that some types of cooperative behaviour do not represent the solution of a Prisoner's Dilemma. Lions can lead and crabs can collude whether or not it is reciprocated, so some forms of cooperation may not be undermined by cheaters who do not cooperate. The reasons are various, but just because we see an organism benefiting another does not, in itself, mean that the behaviour is in some way open to exploitation.

Conclusions

Cooperation is a fascinating area of enquiry because it touches so many different disciplines: biology, economics, sociology, psychology, and even theology. As we have seen, there are several different ways in which cooperation can emerge from sheer self-interest, yet many examples of cooperation in the natural world remain open to several interpretations. Kin selection may be at the heart of much of the intraspecific cooperation we see in the natural world, but it is not the only mechanism, and sometimes cooperation can be maintained by a complex interplay of several different factors including reciprocity, partner choice, and the threat of punishment.

Bernard Mandeville (who penned our opening quotation)[1] observed, 'One of the greatest reasons why so few people understand themselves, is, that most writers are always teaching men what they should be, and hardly ever trouble their heads with telling them what they really are'. Here we have taken Mandeville's advice by trying to understand why animals and other organisms cooperate, rather than make any judgment about how they should morally behave. This is not always easy—terms such as 'defect' and 'cheat' evoke negative reactions, perhaps in part because we are social species. On the flip side, just because self-interest can have the unexpected consequence of occasionally promoting cooperation, then this does not mean that we should encourage self-interest. Mandeville likewise did not believe that vice should be encouraged, but merely that some vices 'by the dextrous Management of a skilful Politician may be turned into Publick Benefits'.

4

Why Species?

Figure 4.1 The natural history store room in the Kendal Museum, UK. Photo: DMW.

Why is not all nature in confusion, instead of the species being, as we see them, well defined?

—*Charles Darwin*[1]

It is quite possible to think of a world in which species do not exist but are replaced by a single reproductive community of individuals.

—*Ernst Mayr*[2,3]

The living world is not a single array of individuals... but an array of more or less distinctly separate arrays, intermediates between which are absent or usually rare.

—*Theodosius Dobzhansky*[4]

In this chapter, we will attempt to address several interrelated questions about species and species formation. First we ask what, if anything, is a species? As we shall see, while most scientists are happy to agree on the essentials, the answer to this question is far from straightforward. We then briefly discuss the range of ways new species can evolve, and provide evidence for these different pathways. Finally, following from our opening quotations, we ask a somewhat more abstract and philosophical question that brings together many of the separate threads we have introduced: why is life not composed of a single species?

What is a species?

The classification of organisms into species is so familiar that it is easy to accept without much critical thought. On reading 'Tiger, tiger burning bright', or headlines such as 'Man bites Dog', we have no problem envisaging who the main protagonists are. Mention a tiger, and one immediately thinks of a large cat with stripes. To most people, species are simply a collection of organisms with a given set of physical traits. All classification systems include elements of personal preference as to how one chooses to classify any group of objects (e.g. by shape, size, or colour). However, there is evidence that 'species' represent categories that are more consistent between observers than the various ways of sorting out one's stamp collection.

The Fore, a highland people of New Guinea, are perhaps best known in the western world for the devastating prion-based disease 'Kuru' that afflicted their population as a result of ritualized consumption of dead family members. However, the people have close links to their natural environment and a remarkably detailed system of classifying the larger animals they see around them. In an early study to test the degree to which species assignations are consistent among peoples with different backgrounds,[5] Jared Diamond compared the Fore nomenclature with that developed by European taxonomists. Birds found regularly in the Fore territory were divided by the Fore into 110 distinct types, and by zoologists into 120 types, with an almost exact one-to-one correspondence between Fore 'species' and taxonomists' 'species'. So, as birds go, the Europeans and New Guineans perceived remarkably similar units. Somewhat surprisingly, the Fore had no detailed classification of butterflies, all of which were lumped together despite striking species differences—quite possibly because the Fore had no practical use for discriminating among these types.[5]

The widespread similarity between folk and scientific designations of species suggests that species assignations tend to be independent of particular cultures, although this in itself does not rule out common perceptual or cognitive biases among humans. Whatever the underlying reasons for the relative consistency of species designations, using species names has proved extremely important for biologists. As we will see, even closely related species can differ markedly in their ecology. Therefore, without some form of label for the organisms studied, it would be almost impossible to convey meaningful results. Yet ironically, a formal definition of species that satisfies everyone has so far not been possible.

To see why the term 'species' has been so difficult to pin down, let us briefly consider some alternatives. First, could we not just formalize the 'look-alike' criterion we invoked earlier to get a definition of the term? Unfortunately, while practical, it does not always fit most people's notion of what a true species is. Great Danes and Chihuahuas look dramatically different from one another, as do Arabian horses and Shetland ponies, but would one really want to consider them separate species[6]? Likewise, male birds of paradise are often brightly coloured yet the females with which they mate are often far less-elaborately coloured: would one wish to use this clear morphological difference to split the two sexes into two distinct species?

Many people would point to the fact that as long as individuals interbreed then they are of the same species. The Swiss physicist and snail enthusiast Albert Mousson had argued much the same thing over a century ago, proposing in 1849 that 'the species is the total of individuals, interconnected by descent and reproduction, maintaining unlimited reproductive capabilities'.[6,7] Indeed, one of the most celebrated and popular definitions of a species, formulated by Ernst Mayr, takes much the same approach. Mayr, who was to become one of the twentieth century's leading evolutionary biologists (and longest lived—he died at age 100 and was publishing books and articles until the end) proposed, 'Species are groups of actually or potentially interbreeding natural populations, which are reproductively isolated from other such groups'. The definition, emphasizing the biological importance of reproductive isolation, is called the 'Biological Species Concept' (BSC).[8,9]

So, while the 5-year-old twin children of one of the authors can confidently distinguish at least 10 types of whale based on their appearance, it is the fact that members of each type of whale are reproductively isolated from one another that makes them species? Well, err, not quite. For example, blue whales and fin whales occasionally hybridize. One such hybrid, born in 1964, was notoriously tracked down to Japanese meat markets in 1993.[10] Similarly, lions (*Panthera leo*) and tigers (*Panthera tigris*) have been known to interbreed (almost always in zoos, although their ranges may overlap in western Asia[11]) creating 'ligers' (male lion × female tiger) and 'tigons' (male tiger × female lion), both with intermediate lion and tiger characteristics. It has been estimated that on average about 10% of recognized animal and 25% of recognized plant species hybridize with at least one other species.[12] In birds, 895 (9%) out of 9,672 described species are known to have produced at least one hybrid with another recognized species.[13] Clearly, *if* one were to treat animals such as lions and tigers as separate species (as well as many other similar cases), then one cannot take too hard a line in the application of the above definition—alternatively, one needs to use a rather different one. Recent reviewers (and, to varying degrees, the originators of the BSC themselves) have argued that the term 'reproductively isolated' should not be taken too strictly, allowing for limited gene exchange.[9] Of course in so doing, one's cut-off for how much gene flow can be tolerated immediately becomes fuzzy, but that may well reflect the continuous nature of speciation itself. New species do not tend to arise overnight (although see our forthcoming sections on hybrid speciation and polyploidy), and cases with a limited degree of hybridization could arguably be considered incipient species.

How can what we think of as a 'species' continue to remain distinct despite the potential for genetic exchange between them? The relative rarity of hybrids at the population level in the wild suggests that crossings between recognized species do not occur very frequently, and/or that the hybrids are themselves relatively unsuccessful at producing viable offspring. Perhaps the most dramatic examples of hybrid formation come from the study of 'hybrid zones', the geographical juncture, sometimes a matter of a kilometre or less, where two species meet. Formally such species are known as 'subspecies' because of the incontrovertible evidence of genetic exchange. The hybrid zone between the all black carrion crow *Corvus corone corone* (Fig. 4.2) and the grey and black hooded crow *Corvus corone cornix* (the third name to denote subspecies status) extends throughout Europe—from Scotland to Italy.[14] The zone is typically 50–150 km wide, with hooded crows populating the north, and carrion crows populating the south. The two subspecies are thought to have arisen following an earlier period of population

Figure 4.2 An inquisitive carrion crow (*Corvus corone corone*) on Dartmoor, UK. Photo: DMW.

isolation and the zone therefore represents an area of contact between two previously isolated populations. Hybrids are common within the zone and have a more or less intermediate appearance between the two types.[15] If hybridization is reasonably frequent, then why does it not gradually erode the differences between these (sub)species? The fact that the crows do not disperse widely may have something to do with it. However, a much more important factor helping to maintain the differences may be 'assortative mating' in which carrion crows prefer to mate with carrion crows over hybrids or hooded crows, and *vice versa*[15] (more on this phenomenon later).

Note that when hybrids form they typically involve the combination of genetic material from two very different sources. Therefore, hybrids, with little history of adaptation and a combination of two genetic systems, are often less viable than either of the 'pure' stocks. A dramatic example of this, again drawn from the study of hybrid zones, comes from the fire-bellied toads *Bombina variegata* and *Bombina bombina* (taxonomists cannot quite bring themselves to give them subspecies status in this case). Their levels of genetic and morphological differentiation suggest that the *Bombina* have evolved separately for many millions of years, yet they interbreed readily across a hybrid zone ('a genetic no-mans' land'[6]) less than 10 km wide but about 5,000 km long, running from eastern Germany all the way to the Black Sea.[16] It appears that the zone is maintained because there is strong selection against hybrids due to their relative inviability, but also because of differences in the two species habitat preferences: *B. variegata* is a puddle breeder, while *B. bombina* lays its eggs in larger semi-permanent ponds.[17]

With all this complexity, one might wonder how taxonomists sleep at night, wondering whether the type they have just labelled a species is actually a true species according to the BSC or some other related definition. Sometimes taxonomists get it wrong. For example, the father of modern scientific nomenclature, Carl Linnaeus, originally classed different-looking male and female mallard ducks as different species, correcting this later.[18] Only recently, researchers have begun to recognize six species of giraffe when previously they believed there was only one (*Giraffa camelopardalis*),[19] although many subspecies have been described in the past. Particularly problematic are cases in which types are similar in appearance but do not occur together, so it is difficult to know whether they would be capable of successful reproduction if they ever met. The answer to how taxonomists sleep at night is that the species label is simply a working hypothesis proposing that should this type meet another type then they will not regularly and successfully interbreed.

Just before one starts to get comfortable, the species concept gets weirder. Sometimes (sub)species A can hybridize with (sub)species B, which can hybridize with (sub)species C, but subspecies A and C cannot hybridize—collectively, such species are called 'ring species'. Therefore, A and C may be fully fledged species with respect to one another, but not to their intermediates. One recently researched example is the greenish warbler *Phylloscopus trochiloides*, which has an almost continuous circular chain of intergrading forms around the Tibetan plateau in central Asia—a gap has subsequently opened up, most likely as a result of human-induced habitat changes.[20,21] There are six recognized subspecies (including the dull-green warbler, the green warbler, and the bright-green warbler), yet DNA analyses (coupled with behavioural observations on song) suggest that the 'terminal' populations of *Phylloscopus trochiloides viridanus* and *Phylloscopus trochiloides plumbeitarsus* coexist without interbreeding in Central Siberia, raising them to the rank of genuine species according to the BSC.

One of the major strengths of the BSC is its tractability—under it, the study of speciation becomes that of understanding mechanisms by which reproductive isolation evolve. As Coyne and Orr have noted,[9] pretty much every paper concerned with the mechanisms or genetics of speciation implicitly or explicitly uses a form of the BSC. Nevertheless, the BSC does not always identify cases of what we might think of as species, most especially if we interpret the phenomenon of reproductive isolation too rigorously. As noted previously, these grey areas are not necessarily problems with the underlying definition, but may reflect the biological reality that speciation is a continuous process and that present-day populations may be at intermediate stages. That said, and partly as a consequence of these perceived limitations, other definitions (grandly called 'Concepts') have been proposed,[22] with alternative leading contenders including the Genotypic Cluster Species Concept (GCSC), which considers species as a (morphologically or genetically) 'distinguishable group of individuals that has few or no intermediates when in contact with other such clusters'.[23] Of course, we are still left with the inevitable fuzziness (how few is few?), but some see this competing concept as being somewhat less restrictive—principally because it avoids mentioning one of the key factors (reproductive isolation) that may bring these genetical differences about. The approach has its own problems, but it may prove a better guide to 'species' in those groups that regularly reproduce without sex, to which we now turn.

Species concepts in taxa with little or no sex

As noted in Chapter 2, many species can reproduce without sex simply through the production of unfertilized eggs (parthenogenesis), budding, and a range of other mechanisms. How on earth does one define species in these so-called 'uniparentals' or 'asexuals'? First, it is important to note that not all recognized asexual species are obligatorily asexual (Chapter 2). For example, in most well-known asexual plants (such as dandelions and blackberries) and animals (such as some lizards and aphids), the recognized species either have sexual forms at some times of year or in some places, or they have arisen as a consequence of crossing two different sexual species. Therefore, in these particular cases, the traditional BSC approach may continue to help define what a species is.

Many bacteria, although traditionally regarded as asexual, occasionally exchange genes with close and (less likely) more distant genetic relatives, through processes that include recombination following conjugation,[24] and by the uptake of 'naked' DNA from the environment.[25,26] For example, penicillin resistance may have been passed between *Neisseria* and *Streptococcus* by recombination of elements of the two bacterial species genomes.[27] The preferential recombination of bacteria with close genetic relatives can, in theory, help maintain morphological or genetic clusters in much the same way that more frequent sex can. Another important force maintaining the clustering may be direct natural selection, with more extreme variants effectively purged. Whatever the mechanisms, it turns out that the variation within bacteria is far from continuous, so that it is possible to identify clusters in shared morphology ('phenotype'), or shared DNA, we might think of as 'species'.[25,26]

One particular working definition of a prokaryotic species that adopts the cluster approach is 'a group of strains that have some degree of phenotypic consistency, exhibit at least 70% DNA–DNA hybridization, and greater than 97% 16S [ribosomal] rRNA sequence similarity'.[28,29] The definition identifies species through quantitative rules, but it is worth noting that under these criteria humans and chimpanzees would probably be classed as the same species. Some have argued for an even more subtle approach to identifying species in microorganisms that relates more closely to the BSC, noting that despite relatively free incorporation of novel genes, there are some 'non-exchangeable' genetic sequences between lineages since they would kill or harm the recombinant form.[30] Therefore, one way of thinking about a species from a bacterial perspective may be to consider 'reproductively isolated sequences',[30] although the validity of the approach needs further evaluation.

With the application of molecular genetics to a variety of biological problems, there is likely to be a growth in definitions and analytical methods to identify asexual species, but it is of interest to see how researchers currently identify species of bacteria. Typically the designation of species has been based on sharing common appearance features; increasingly frequently it is based on sharing a high proportion of genetic material and yet occasionally it is based pragmatically on shared ecology or mode of action. For example, it seems that bacteria that are very different genetically have nevertheless been lumped together as *Legionella* simply because they have something important in common: they all cause Legionnaire's disease.[31]

One example of the application of a number of different rules to identify species of microorganism is seen in the eukaryotic testate amoebae, so called because of their characteristic shell (a 'test')—see Fig. 4.3. The testate amoebae are widely considered asexual,[32] but there is some evidence for occasional sex[33] with many observations of apparent conjugation dating back to the nineteenth century when people seemingly had more time and leisure for observation.[34] As a predominantly asexual taxonomic group, there is the usual question about what constitutes a species. Some limited molecular work is starting to be done, but currently their taxonomy is based mainly on morphological characteristics—their shells provide more visual characters to use than most microbes, although shell shape can change somewhat with environmental conditions.[35] Even with more molecular data to inform species designations, there is still the thorny and somewhat arbitrary question of how much difference justifies a new species. This in itself brings the familiar challenge of reconciling species based on DNA with the more traditional morphospecies concepts, since the vast majority of

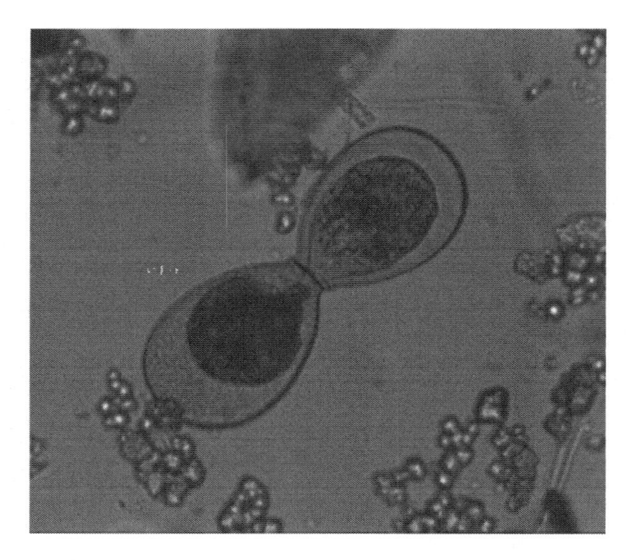

Figure 4.3 Division by fission in the testate amoeba *Nebela tincta*. The offspring takes a newly synthesized shell, while the parent keeps the old shell. Photo courtesy of Edward Mitchell.

ecological information that has been collected applies to morphospecies not molecular species.

In summary, one cannot escape the fact that the scientific community has evolved a range of different species definitions for both sexual and asexual groups, with different researchers preferring the definition that most satisfies their aims and which most suits the patterns of variation they need to explain. One such group is palaeontologists, who cannot evaluate whether groups of fossils might have been capable of interbreeding, and have therefore adopted the 'morphological species concept' (an example of the GCSC) in which species are defined and distinguished on the basis of variation in their morphology.[36]

Writing in the second chapter of *On the Origin of Species,* Darwin summed up his position which might apply equally today[37]: 'No one definition has as yet satisfied all naturalists; yet every naturalist knows vaguely what he means when he speaks of a species'.[1] Indeed, in the case of large multicellular sexual species it is increasingly clear that the different definitions tend to produce more or less the same outcome when it comes to recognizing individual species. For example, a recent survey of the plant literature, a group well known for its taxonomic challenges, found that 70% of recognized taxonomic species and 75% of phenotypic clusters (based on morphological appearance) also constitute species based on the BSC approach (i.e. reproductively isolated lineages).[38] Most heated disputes arise exactly where one would expect to see them—where putative species do not differ much in appearance or genetic make-up, and/or when there is a degree of genetic exchange.

A river runs through it—allopatric speciation

So that is how species are defined; but how do new species form? It is abundantly clear that genetic differences among populations, along with the potential for reproductive isolation, can be enhanced by long periods of geographical separation—'allopatry'. An early advocate of the importance of significant geographical barriers in driving speciation was German naturalist and explorer Moritz Wagner, who, working before the publication of *Origin of Species*, was struck by the distribution of different species of wingless darkling beetles on different sides of rivers.[39–41] Likewise, Alfred Russel Wallace noted that the geographic boundaries of primate and bird species in Amazonia tended to coincide with major rivers.[42,43] These ideas have become incorporated into a rule that has become known as Jordan's law, although its namesake David Starr Jordan was quick to recognize the priority of his predecessors[44]:

> Given any species, in any region, the nearest related species is not to be found in the same region nor in a remote region, but in a neighboring district separated from the first by a barrier of some sort or at least by a belt of country, the breadth of which gives the effect of a barrier.[45]

Several different versions of allopatric speciation have been presented, and all, by definition, involve long periods with no gene flow between the populations of interest due to some form of spatial separation, or temporal separation of reproducing forms. For example, 'vicariant speciation' (derived from the Latin for 'interchange') is said to occur when a single reasonably large population becomes split into two or more populations due to the appearance of a significant barrier to dispersal such as an ice sheet, a desert, a river, or mountain chain. Many examples of vicariant allopatric speciation have been proposed, including the different species of snapping shrimp in the Caribbean and Pacific that arose when the population was cleaved apart following the gradual formation of the Isthmus of Panama about 3–10 million years ago[46,47] and the evolution of related seaweed species in the North Pacific and North Atlantic Oceans.[48] These examples and others clearly show how new species can evolve following sufficient periods of isolation of separated populations.

In the 'peripatric' version of allopatric speciation, a few dispersers by chance overcome a major geographical barrier such that at least one of the separated populations is small. A fascinating recent example comes from the study of lupins (genus *Lupinus*) in the high elevation Andes in South America.[49] Over 80 related species of *Lupinus* are found in this area and almost certainly arose as a consequence of the uplift of the northern Andes, effectively creating competitor-free islands that could be colonized by lupins, with populations gradually diverging from one another over time.[49] Perhaps the best-known examples of peripatric speciation, however, come from oceanic islands. Textbook cases include the radiation of Darwin's finches in the Galápagos and the radiation of drosophilid (fruit) flies in the Hawaiian Islands, both of which are thought to have been facilitated by geographical isolation.[9] On-going collaborative work in one of our own research groups has looked at the distribution of about 30 endemic species of damselfly within the Fijian archipelago, with each species, based on both morphology and genetic analyses, classed within the genus *Nesobasis* (Fig. 4.4). What is remarkable is that while individual species are relatively broadly distributed within the islands where they are found, no species is found on *both* the main islands Viti Levu and Vanua Levu. Moreover, several closely related (sister) species are found on different islands,

Figure 4.4 An array of Fijian damselfly species, all from the genus *Nesobasis*. *N. malcolmi* (left upper panel), *N. unds2* (as yet undescribed species 2, left middle), *N. erythrops* (left lower panel), *N. rufostigma* (right upper panel), *N. brachycerca* (right middle), and *N. heteroneura* (right lower panel). Photos courtesy of Hans Van Gossum.

further supporting Jordan's law and indicating a role of geographical isolation in species formation.[50]

What is it about isolation that facilitates speciation? When populations become separated, then different mutations in these populations can arise by chance and spread by selection and/or by genetic drift. First, it is possible that the form of natural selection is similar in both populations, but by chance they evolve different solutions to the same challenges. Second, it is possible that the isolated populations experience subtle differences in their environment, so that factors such as climate, salinity, or competitors cause them to experience slightly different forms of natural selection and that this helps drive the two populations apart. Third, it is also possible that the two populations experience rather different forms of sexual selection—for example, females in the two populations might evolve a preference to mate with a rather different type of male, so that differences between populations can be driven by mate choice. Note, however, that invoking sexual selection only pushes the problem back a step to asking how such differences in preference arose in the first place.

The small size of populations involved in peripatric speciation might lead one to suspect that the chance genetic make-up of the colonists ('founder effects'), coupled with their subsequent chance drift, plays some role in facilitating speciation. This founder-effect mechanism of speciation (often called the 'peripheral isolate theory') enjoyed considerable popularity for several decades of the past century, advocated by Ernst Mayr and influenced by his comprehensive observations of New Guinean birds (including paradise kingfishers that show striking differences in appearance when they are found on small islands, compared to their relatives on the mainland[51]). Here is not the place for a detailed critique of the theory, observations, and experiments to evaluate the role of founder effects and drift in speciation. Suffice to say, many researchers are currently sceptical that founder effects *per se* play an important part in speciation,[9] in part because the very traits that facilitate reproductive isolation are typically under strong selection. Theory indicates that small populations do not readily become reproductively isolated through chance drift, and several experiments, notably those with replicated populations of fruit flies started from small numbers of founders,[52,53] fail to show any incipient preferences of members of isolated populations for their own kind after many generations of being kept apart. This is perhaps unsurprising as such experiments have been performed unconsciously many times when researchers establish a laboratory stock of their favourite organism from a small collection from the wild or a shipment from a stock centre. In no case has a new species been reported to arise.[9]

Mate choice and speciation

As one might expect, there are numerous barriers that continue to separate incipient or actual species today, including oceans and high mountains. However, when previously separated populations do meet again then they often fail to mate successfully.

The cause of isolation can be 'pre-zygotic', in which members of the two populations have no opportunities to mate (perhaps because they are common at different times of year, or arise in different habitats, or are pollinated by different insects), or they have no inclination to do so. However, the isolation can also be 'post-zygotic' in which individuals from the two populations can mate, but they produce no viable offspring. Many of these same general mechanisms for reproductive isolation of reunited allopatrically evolving species apply equally to various stages in speciation that do not involve geographic barriers, so-called sympatric speciation, which we cover later.

One recent example of pre-zygotic isolation comes from the closely related and co-occurring damselfly species *Nehalennia irene* and *Nehalennia gracilis* in Canada.[54] In a recent experiment, females of the two species were presented to males of the two species. It turns out that male *N. irene* are relatively indiscriminate in their sexual preferences, showing little if any difference in their propensity to form a 'tandem' (a necessary prelude to copulation that involves the male grasping the female just behind her head) with female *N. irene* and female *N. gracilis*. In contrast, *N. gracilis* are considerably more discriminating, preferring to attempt to form a tandem with members of their own species.[54] One explanation for this difference is that *N. gracilis* is rather rare, so that most females a male *N. irene* meets will be of its own species and the chance of a male *N. irene* making a mistake is correspondingly low; conversely, the chance of a male *N. gracilis* making a mistake is very high, so it pays to be choosy. In both cases, however, there was an even more significant barrier to hybridization—repeated trials clearly showed that males could not complete the latch on to females and form a successful tandem with females of the opposite species, because their appendages used for clasping the female simply could not make the connection[54]—their keys simply did not fit the lock (Fig. 4.5). Whether the two species persist as independent entities because *by chance* they have developed well-differentiated male and female reproductive structures in isolation, or whether natural selection has actively pushed them this way, is an open question. However, the observation that mate discrimination has arisen predominantly in the species that needs it most is indicative of the latter.

Laboratory experiments, particularly those using fruit flies reared for multiple generations in different environmental conditions (such as media with starch or maltose), regularly show that mating preference for one's own type can arise simply as an incidental chance consequence of selection for reproductive success in the two rearing environments.[55,56] Moreover, it has long been recognized that if hybrids are somehow dysfunctional or infertile, then there may be selection within the population to evolve pre-zygotic barriers, a process known as 'reinforcement'[9] (see Glossary). The idea of reinforcement is fascinating, because it suggests an active role for natural selection in generating new species, at least from the stage at which hybrids start to become less viable than pure crosses. One of several reasons why reinforcement may not be particularly widespread as an adaptation, however, is that there will only be selection for pre-zygotic isolation at the juncture where populations meet, unless the two lineages are heavily intermixed.

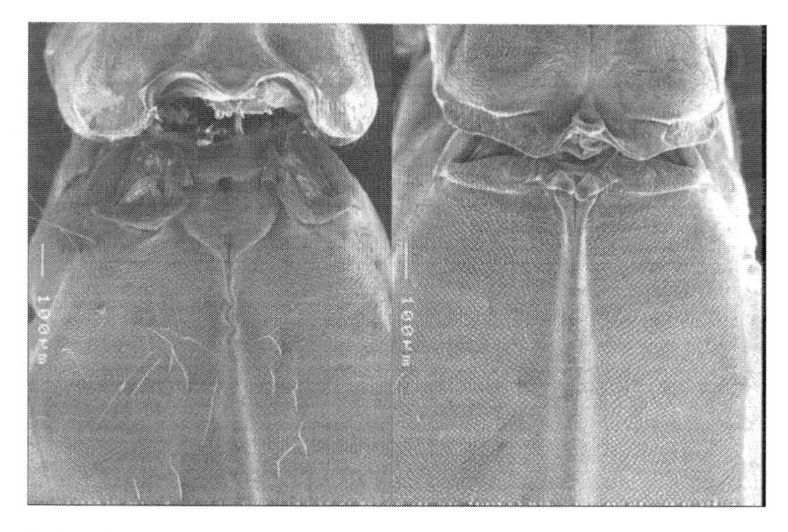

Figure 4.5 The hind margins of the pronota and proximal parts of the thorax in the female damselfly *Nehalennia irene* (left) and its sister species *Nehalennia gracilis* (right), showing significant differences in shape and structure. While males of the same species can readily grasp the female as a prelude to mating, males of the closely related species cannot. Through this simple lock and key mechanism, the two species are reproductively isolated. Photo courtesy of Mark Forbes.

Good evidence for reinforcement is hard to come by, but a fascinating recent example comes from the green-eyed tree frog (*Litoria genimaculata*) in northeastern Australia[57] (Fig. 4.6). The species (singular or plural, depending on one's perspective) comprises two highly divergent lineages, northern (N) and southern (S), which probably arose allopatrically and now meet across two spatial contact zones: 'A' is the primary contact, but there is also a remnant population of S surrounded by N, forming another much smaller contact zone 'B'. Hybridization does occur in the field but it is more common in contact zone A than B. In rearing experiments, the offspring of all crosses of female S with male N died in tadpole stage, but the offspring from crosses of male S with female N produced somewhat more viable offspring—this interpretation was supported by genetic analysis of field-caught specimens which showed that all sampled hybrids had mitochondrial DNA that could only have come from female N parents. Under the above conditions, one might expect that there would be selection against hybridization in the contact zones, but more selection on S females to choose their mates correctly than N females because their hybrid offspring are much less vigorous. As predicted, while it is clear that both types were under selection to avoid hybridization, females of the southern lineage were significantly more inclined to choose mates of their own lineage than northern females.

What explains the lower rates of hybridization of the frog in contact zone B? It turns out that mate choice in the tree frog is influenced primarily by female preference for

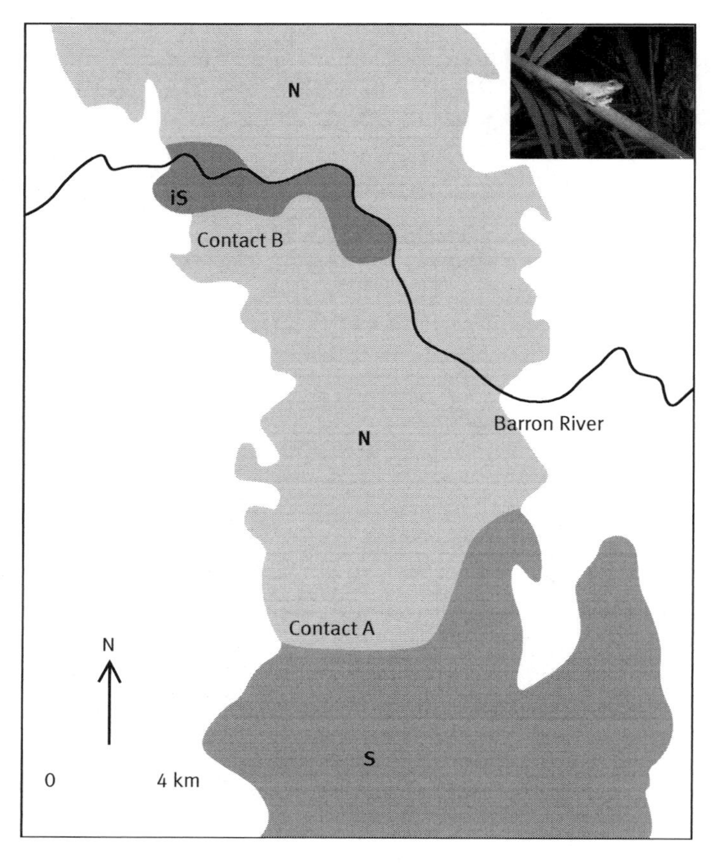

Figure 4.6 Distribution of the Northern (N, pale shading) and Southern (S, dark shading) lineages of the green-eyed tree frog *Litoria genimaculata* in their area of overlap in northeastern Australia. Inset shows the frog. Reproduced with permission from *Nature*. Map and photo courtesy of Conrad Hoskin.

male calls, and in zone B there is greater divergence in the calls of N and S males. This divergence in turn may arise because the individuals in the isolated S population are surrounded by N individuals and so have more to gain by choosing their mates even more carefully. As such, the results mirror the case of the damselfly *N. gracilis* considered earlier, and support the argument for reinforcement as a facilitator of speciation.

Character displacement

An idea closely connected to reinforcement is 'character displacement', where differences among closely related species are accentuated in regions where the species co-occur, yet are small or absent where the two species' distributions do not overlap.[58,59] Potential examples include the nuthatches (*Sitta neumayer* and *Sitta tephronota*) in Eastern Europe and Asia—allopatric populations can only be identified by experienced taxonomists, yet where they co-occur they can readily be distinguished using bill length and facial stripes by anyone familiar with the diagnostic traits. Likewise, the ants *Lasius flavus* and *L. nearcticus* are extremely similar in appearance when found separately in North America, but when found together they differ markedly in traits such as antennae length, head shape, and mouthparts.[58]

While distinguishing traits can conceivably evolve by natural selection as a means of facilitating species recognition (and thereby reduce the likelihood of less-viable hybrids being produced), character displacement is generally seen (and defined) as a consequence of competition for limited resources. In essence, natural selection will favour, in each species population, those individuals whose morphology allows them to exploit those resources less frequently utilized by members of the opposite species. Good proof, which fulfils a variety of formal conditions, of character displacement via competition is rather hard to come by,[60] but one recent candidate example comes from the Grants' celebrated studies of 'Darwin's finches' on the Galápagos Islands.[61] In 1982, a breeding population of the seed-eating large ground finch (*Geospiza magnirostris*) established on the island of Daphne Major. As their numbers increased, which coincided with times of little rain, they began to compete for seeds with their relative, the medium ground finch (*Geospiza fortis*). As a result, those members of the medium ground finch with bill sizes that allowed them to concentrate on smaller seeds tended to survive relatively better, and the population of medium ground finches evolved these divergent characteristics.[61] The exact time at which incipient or established species might begin to show signs of such character displacement is open to debate, and may vary, but as we will now see, selection to exploit novel resources with less competition may be one factor driving speciation in the first place.

Species without frontiers—sympatric speciation

Allopatric speciation is widely considered such an obvious process that it is 'theoretically trivial'.[62] One can even make the rather extreme case that it should be considered the 'default' method of species formation.[9] However, a crucial question is: can new species form without geographical isolation, that is, within a population that can initially freely interbreed? It is a fascinating possibility, but highly controversial. One of the doyens of genetics and evolutionary biology Theodosius Dobzhansky, who penned the now-famous polemic 'Nothing in biology makes sense except in the light of evolution',[63,64] is alleged to have quipped 'Sympatric speciation is like the measles. Everybody gets it, but they all get over it'.[65] Eminent evolutionary biologists including Dobzhansky and Mayr, and a new generation of researchers including Coyne and Orr, argue that there is little good evidence for sympatric speciation in nature, and there are some fundamental theoretical objections,[9] although all have recognized that there are some very likely cases.

The key problem, as we will see, is whether any sexually reproducing population can split into two when there are the homogenizing effects of gene flow.

Consider this: there are about 1 million named species of insect and a lot more waiting to be documented.[66] Did the vast majority arise out of repeated geographical partitions over evolutionary time? It is conceivable, but it is also important to consider alternatives. One of the most plausible candidate routes to sympatric speciation arises when a population feeds on two or more food resources and individuals mate near or on their food resources. One such example comes from the apple maggot fly *Rhagoletis pomonella*, which has become a celebrated example of incipient sympatric speciation thanks to the dedication and inspiration of Guy Bush.[67] This fly, now a pest in orchards in North America, lays eggs on fruit, from which maggots hatch and begin consuming the fruit. Hawthorn is the native host for apple maggot fly (*R. pomonella*) in North America, but in the mid-1800s individuals began to parasitize the co-occurring introduced domesticated apple.[68] It turns out that male and female flies show a strong tendency to return to the host from which they derive (known more generally as the 'Hopkins host-selection principle'[69]). Since the maggot flies mate exclusively on or near the fruit of their host plants, then the differences in host preference (coupled with the fact that apples and hawthorns fruit at different times, with each race tending to be synchronized with their particular host plant) can result in the virtual isolation of races.[70] The keyword is 'virtual' because mark-recapture experiments suggest some 6% of flies were found on the 'wrong' host plant,[71] so that host fidelity only partly restricts gene flow between the hawthorn and apple races. However, another important factor maintaining racial distinctiveness may be active selection on the maggots following egg laying, since the wrong developmental schedule in the wrong host can be catastrophic.[68]

So, do we have solid evidence for sympatric speciation in the apple maggot fly? The system is indeed beginning to look like a good case of incipient sympatric speciation although, as one might expect, with less than 200 generations of diversification, the genetic differences between the two host races are relatively small. Interestingly, it now appears that the original source of genetic variation in developmental schedules exploited by the apple maggot flies to colonize different co-occurring host plants may have come originally from populations in different geographical areas of North and Central America.[72] All genes have 'history', and those from these flies are no exception. In this case, allopatric processes may have furnished the genetic variability, which have allowed incipient sympatric speciation to emerge.

More generally, it does seem that host-shifting is a plausible cause of speciation in a range of plant-feeding ('phytophagous') insects. Many such insects are extreme specialists, utilizing only one or a small range of host plants. For example, in Chapter 7 we note that many butterflies are restricted to a single host plant. Likewise, there are about 750 species of fig wasp and each species breeds on its own species of fig.[73,74] Indeed, figs themselves are among the most species rich of any plant genus—with currently some 850 described species.[75] It is possible that these distributions of wasps on figs arose post-speciation as a consequence of competition, but the well-documented maggot fly story suggests that speciation and specialization can go hand in hand. It therefore seems reasonable to suggest that phytophagous insect speciation is to a degree influenced by host plant diversity, and possibly *vice versa*.

Another intriguing example of sympatric speciation has recently come to light, that of palms on Lord Howe Island.[76] Lord Howe Island, named after Richard Howe, first Lord of the Admiralty, is a small (about 12 km^2) crescent-shaped volcanic island situated about 600 km east of Australia. About half of the island's native plants are unique to the island, with one of the best known endemic groups being palms of the genus *Howea*. You may not think you know these palms but you have probably seen one—*Howea forsteriana*, the kentia palm, is one of the most widely sold house plants in the world[76]—even though it can grow to 18 m. Both *H. forsteriana* and its relative *H. belmoreana* occur widely throughout the small island, often together (although they have different soil pH preferences), yet hybrids have only rarely been reported. Much of the isolation may well be temporal, because *H. forsteriana* flowers a few weeks before *H. belmoreana*. One plausible scenario is that *H. forsteriana* is a species that has adapted to colonizing soils bearing the widespread lowland calcarenite deposits (with high pH), and that the difference of flowering time arose simply as a physiological consequence of adapting to the different soil conditions, reminiscent of the copper-tolerant monkey flower *Mimulus cupriphilus* found on contaminated mine soils in California, which flowers earlier than its relative *Mimulus guttatus*.[77] Once established, any temporal difference in flowering time can readily lead to reproductive isolation and hence distinct species.

Darwin's dream ponds

Even a short review of speciation such as this would be deficient if it did not consider speciation in lake-dwelling fish, which have provided some of the most celebrated and well-researched examples to date, as well as some excellent candidates for sympatric speciation.[78] We now take a short tour through two such examples—the speciation of cichlid (pronounced 'sick-lid') fish in African lakes, and, even more briefly, the speciation of sticklebacks in glacial lakes in Canada.

The Cichlidae are among the most species-rich families of all vertebrates. An estimated 3,000 species are found worldwide[79] from South America to southern India, and in each area they are highly diverse.[80,81] However, it is the radiation of cichlids in the African Great Lakes that has captured most attention. Lake Victoria (at about 70,000 km^2, the world's largest tropical lake), Lake Nyasa (also known as Lake Malawi), and Lake Tanganyika (both approximately 30,000 km^2) are three large freshwater lakes created by tectonic movements in the African Rift Valley of East Africa. Collectively, they are home to an estimated 1,500–2,000 species of cichlids, although it is difficult to say precisely how many because of the obscurity of the species boundaries and the extremely wide range of appearances.[82] Indeed, the rapidity of the radiation has made it challenging to work out evolutionary relationships among the multitude of fish species.[83]

Locals call the cichlids by several names (including 'mbipi' and 'mbuna' for the colourful rock-dwelling fish in Victoria and Malawi, respectively[82]), but unlike the Fore people of New Guinea, a more sophisticated nomenclature has not developed to identify the bewildering variety of forms of these fish.[84] The radiation has been nothing short of explosive: Victoria is somewhere between 15,000 and 750,000 years old, Malawi between 1 and 2 million years old, and Tanganyika between 5 and 12 million years old, with researchers referring to the highly diverse species collections as 'flocks'.[79]

In each of the lakes, different species of cichlid have specialized in feeding on different types of prey including algae, zooplankton, fish scales, molluscs, and other fish.[85] There are even predatory cichlids that feign death, luring smaller fish to investigate. As a consequence, the cichlids range dramatically in their habitat preferences, size, jaw and tooth morphology, body shape, and colour. All the same, closely related species within the same lake all tend to have the same feeding specialization. The mating systems of cichlids are also diverse in form. Owing to intense predation, most female cichlid species in the lakes provide parental care either through mouthbrooding (incubating fertilized eggs for several weeks in their mouths) or through nest guarding. Males do not engage in these activities, but they are frequently highly territorial, defending areas that females might find attractive to spawn in. There is clearly intense competition among males to secure mating opportunities, and this selection, an example of 'sexual selection', appears to have driven bright coloration in males of many species, either permanently or during times of breeding. Some males even change colour according to their 'mood'.[82]

What explains the high species diversity? On one level, morphological features such as pharyngeal jaws may have allowed the flexibility to evolve a range of different specializations.[86] However, the bright coloration of many males has led to suggestions that sexual selection is an important driver of speciation in this group.[9] The theory, which has been formalized with a variety of mathematical and computer models,[87,88] goes as follows. Imagine a population of females that prefer one type of male (e.g. red ones), and a mutant form of female that happens to prefer another type of male (e.g. blue ones). If there were any blue males in the population then the blue-liking females will be more likely to mate with them, generating a degree of reproductive isolation in which blue-liking females and blue males form one group, and red-liking females and red males form another. This process is called 'assortative mating'—we saw a spatially mediated example earlier in the case of apple maggot flies and also in the case of the crows, and it may seem, on first reflection, to help to cleave the population into two. However, there is a serious complication: the genetic exchange between the two incipient sexual populations. Recall from Chapter 2 that sexual recombination can scramble genes in meiosis, allowing genetic contributions from mother and father to be spliced together in the next generation. Thus, depending on the precise genetic details, sexual reproduction could lead to some females with a preference for one trait, yet a tendency to produce offspring with the opposite trait: not exactly a good set of rules for facilitating a clean break between the 'blues' and the 'reds'. Despite this complication, sophisticated mathematical models involving multiple genes for traits and preferences[89,90] do indicate that speciation could be driven in this way, particularly if there is initially relatively wide, and symmetrical, variation in preferences and traits for sexual selection to work on,[91] and especially if the assortative mating evolves in conjunction with some form of ecological specialization such as feeding on particular food resources.[83,92]

Evidence for a role for sexual selection in facilitating sympatric speciation in cichlid fish is regularly presented, but remains somewhat circumstantial. First, males of related species in the same lake often differ in colour, suggesting that they have been subject to some form of selection based on their physical appearance.[85] Second, males sometimes show variability in colour within species, while females in the same species show significant (and consistent) variation in preference for particular types of male, suggesting that there is sufficient variability to allow for incipient speciation.[82] Most importantly, there is now considerable evidence for assortative mating in cichlid fish, such that females tend to choose members of their own species with which to mate. In a well-known experiment, Ole Seehausen and Jacques van Alphen[93] looked at mate choice by females of the Lake Victoria cichlid species *Pundamilia nyererei* (which has reddish males to human eyes) and *Pundamilia pundamilia* (with bluish males)—for consistency, we have used the more recent species designations here rather than the species names used in their original publication. The authors found that, under broad spectrum (white) light conditions, the females strongly preferred males of their own species over the opposite species. However, when the experiment was conducted under monochromatic light, rendering it much harder for females to detect colour differences, then the females did not exhibit any of these preferences. Reassuringly, neither the males' behaviour nor the female response frequencies differed between the two experimental treatments. This suggests that female mate choice was primarily based on visual cues relating to colour and that it helped to maintain species boundaries. Interestingly, the hybrids that formed from species crosses were also fertile and they continued to breed over repeated generations,[93] indicating that the two species were isolated pre-zygotically, but not post-zygotically. As it turns out, although hybrids are relatively rare in Lake Victoria, they are common at several locations with exceptionally low water transparency,[93] suggesting that lake pollution—notably an increase in water turbidity caused by deforestation and agricultural practices—could serve to break down species barriers.[94]

Nevertheless, even if assortative mating is present and can help *maintain* species, we still need to know whether it can *drive* speciation of cichlids itself. While the majority of proposed mechanisms implicitly assume some degree of associative mating, based indirectly either on habitat choice or on mate choice, or both, it may well be that a variety of driving forces is involved. The clearest indication of the multiplicity of factors at play comes from Lake Malawi where the distinct rock- and sand-dwelling clades of cichlid may represent the first, and most fundamental, stage of radiation. The second radiation appears to have been based on different feeding specializations since genera within these clades are distinguished by differences in feeding morphology. The final stage may well have been driven by mate choice, because species within genera are typically distinguished by male breeding colour.[85] Thus, speciation in these fish may first be driven by natural selection for differential use of habitat, then food specialization, and finally sexual selection for attractive male traits. The role of allopatric processes in each of these stages is unclear, but it seems likely to be involved at least in the first stages when fish occupy different habitats and feed on different prey types. Moreover, the history of water-level changes in the lakes indicates that there may have been many isolated refugia, allowing speciation through allopatric means: for example, Lake Tanganyika comprises three discrete basins separated by shallow sills.[82]

Large lakes are complicated systems and for this reason it is hard to say whether any given set of speciation events are sympatric or allopatric. So, it will come as no surprise that the most convincing examples of sympatric speciation in cichlid fish come from much smaller lakes where the opportunities for allopatric speciation are considerably less. The volcanic crater lakes Barombi Mbo (4.15 km^2) and Bermin (0.6 km^2) in Cameroon are uniformly conical in shape and so would not have produced separate basins with water-level changes, and they also (currently) contain no obvious microgeographical barriers.[95] Yet despite this small size and uniformity, Barombi Mbo and Bermin contain 11 and 9 cichlid endemic species, respectively.[95] Mitochondrial DNA analysis indicate that each lake was most likely colonized once only, although Barombi Mbo may have been colonized twice,[82] and subsequently radiated. Interestingly, these mini-radiations seem to have been based on ecological specialization rather than sexual selection; indeed many of the species are not sexually dimorphic. In Barombi Mbo, the basal lineages (as determined by genetic analysis) comprise three ecological groups based on feeding specializations (generalist predators, particle feeders, and specialist predators). In Bermin there are two lineages: open water plankton feeders and bottom feeders.[95] Other examples of speciation of cichlids in small water bodies, including crater Midas cichlids in Nicaragua, are now being investigated,[96] and they all point to sympatric speciation.

More fishy tales

Stickleback fish have long provided sources of inspiration to evolutionary and behavioural ecologists, including Niko Tinbergen who observed the apparent heightened behaviour of males when a red postal van was in the proximity of fish tanks in the window of his laboratory.[97,98] The three-spined stickleback is a case in point. This fish occurs throughout much of northern Europe, North America, and the Asian Pacific Coast and is rather unusual in that different forms of the species can occur in saltwater and freshwater.[99] Six different pairings of morphologically and ecologically distinct forms have been recognized,[99] but here we will concentrate on the benthic (bottom-dwelling) and pelagic (surface-dwelling) forms that have arisen in a series of freshwater glacial lakes around the Strait of Georgia in British Columbia, Canada. The lakes were formed as the ice retreated at the end of the last ice age about 10,000 years ago, and colonized by marine forms of the stickleback.[100] From these marine origins, each lake now contains a pair of freshwater-adapted sticklebacks—a large-bodied benthic form that feeds on macro-invertebrates and a smaller more slender form that feeds primarily on plankton in open water. The two forms have not yet been given the status of distinct species, but they are effectively just that. What is remarkable is that each of the lakes has produced the same combination of benthic and limnetic forms, despite the fact that the different forms within each lake are genetically more closely related to one another than the same forms between lakes.[99] This remarkable repeated pattern, albeit restricted to this part of British Columbia, has been referred to as 'parallel speciation'.[101]

Nevertheless, we should be cautious. Although the two forms of stickleback are radically different in habitat and feeding specializations, and females mate assortatively, at least in part due to their large size differences,[102] the forms can and do occasionally hybridize and this hybridization may play some role in generating the high genetic relatedness among sympatric forms that have been observed. Moreover, it remains possible that the two forms evolved allopatrically, with the benthics evolving after the first wave of colonization and the limnetics evolving following a subsequent colonization in a second marine submergence approximately 1,500–2,000 years later.[103] Indeed, this may explain why the species pairs are seen only in this region. That said, we still need to explain the remarkable parallelism, and once again this may be driven by predictable ecological forces, pushing the latter colonists to a way of life with fewer competitors. In this way, historical contingency and ecological determinism can walk hand in hand.[103]

Sympathy for sympatry

We have briefly reviewed several classic cases of suggested sympatric speciation, ranging from incipient host-based speciation in the apple maggot fly to cichlid speciation in the small crater lakes of Cameroon. In many of these cases, ranging from palms on Lord Howe Island to sticklebacks in British Columbia, we see that speciation may have arisen largely as a by-product of ecological differences. We have also touched upon mathematical and computer models that indicate that the selection for specialization, such as feeding on different food resources, coupled with assortative mating, or even assortative mating based on arbitrary traits, can, under some conditions, be expected to generate reproductive isolation among co-occurring groups. There now seems little doubt, even among the sceptics, that sympatric speciation can and does happen. Indeed, in Ernst Mayr's final book he noted:

> After 1942 allopatric speciation was more or less victorious for some twenty-five years, but then so many well-analysed cases of sympatric speciation were found, particularly amongst fishes and insects, that there is now no longer any doubt about the [non-zero] frequency of sympatric speciation.[104]

Researchers are currently increasingly open to the possibility of sympatric speciation, although many suggest that it is unlikely to be as prevalent as allopatric speciation. For example, a pair of recent reviews concluded that there was little evidence of sympatric speciation in birds.[105,106] In another recent review, Feder suggested that the question now is not whether sympatric speciation occurs, but 'How frequently does sympatry underlie the genesis of new taxa?'.[107] Here is not the place to answer this entirely new question, but recent estimates based on the raw proportion of sister species (from a variety of taxa) that share much of the same distributional range indicate that less than 10% of pairs occur together, and are therefore plausible candidates for sympatric speciation.[108] Of course, this whole argument is based on the assumption that current distributions reflect past distributions, and allopatric sister species are genuine species.[108] Another way to get at the question, or at least a subset of routes that might lead to sympatric speciation, is to ask whether the number of species or rates of speciation are greater in species that show signs of sexual selection, such as sexual dimorphism. A variety of researchers have just done this on groups ranging from birds[109] to insects,[110] each carefully controlling for phylogenetic relationships, but there is a variety of pitfalls in deducing mechanisms from correlative studies, and evidence for and against a role of sexual selection has so far been rather mixed.[83]

Sympatric speciation is attractive as a subject in part because one has to work slightly harder to understand how a population might spontaneously break into two. Yet in other ways, the whole allopatric–sympatric debate may have distracted evolutionary biologists from other important challenges, which include why certain groups (such as cichlid fish) are so capable of radiating, when others show such limited diversity? Before we get on to these questions, there is one more route to sympatric speciation that we must briefly deal with and what may be another important source of biological diversity, that of hybridization.

Speciation through hybridization and related mechanisms

There are particular cases of sympatric speciation that are uncontroversial, because the mechanism is reasonably obvious. One such example is speciation by hybridization, in which members of one species, or two species, combine to produce a reproductively isolated third species. In fact, we have already met one such example—of the 11 species of cichlid found in Barombi Mbo in the Cameroon, at least one was generated by the hybridization of two other species.[107]

Pioneering researchers in the early part of the past century, notably Öjvind Winge (see the humorously titled 'On Ö. Winge and a prayer'[111] for some early history) observed that the number of chromosomes carried by plant species was frequently even, and that certain groups of closely related species were interrelated by simple arithmetic series (2, 4, 6 sets of chromosomes), leading him to suggest that speciation could arise simply through combining whole genomes. The possibility of speciation via hybridization must have come as something of a shock to many researchers as more examples came to light in the first half of the past century. Indeed, the evolutionary biologist J.B.S. Haldane argued that this almost instantaneous ('quantum') mode of speciation represented: 'the most important correction which must be made to his theory of the origin of species'.[112]

Several types of speciation through hybridization are now recognized and each is particularly well known in plants, although, as we will see, occasional examples occur in animals.[12,113,114] Some of the most familiar examples of polyploids come from commercial crops, which tend to have larger cell sizes due to the high number of chromosomes ('genomic obesity')—and, by design, higher yields than their natural relatives. In polyploidy speciation, the number of chromosomes of the newly formed species is greater than either of its parents. One reason, among others, why polyploidy may be much more prevalent in plants than most other groups is asexual reproduction and hermaphrodism (modes of reproduction that are particularly common in plants), allowing populations of polyploids to build up from extreme rarity, thereby allowing newly formed polyploids more opportunity to eventually mate with their own kind.[12]

Allopolyploids (allo—different) are a class of polyploids formed by the hybridization of two different species, and so contain some or all of both parent species' chromosomes. Here, two species effectively combine to generate an additional one, rather than one species somehow splitting into two. For example, triticale (with six sets of chromosomes) has been produced by combining particular strains of wheat (contributing four sets of chromosomes) and rye (contributing two sets). Likewise, three different *Brassica* species (cabbage, black mustard, and turnip) have been hybridized in all possible ways to generate oilseed rape, Abyssinian mustard, and leaf mustard.[115] The phenomenon is also common in nature. For example, the salt marsh plant, common cord grass *Spartina anglica* (61 pairs of chromosomes), arose in the United Kingdom some time after 1870 when the native small cordgrass (*Spartina maritima*, 30 pairs of chromosomes) hybridized with the accidentally introduced, probably through ship's ballast, smooth cordgrass (*Spartina alterniflora*, 31 pairs of chromosomes).[116] This particular hybrid grass species has proved to be highly invasive due to its ability to colonize and bind together unstable sediment. In fact, this same property has led to the hybrid being purposefully introduced to North America and Australasia, where it has causing serious disruption of salt marsh ecosystems.

As might be expected, testing whether particular species have arisen through allopolyploidy is not always easy, and researchers have occasionally resorted to laboratory attempts to recreate wild plants with suspected hybrid origins. For example, as early as 1930, Muntzing hybridized the diploid mint species *Galeopsis pubescens* and the diploid *Galeopsis speciosa* to make the tetraploid *Galeopsis tetrahit* which was similar in appearance to the wild plant and could interbreed with it.[9]

Autopolyploids (auto—self or same), in contrast, involve combining chromosomes from members of the same species. Artificial examples include crops such as potato, sugarcane, and banana. Natural examples include the perennial plant *Galax urceolata* (beetleweed) found in the mixed deciduous forests in Eastern North America where it occurs in diploid (regular two sets of chromosomes), triploid (three sets), and tetraploid (four sets) forms.[117] To see how an autopolyploid might be generated, and subsequently reproductively isolated from members of the population that generated it, imagine a case in which parents of the same species occasionally omit chromosome reduction in their meiosis, producing diploid rather than haploid gametes. When two diploid gametes happen to fuse, the resulting offspring will have four copies of chromosomes (i.e. they will be tetraploid). Any hybrid species with four times the copies of chromosomes is unlikely to be able to produce viable offspring with a diploid because the resulting progeny would have three sets of chromosomes (triploid) which themselves could not divide equally to produce gametes in the next generation.

Just how common is polyploidy as a route to speciation in plants? Some plant species have an extraordinarily high number of chromosomes, which almost certainly arose from combining genomes: for example, the stonecrop, *Sedum suaveolens*, has an impressive 320 pairs of chromosomes.[118] It has been estimated that somewhere between 40% and 70% of all plant species are polyploids, and that up to 95% of ferns are polyploids.[119] Yet these are just estimates of the number of species with a polyploidy history, not the proportion of speciation events involving polyploidy. More conservative estimates, excluding complicated cases where the chromosome number varies within species, put the relative role of polyploidy in generating new species of flowering plants and ferns in the region of 3% and 7%, respectively (much smaller than some texts imply, but still impressive).[118]

Sometimes, the hybrid routes to new species do not involve a change in chromosome number—these examples come under the umbrella of 'homoploid' hybrid speciation (or recombinatorial speciation). In a manner reminiscent of Muntzing's experiments on mint, Mavarez and colleagues have recently artificially recreated a butterfly species, which occurs in the wild, and which had suspected hybrid origins.[120] The butterfly *Heliconius heurippa* (see Fig. 4.7) has long been recognized as having wing patterns intermediate between that of *Heliconius melpomene* and *Heliconius cydno*, all of which have the same number of chromosomes. Within three generations of crossing and backcrossing of the parental *H. melpomene* and *H. cydno* species and their progeny in the laboratory, the authors were able to produce a fertile hybrid with almost identical patterns similar to that of *H. heurippa*. Moreover, these *H. heurippa* with their characteristic wing patterns were undesirable as mates for members of their parent species, but attractive to each other, emphasizing a role for assortative mating in pre-zygotic isolation.

Another way of generating new species may be through chance chromosomal rearrangements, such as reciprocal translocations where different chromosomes nudge up to one another and swap genetic material, similar to the way paired sister chromosomes can recombine. Offspring with such rearrangements may be genetically compatible with their own type, but incompatible with other members of their own population. A fascinating example of this has recently been investigated in yeast. The *Saccharomyces* '*sensu stricto*' yeasts comprise six species based on the sterility of their hybrid crosses. Yet the genomes of specific strains of *Saccharomyces cerevisiae* and *Saccharomyces mikatae* are very similar, except for the fact that a certain series of genes is found on different chromosomes in the two species. To test whether this reciprocal translocation was the primary cause of the post-zygotic isolation, Delneri and colleagues[121] craftily

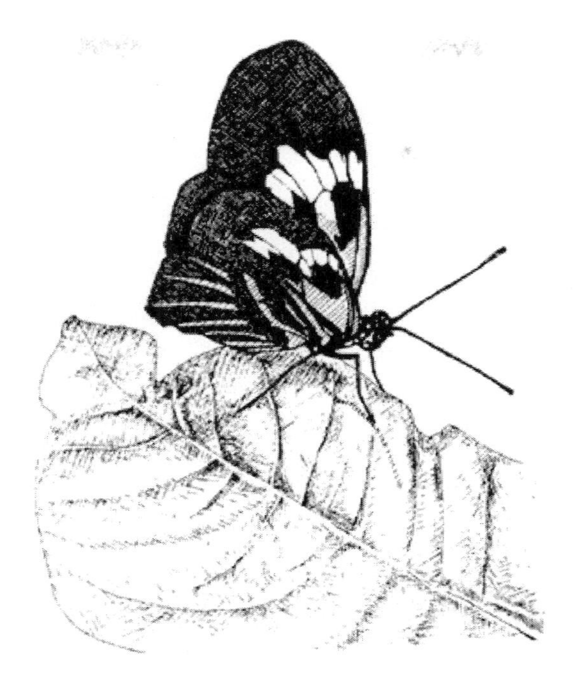

Figure 4.7 A heliconid butterfly. Drawing by Fiona Burns, University of Glasgow.

engineered the *S. cerevisiae* strain so that it matched that of *S. mikatae* and found that hybrid offspring were much more viable (as were genetically engineered *S. mikatae* when crossed with standard *S. cerevisiae*). In this way, it has proved possible to test theories of speciation not just by recreating new species (as in the case of *Heliconius*), but by reversing speciation altogether.

So, sympatric speciation can occur overnight through combining genomes of different species, or the genomes of members of the same species. Earlier, we have seen that a variety of phenomena ranging from steady differentiation following geographical isolation, to assortative mate choice, can help generate and maintain species integrity. We now end this chapter with a brief answer to the question we have been building towards: why is life not composed of a single species?

Why species?

Taxonomically speaking, the world is clumpy, and we have identified many potential mechanisms through which clumps arise: the 'transformation of the homogeneous to the heterogeneous' as English philosopher Herbert Spencer put it.[122] But are such clumps inevitable, and if so, what makes them inevitable? In several reviews of the phenomenon of speciation, Jerry Coyne, Allen Orr, and colleagues have argued that the question of 'why are there species' has been almost entirely neglected,[9,87,123] yet have suggested that it represents 'one of the most important unanswered questions in evolutionary biology—perhaps *the* most important question about speciation'.[9] Putting the question in another (and rather different) way, if we were to find life on another planet, would there be many recognizable 'species' or just one?

We have seen already that geographical isolation can readily lead to speciation in sexual groups because over time the populations develop differences that can subsequently be maintained in sympatry due to some form of pre-zygotic (e.g. mate choice) or post-zygotic (e.g. non-viable offspring) isolating mechanism. Since geographical barriers are commonplace, it seems highly likely that any complex sexually reproducing life would eventually speciate into reproductively isolated populations. Here, natural selection plays a relatively incidental role in speciation, serving to maintain species' integrity through mechanisms such as adaptive mate choice (reinforcement) if the hybrid is less viable.

The observation that geographical barriers will frequently lead to species via prolonged reproductive isolation necessarily applies to sexually reproducing groups. Yet, as we have seen, there is considerable evidence that discrete clusters (species) can arise even in well-mixed taxonomic groups that engage in little or no sex. For example, in Chapter 3 we described the way in which a single clone of the bacterium *Pseudomonas fluorescens* can diversify through mutations into morphologically and genetically distinct forms that include 'wrinkly spreaders' (which bind together on the surface of the broth, causing a film) and smooth forms (which occupy the liquid media), each with their ecological 'niche'.[124] This separation is itself reminiscent of the sympatric speciation observed in crater lakes with the open water and bottom-dwelling cichlid fish, each with their own feeding specializations. These observations point to a strong role for ecology in shaping the type of incipient species that can, and cannot, be wedged into the existing community. Perhaps, therefore, separate species would arise and be maintained for purely ecological reasons.

Ecological factors may be important in two ways. First and foremost, they may facilitate speciation directly. For example, if there are distinct ways of making a living (such as feeding on the surface, or on a particular type of fruit, on a particular-sized seed, or on a particular type of sugar), then any mutant forms that exploit a rather different resource from their conspecifics (an 'innovation') are likely to do relatively well. If these distinct ways of making a living were spatially separated, then this would naturally lead to assortative mating. Moreover, one can imagine that crossing one specialist, such as a surface feeder, with another, such as a bottom feeder, is unlikely to produce a particularly successful hybrid. In this way, ecological 'trade-offs'[125] (in which being suited for one way of living makes one less suited for other ways of living) helps to keep populations true.

Second, ecological factors may also have an influence on which newly formed species do or do not persist, an equally important consideration when asking why there are species. This may be particularly significant in species that have evolved allopatrically yet meet again, where ecological pressures such as close competition with related species will have originally played less of a role. We have already seen that competition between co-occurring related or incipient species, such as seed-eating finches, can drive the species populations to evolve traits that reduce the competition between them (character displacement). It, therefore, stands to reason that incipient species must somehow develop traits that make them capable of carving a distinct niche if they are to avoid being outcompeted in sympatry.

So, we strongly suspect that any planet found to have life will have recognizable species. There will be multiple species because geographical barriers can prevent intermixing for long enough to make many potential sexual populations incompatible with one another. Yet even asexual organisms should be classifiable into discrete clusters, simply because there are a number of incompatible ways of making a living. When jack of all trades truly means master of none, then natural selection will have promoted specialization.

Cichlid fish are a group well known for their high species diversity and provide some of the best candidate examples of sympatric speciation. Photo: TNS.

5

Why are the Tropics so Diverse?

Figure 5.1 A tropical bromeliad (a group that includes the terrestrial pineapple) in Trinidad, West Indies—just one of many species that conjure up a lush tropical environment. Many bromeliads (such as the one shown) are epiphytic, using plants for support but not nutrients, a fact supported by the observation that they can be seen growing not only on telegraph poles, but also sometimes on the wires. Photo: TNS.

> *The day has passed delightfully. Delight itself, however, is a weak term to express the feelings of a naturalist who, for the first time, has been wandering by himself in a Brazilian forest. Among the multitude of striking objects, the general luxuriance of the vegetation bears away the victory. The elegance of the grasses, the novelty of the parasitical plants, the beauty of the flowers, the glossy green of the foliage, all tend to this end.... To a person fond of natural history, such a day as this, brings with it a deeper pleasure than he ever can hope again to experience.*
>
> —*Charles Darwin (1839), Journal of researches into the geology and natural history of the various countries visited by H.M.S. Beagle.*[1]

Our opening quotation describes Charles Darwin's first experience of tropical forest on 29 February 1832. He had been looking forward to this moment for several years. While completing his studies at the University of Cambridge he had read Alexander von Humboldt's accounts of tropical natural history and resolved that he too must experience the luxuriant vegetation and diversity of tropical species at first hand. Initially, Darwin planned to visit the subtropical island of Tenerife; however, this plan was superseded by the opportunity to join H.M.S. Beagle's circumnavigation of the Earth—to his great disappointment Darwin never did get to land on Tenerife, although he saw it from the sea as the Beagle passed close by.[2]

Since Darwin's time we have learnt much about the nature of biological diversity, both in the tropics and at higher latitudes. In this chapter, we review current knowledge of tropical diversity and how it compares with diversity at higher latitudes, before going on to discuss the various explanations that have been put forward to explain why the tropics have so many species. Here we define the tropics as the area between the Tropic of Cancer (23°28′ N) and the Tropic of Capricorn (23°28′ S) when we are discussing the modern world. In discussions of past climates, we refer to areas as 'tropical' if their reconstructed climates are similar to those currently experienced in the modern tropics. While we describe below how diversity changes with latitude, it is obvious that latitude itself is only part of a grid system that allows us to define the location of a point on the Earth's surface, so it cannot itself have a *direct* effect on the number of species. However, many variables such as climate and land or ocean area are correlated with latitude and may provide an explanation for tropical diversity. Indeed, latitude itself is defined by the rotation of the Earth about its axis—a fundamentally abiotic (i.e. non-biological) planetary event. It follows that the ultimate cause of the gradient in diversity over latitude must be attributable to abiotic factors that are correlated with latitude, even if biological factors subsequently play a role in maintaining or promoting this diversity.

Arguably the way we, and most other scientists, have phrased this question may be in part an artefact of the way of working in the pre-Internet age. As Hawkins and Diniz-Filho[3] have pointed out, because 'latitude' is easily read off a map, then it has been effectively treated as a 'variable' for over 200 years in studies of large-scale differences in global diversity. Today, one can easily download extensive data on climate (and many other environmental variables) to compare directly with species richness. Hawkins and Diniz-Filho[3] suggest that there is no longer any need to use latitude as a surrogate for climate (or whatever variable we are interested in) when the real environmental data are so readily available. They have a point; one of us can remember the time-consuming job of assembling a climate data set for remote parts of the southern hemisphere from a diverse array of obscure publications, during a study of microbial biodiversity in the early 1980s[4]—a decade before the birth of the Internet. However, the idea of latitudinal gradients does nicely summarize the empirical observation that, at least to a first approximation, diversity of species declines as one moves away from the equator. In addition, some historical factors (such as past climates) that have varied with latitude are still difficult to use directly in a quantitative manner—so latitude as a surrogate still has its uses.

Since the extent of tropical diversity became apparent during the nineteenth century, a large number of different factors have been suggested as possible explanations. In a well-known paper published in 1966, Eric Pianka[5] suggested that there were six main classes of explanation, most of which were not mutually exclusive. None of these candidate explanations have subsequently been ruled out as definitely wrong, and some new possibilities have been added. For convenience, we classify these explanations into three main types: (1) null models, which make use of aspects of geometry rather than biology to explain tropical diversity; (2) historical explanations, which use the geological past to explain present distributions; and (3) explanations based on ecological processes operating today. Null models of various kinds have attracted particular interest from ecologists in the past decade, particularly because of the work of Stephen Hubbell[6] in arguing that they are capable of explaining many of the patterns seen in global biodiversity. However, we argue below that the various 'null' hypotheses do not by themselves provide satisfactory answers to this chapter's question. In reality, the 'true' explanation may well be a mix of several processes and we will attempt to suggest how these various theories fit together in the latter part of this chapter.

The nature of tropical diversity

While ecologists rightly associate many areas of the tropics with both high biomass (pure weight of life) and high species diversity, not all of the tropics are as rich in life as Darwin described in his accounts of his initial wanderings in Brazil. In Chapter 8, we describe the low-nutrient status of many tropical oceans, which leads to low biomass of organisms living in them. Some tropical areas on land also have little life because they are hot and arid (Fig. 5.2). Other areas of the tropics, such as the area east of the central cordillera through Central America, are relatively low in species richness for reasons that are less certain. Yet the tendency for the tropics to be rich in species has been described as 'one of the most venerable, well-documented, and controversial large-scale patterns in macroecology'.[7] For example, the rain forest in the Gunung Mulu National Park, Sarawak (just north of the equator on the island of Borneo) has at least 223 species of tree per hectare[8]; this can be compared with the entire tree flora of the United States and Canada, which comprises only around 700 species.[9] Similarly with freshwater fish,

Figure 5.2 An arid part of Ascension Island. When Darwin visited the island he was not impressed by this arid landscape writing about its 'naked hideousness'[1] compared to other tropical locations covered with lush forest. Photo: DMW.

a study of just one relatively small section of Amazonian flood plain recorded 286 fish species,[10] more species than the whole of Europe (c. 215 freshwater fish species).[11]

A good example of a diversity gradient where species richness has a maximum in the tropics is provided by invertebrates living in estuarine sediments. Martin Attrill and colleagues[12] compared data on invertebrate species diversity from studies of 20 estuaries around the world (chosen to have similar salinities, sediment particle size, and invertebrate sampling methods), and found a significant relationship between latitude and diversity, with higher diversity at low latitudes. Reptiles are a very different animal group from estuarine invertebrates; however, when Gaston and colleagues[13] looked at global variation in the diversity of reptile families (using family as the taxonomic unit for comparison at least partially sidesteps the problem of incomplete data on species richness, especially in areas of the tropics which are less well studied), they found that the top areas for diversity were all in tropical America; with southern Mexico highest, followed by Nicaragua, southern Colombia, then a tie between central Venezuela and central Columbia.

Another example comes from a classic data set, repeated in many textbooks, that shows increasing diversity towards the tropics in the numbers of breeding land bird species for Canada, the United States, and Mexico.[14] Many similar examples are given in both introductory textbooks[15,16] and more advanced texts,[17,18] which provide long lists of taxa that show a gradient of declining species richness as one moves away from the tropics; these include global tree species, freshwater fish in the rivers of the world, marine bivalves, termites, and African primates. Even aspects of human diversity have been shown to exhibit similar patterns. For example, Collard and Foley[19] analysed a data-set on 3,814 human cultures and showed that both the density and diversity of cultures decline as one moves to higher latitudes—although it is not clear if this anthropological pattern can be explained by processes similar to those applied to gradients in species richness.

Exceptions to the general pattern

There are many examples of groups where diversity peaks somewhere in the tropics, and indeed there are far more documented examples of groups that follow this pattern than ones that break this ecological generalization. However, there are a few examples that buck this general trend, and it is worth briefly discussing them. Consider penguins; a family of birds usually associated with Antarctica although, in fact, penguins do sometimes get into the tropics. For example, L. Harrison Matthews—one of the last generation biologists to start his career in a Darwinian manner with major expeditions on sailing ships—reminisced in a book, written in old age, of the first time he ever saw wild penguins as a young man. This, he fondly remembered, was while swimming naked with a strikingly attractive woman on a tropical Brazilian beach.[20] These birds were probably Magellanic penguins somewhat north of their normal range, but one species of penguin does breed on the equator—namely, the Galápagos penguin. Nevertheless, most of the 17 current species are restricted to much colder polar waters.[21] Another bird example is provided by the Procellariiforme seabirds (which include albatrosses, petrels, and shearwaters): these also peak in diversity away from the tropics. Indeed, Steven Chown and colleagues[22] used data compiled for the first volume of the *Handbook of the Birds of the World*[21] to show that Procellariiforme diversity peaks between 37° and 59° South; although in addition there were concentrations of endemic species in the northern hemisphere, north of the Tropic of Cancer (Fig. 5.3).

The above bird examples are by no means unique. A range of other groups that show a peak in diversity away from the tropics has been described, including parasitic ichneumon wasps, which reach maximum diversity in temperate areas;[23] free-living (non-parasitic) soil nematodes, which show lower diversity at polar latitudes but little difference between temperate and tropical diversity;[24] and soil-living oribatid mites, which show a pattern similar to that of the non-parasitic soil nematodes.[25] In the marine realm, we are used to thinking of tropical coral reefs as spectacularly diverse. However, it has become apparent in recent decades that the Southern Ocean around Antarctica is also surprisingly species rich—although requiring a hardier breed of marine ecologist to study it. For example, recent studies of deep waters (748–6,348 m) off the Antarctic Peninsula have emphasized their biodiversity, including 674 species of isopod crustacean—of which 585 were new to science.[26] Therefore, while there appears to be a steady decline in the diversity of marine species from tropics to pole in the northern hemisphere, the richness of the Southern Ocean means that this pattern is less clear, and possibly completely absent, in the southern hemisphere oceans.[27]

Figure 5.3 Cory's Shearwater off the coast of Madeira in the North Atlantic. Although they have to breed on land, Procellariiformes spend most of their time at sea, feeding over the open ocean as in this photograph. Their diversity peaks away from the tropics especially in the southern hemisphere. In the north, Madeira and the Canary Islands are areas of high endemism for this group—with several species of very limited global distribution breeding there.[22] Photo: DMW.

The apparent difference in marine species richness with latitude between the North and South Atlantic highlights another point that has been raised recently. Patterns may differ between the northern and southern hemispheres—for example, groups such as New World birds and the number of mammalian families show a steeper decline in species richness with latitude in the northern hemisphere as one moves away from the tropics compared to the southern hemisphere.[7] One potential explanation for these different patterns is that, latitude for latitude, land in the southern hemisphere is usually warmer, being buffered by heat from the more extensive southern oceans. While there are less-pronounced hemispherical differences in ocean temperatures, the nearly landlocked Arctic Ocean has more variable annual temperatures than those of the seas surrounding the Antarctic.[7] This is a complication that is overlooked by many studies, such as the one on oribatid mites we described earlier,[25] that plot latitude against species richness without distinguishing between hemispheres (in the mite case, the majority of the data were taken from the northern hemisphere). However, given that latitude itself cannot directly affect species richness it is perhaps not surprising that different patterns are sometimes observed in the two hemispheres.

Patterns in microorganisms

All the studies we have described so far have been of multicellular organisms. However, this ignores some of the most significant contributors to biodiversity and ecosystem functioning. Microorganisms play a crucial role in the working of most ecosystems—breaking down organic matter, fixing nitrogen, and producing oxygen, amongst many other roles. One important unknown in ecology is our lack of knowledge of the basic patterns of microbial diversity on Earth; this is illustrated by our ignorance of latitudinal patterns in microbial species richness. In an early attempt to address this problem some 25 years ago, Humphrey Smith[28] plotted species richness for testate amoebae (the eukaryotic microorganism shown in Fig. 4.2) from a range of sites in the Antarctic and sub-Antarctic. He found that higher-latitude (i.e. in Antarctica) sites tended to have fewer species than lower-latitude sites, and pointed out that if one extended the regression line, fitted through these data, to the tropics then it predicted around 100 species—which was close to that described in the limited number of tropical studies then available in the literature. However, recent work has greatly increased the number of species known from some of the sub-Antarctic Islands and their diversity now rivals that described for many parts of the tropics.[29,30] Currently, it is highly uncertain if testate amoebae show any general diversity gradient with latitude.

What about other groups? One of the best-known phyla of eukaryotic microorganisms is the diatoms. Since diatoms have highly distinctive and beautiful shells they can be identified to species level using morphological criteria with relative ease, and so we have better data on their ecology than is the case for many microbial groups. Using data from an impressive 179 studies, Hillebrand and Azovsky[31] failed to find any convincing evidence for a general latitudinal effect on diatom species richness (there was a significant relationship for just the southern hemisphere data; however, they had some concern that this may have been a product of a small sample size in the studies from the Antarctic). More strikingly, a recent study of the diversity of a community of eukaryotic microbes in a tidal mud flat in Greenland (using ribosomal RNA) found that this site had greater diversity than any other comparable site yet studied—suggesting that diversity may peak in the Arctic,[32] a very different pattern to that which we see in most large organisms. Only one group of eukaryotic microbes appears to exhibit a completely unambiguous increase in diversity in the tropics, namely the benthic foraminiferans (forams for short)[33]—bottom-living marine microbes that build characteristic shells.

To sum up, there appears to be only limited evidence for latitudinal gradients for most eukaryotic microbes—although it is currently difficult to be sure if this is because these gradients are generally absent, or because we have only limited data of variable quality to use in these analyses. Indeed, the Greenland mud flat study[32] suggests that if there is such a gradient, the high diversity may be in polar latitudes.

Ecologically, the most important microbes are the prokaryotes—the 'bacteria' in the old sense of the term, although many biologists now split this group into 'true' bacteria and archaea. Do prokaryotes show a latitudinal gradient in diversity? Until recently this has been a very difficult question to even try to answer, since species concepts are not readily applied to many microbial groups (see Chapter 4). For example, bacteria tend to show little morphological variation when viewed under the microscope—unlike the shell-building testate amoebae, diatoms, and forams we described earlier—so it is very difficult to compile lists of species from soil or water samples. The methods of molecular ecology (using analyses of DNA or RNA) are now allowing a start to be made on addressing this important question. Before the rise of these molecular methods the main approaches researchers were using to quantify bacterial diversity included differential growth in culture media and/or lipid analyses along with direct observation by microscope.[34]

One important recent molecular study[35] looked at bacterial diversity in 98 soil samples from North and South America. They found no relationship with latitude but did identify soil pH as a good predictor of bacterial diversity, with low diversity in acidic soils. This is one of the most detailed attempts to look for a latitudinal gradient in bacterial diversity of which we are aware; however, it still has some problems that prevent it from being a fully conclusive demonstration of the lack of such a gradient in bacteria. As with the mite study we described earlier,[25] this study did not look for separate northern and southern hemisphere patterns. More importantly, it failed to find any strong relationship between the geographical distance between sites and the specific composition of the bacterial communities. This result runs counter to the conclusions of several other studies of prokaryote diversity,[36] which appear to show increasing differences in bacterial communities as sampling sites get further apart, and until we have an understanding of the reasons behind these apparently conflicting patterns, it is difficult to know how to interpret these results. So, as with eukaryotic microbes, there is currently a shortage of convincing evidence for latitudinal gradients in prokaryotic diversity, but also no really convincing proof of their absence.

To summarize the mass of observational studies: the basic pattern we are trying to explain in this chapter is that the tropics tend to be more species rich than higher latitudes. As demonstrated by the examples we refer to earlier, this is now widely established for many groups of macroscopic multicellular organisms, although some exceptions are known. It is less clear if such patterns are common in microorganisms, although forams certainly show this pattern. Indeed, in general, there seems to be an increased likelihood of a group showing a strong latitudinal gradient with increasing body size: in a statistical study of nearly 600 examples, assembled from the scientific literature, Helmut Hillebrand[37] found that both the strength and the slope of the relationship increased with body mass. The vast majority of the studies analysed by Hillebrand were of multicellular organisms. All the microbial studies he used were of eukaryotes and these mainly came from just three groups; the forams (which tended to show a tropical peak in diversity), diatoms, and ciliate protozoa (both of which mainly showed an absence of the classical tropical latitudinal diversity gradient).

How old is the latitudinal gradient?

An obvious question, which follows from the observation that this relationship is widespread amongst modern organisms, is to ask about its geological history. Clearly, the type of explanations we use to explain this pattern could be different if it has been common for tens or hundreds of millions of years compared to being a product of particular conditions on the modern Earth. It turns out that there is good evidence that, in at least some groups, this pattern has a long geological history.[38] For example, foram fossils extracted from marine sediments suggest that the tropics have had more species than temperate regions for at least 10 million years; interestingly, they also suggest that the difference has become more pronounced over this period, with tropical diversity increasing much more than temperate diversity over time (although this conclusion is based on rather limited data).[33]

There is also evidence of a long-established (geological) latitudinal gradient in flowering plants (angiosperms). Ever since the origin of this group in the tropics approximately 145 million years ago,[39] the angiosperms have remained a predominantly tropical group. Again it is interesting that this gradient appears to have become more pronounced during the past few million years.[39,40] Clearly, any explanation for latitudinal gradients in species richness needs to be applicable to the geological past as well as the present, although it is possible that one or more of the relevant processes may have become more pronounced over the past few million years than it was over longer spans of geological time.

A null model—the mid-domain effect

So what of the proposed mechanisms to explain latitudinal gradients in diversity? The vast majority of scientists who have discussed the question of tropical diversity have assumed that some combination of ecological, evolutionary, or geological processes will provide the explanation and, that in the absence of these processes, all latitudes would have similar species richness. This was challenged during the 1990s by Robert Colwell and others with models of the so-called mid-domain effect—which suggested that we could expect to see peaks in some areas of the tropics for purely geometrical reasons.[41] These models were conceptually similar to ones that had previously been used by two distinguished theoretically inclined ecologists (R.H. MacArthur in the 1960s and E.C. Pielou in the 1970s) to try to explain the relative abundance of species based on distributions of sizes of ecological niches. MacArthur, Pielou, and later Colwell all produced models that partitioned one-dimensional space (a line or 'stick') between species using simple rules and then compared their results with what was seen in nature.[41]

The mid-domain effect applies to any area of land or ocean that is bounded at either end. Consider Africa, which is bounded to the north by the Mediterranean Sea (or the Sahara for organisms that are not drought tolerant), and the Southern Ocean off South Africa to the south—with the equator running approximately half way between these two points. If species were distributed randomly within Africa, then one might at first expect similar species richness at all latitudes. However, this may not be the case. In the simplest version of a mid-domain model[41] Africa (or South America, or the Atlantic Ocean, or any other area of interest) is represented as a one-dimensional line. Species are randomly assigned plausible range sizes (from very small ranges, to ones that would cover the whole continent) and these are then randomly positioned along the line. Species with small range sizes can be positioned almost anywhere, yet ones with larger ranges (spanning a non-negligible fraction of the continent length) are likely to be placed in such a way that they overlap with the middle of the line simply to fit their distribution in—as it is assumed that these are terrestrial species whose ranges cannot overlap into the sea (or *vice versa*). As shown in our graph of such a model (Fig. 5.4a) the result is a peak in species richness in the middle of the line (at 'mid-domain'), which in the case of Africa would lead to a peak of species richness in the tropics. Similar results can be achieved with more complex two-dimensional models.[42] So the pattern we described earlier, of species peaking in the tropics can, in principle, be produced by nothing more than simple geometry—without any need for ecological or evolutionary processes. Analogous arguments can be applied to east/west-bounded continental gradients, or ones that change with altitude or depth.

While it is clearly useful to realize that non-random patterns do not necessarily require complex explanations, there are serious problems in trying to use the mid-domain effect to explain tropical diversity. Similar to all 'null models', the model has no evolutionary component and there is no underlying biological basis for the idea that species ranges are predetermined and 'placed' on the earth as one would place pieces of a jigsaw. Most obviously, while some continents and oceans bestride the equator (e.g. Africa and the Atlantic) it would be more difficult to use the mid-domain explanation for tropical diversity in regions such as Southeast Asia. Thus, while there are potential examples of mid-domain peaks of richness as a function of latitude in cases where the equator is in the middle (e.g. the Americas), crucially the 'mid-domain effect' is not observed on continents where the equator is not in the middle (e.g. Asia or Australia). Likewise, mid-domain effects should also occur with respect to longitude, but we know of no published examples.

Some of the assumptions behind the standard mid-domain models also seem overly simplistic.[43] Species are not born with fixed distributional ranges, and for many species they are in flux. Indeed in some extreme cases, species ranges can be so strikingly discontinuous as a consequence of past history ('disjunct' in the jargon of biogeographers) that the gaps are obvious even when looking at a summary distribution map of the type reproduced in many identification field guides. A classic example of this is the azure-winged magpie, which in Europe is restricted to Iberia but also occurs over much of China—but nowhere in between.[44] The nature of the boundaries of domains is also somewhat problematic. In some cases, such as where the land ends and the ocean starts, the boundary is clear but in many other cases it is less obvious.

As discussed earlier, it will come of no surprise that attempts to compare the predictions of mid-domain models with real data have usually suggested that the model is insufficient to fully explain the observed pattern.[42,43] Some recent detailed analyses of real-world data suggest that the mid-domain effect is most likely to predict the distribution of groups that have large range sizes.[45] This makes geometric sense as small range sizes are less likely to be constrained by hitting the edge of a domain. However, a given

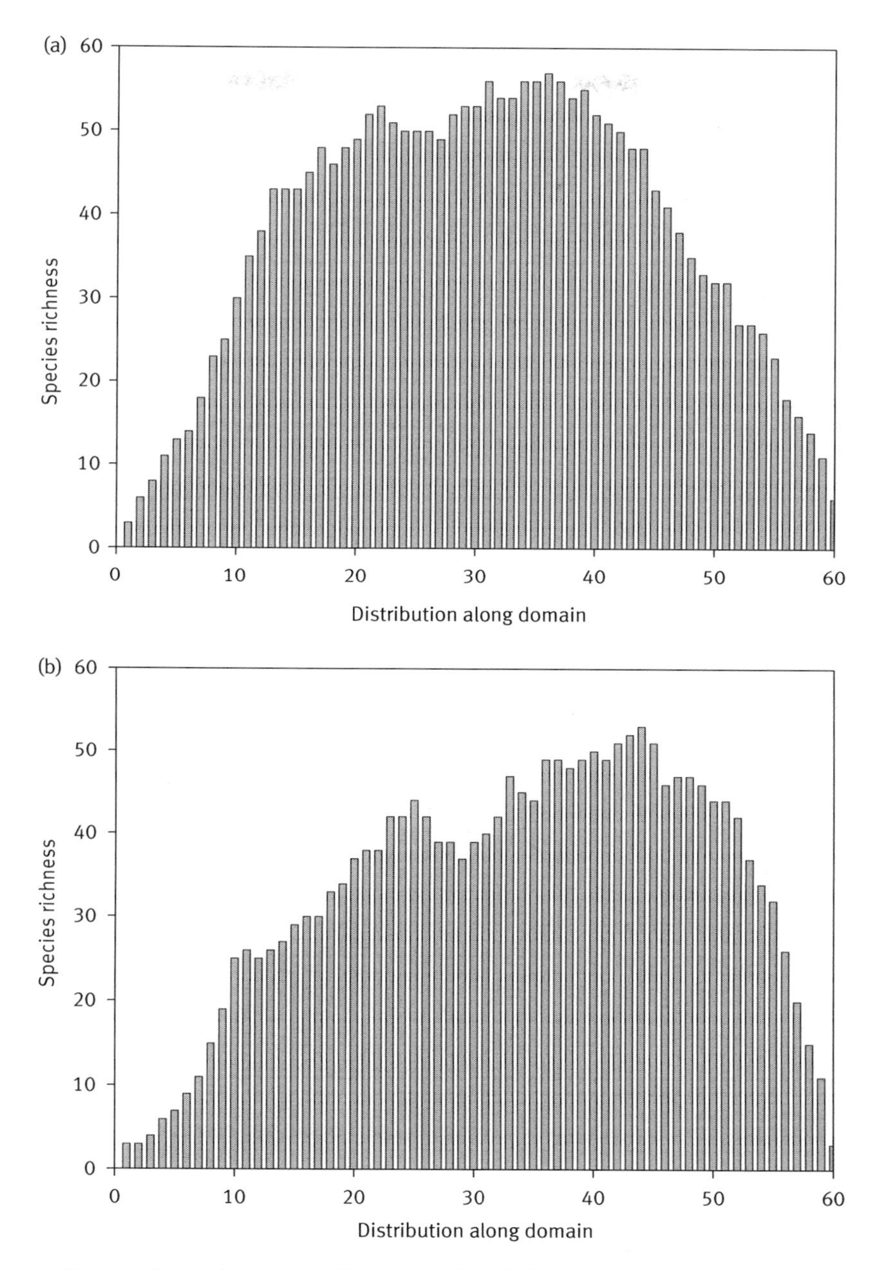

Figure 5.4 Outputs from simple one-dimensional mid-domain models where 100 species with randomly generated range sizes have been located randomly along a line ('domain') using Colwell's RangeModel software.[99] (a) Output from a typical run of this model. (b) The most asymmetrical result from 100 consecutive runs of the model illustrating the possibility that some of the asymmetries in species richness gradient could be the product of random processes—although in most cases, there are probably ecological explanations (such as hemispherical asymmetries in climate).

number of large ranges must necessarily overlap more than the same number of small ranges in a fixed area, making this a poor test of the mid-domain effect.[46]

What are we to make of the mid-domain explanation for tropical diversity? It is certainly valuable in showing that under some circumstances pure geometry can lead to a peak in tropical diversity, indeed, other patterns relevant to this chapter, such as hemispherical asymmetries in species gradients, can also on occasion be produced by a mix of only geometry and random processes (Fig. 5.4b). Beyond these theoretical insights it may actually contribute towards explaining the pattern of tropical diversity for some continents and oceans that lie across the equator (especially for groups of species with large range sizes)—however, not even its supporters suggest that the theory provides a complete answer to this chapter's question,[47,48] and a recent attempt to test the idea by a statistical analysis of 53 relevant published studies was not encouraging.[46]

The ins and outs of historical explanations

In trying to explain latitudinal variations in species richness, we are really trying to answer the question of why there are a certain number of species at a particular place but more (or less) species at another place. At its most basic, a species is present at a particular site either because it evolved there or because it evolved elsewhere and has dispersed to that site. Conversely, if a species is absent from a site then either it has never been present at that location or it was there in the past but has become extinct at that site.[49] This means that rates of speciation, extinction, and dispersal will necessarily play a fundamental role in any explanation of why the tropics are species rich, and indeed most of the hypotheses we will subsequently consider can be traced back to some latitudinal disparity in the relative rates of these processes. Of course they are all processes that happen over time, and for this reason they can be considered under a general umbrella of historical explanations for tropical diversity. We now consider several of them in turn.

Area effects

A potentially important mechanism to explain latitudinal gradients did not feature in Pianka's[5] classic list of the mid-1960s—as it was first suggested by John Terborgh,[50] seven years after Pianka's paper, and brought to prominence by Michael Rosenzweig in his influential book *Species Diversity in Space and Time*[17] in 1995. Both Terborgh and Rosenzweig pointed out that the tropics cover more of the globe than any other region. This is not immediately obvious to most people, many of whom were taught geography at school with maps of the world that utilize the Mercator projection, which exaggerates the size of areas away from the tropics—for example, making Greenland appear relatively much larger than it actually is (in fact it is only 1/14th the size of Africa).

To appreciate the point about the size of the tropics, it is better to look at a globe than a two-dimensional map. While looking at a globe, note that at the tropics the northern and southern hemisphere meet (doubling the area of continuous tropics) while the temperate or polar areas of both hemispheres are isolated from each other. It has long been known that the number (in total, and per unit area) of species on islands (be they oceanic islands, lakes, or mountains) increases as the area of the island size increases.[14] Several complementary explanations for the phenomenon have been put forward, including greater habitat diversity on larger islands, and some form of equilibrium between immigration and extinction. What if the same were true on a much larger scale?

Note that we are generally asking questions about the density of species in the tropics compared to the temperate areas, rather than the total number of species. That is, why does a kilometre squared of tropics have more species than a similar sized area of land away from the tropics? Borrowing from island biogeography, the proponents of the importance of area to tropical diversity go on to suggest that the area of the tropics, and relative isolation of the two temperate regions, indirectly influence the relative speciation and extinction rates in these regions. At its simplest, the idea is that larger areas contain more geographical barriers in total, so promoting allopatric speciation (as described in Chapter 4). Similar to oceanic islands, larger areas are also likely to have a wider range of habitats fostering species diversity. In contrast, smaller areas tend to have smaller populations, which are more prone to extinction, and also fewer habitats or geographical barriers. So, the theory goes, increased species richness in the tropics per unit area is explained by historical differences in speciation and extinction rates in this region, which are in turn mediated by area; this has allowed the tropics to diversify faster than other parts of the world.[17,50]

This standard formulation of the idea immediately runs into a complication, since it claims larger population sizes in the tropics, because of larger range sizes, but also more chance of such ranges being split by geographical barriers—so simultaneously claiming larger range sizes and smaller (more subdivided) ranges in the tropics. There are several other problems with this idea, which have recently been reviewed by Mittelbach and colleagues.[51] First, as species diversified it would seem likely that species population sizes would have had to decrease, so potentially increasing their extinction rate. The assumption that large areas increase the likelihood of speciation can also be questioned, since species with large ranges may have wide environmental tolerances and high dispersal abilities, making allopatric speciation less likely. Moreover, it is worth remembering that the tropics are not uniform, but made up of various biomes such as tropical rainforest, savannah, and desert—so just considering the total area of the tropics may be misleading.

Accentuating the difference through niche conservatism?

Another historical theory for the latitudinal gradient in species richness, at least in terms of accentuating an asymmetry that was already there, is based on the idea of 'niche conservatism', the tendency of species to retain many of their ancestral ecological characteristics.[52] This theory, dubbed the 'tropical conservatism hypothesis', has been presented in a variety of forms but it has been developed and explored most recently by John Wiens and colleagues.[52,53] The hypothesis is built on just a few simple premises. First, it begins by assuming that there was an initial asymmetry in species diversity for some reason—for example, disproportionately more species start out in the tropics because the area is bigger (see earlier) or because they have had more time to speciate. Second, if the niche conservatism argument is valid, then relatively few species would be able to successfully disperse (and thereby speciate allopatrically) from tropical to temperate areas because they lack the adaptations to survive the temperate climate, such as cold winter temperatures. Therefore, if new species are going to evolve in the tropics, then they are likely to remain in the tropics, maintaining and even accentuating the disparity in diversity.

We have already described the palaeontological evidence that flowering plants evolved in the tropics and the modern observation that the majority of these species are still tropical. Of course, it works both ways (temperate-adapted species should remain in the temperate areas), and we must still strive to explain the initial asymmetry, but the theory makes a good deal of intuitive sense and would benefit from further investigation. Such work is already underway, and exploits both climate data and an understanding of how the species in any given taxa are related to one another.[53]

Palm trees in London?

There is an additional complicating factor affecting the niche conservatism and area arguments, which we must consider head-on because it helps inform related hypotheses. Climatically and geographically speaking, the area of land one might consider the tropics (and hence the size of non-tropical areas) has varied over geological time, both with changes in global climate and, on a longer time scale, with changes in the position of continents. This not only considerably complicates the evaluation of the area and niche conservatism theories, but also indirectly provides new explanations for the tropical diversity gradient and therefore justifies more detailed consideration.

The idea of tropical climatic conditions occurring at high latitudes seems highly unlikely in the context of our experience of the modern Earth. Indeed, it seemed so strange that many talented scientists in the past were slow to fully accept the evidence. Marie Stopes is best known today as the author of a highly influential sex manual[54] and later as an important campaigner for contraception. Earlier in her career (around 1903–1945), however, she was 'among the leading half-dozen British palaeobotanists of her time'.[55] Amongst her botanical publications was a *Catalogue of the Cretaceous Flora* (currently the Cretaceous is dated to 145.5–65.5 million years ago). In the second volume of this publication,[56] she summarizes what the plant fossils appeared to tell us about the climate experienced by some of the dinosaurs. In doing so, she describes some apparently tropical plants from England and even from the Arctic, but urges caution over the climatological implications. Reading her brief comments over 90 years later, we can imagine Stopes telling herself that the climate cannot be that warm, so far north. We now have a far larger array of lines of evidence available to us and know that the world was indeed much warmer with forests extending to within 1,000 km of the poles at that time.[57] More recently, around 45 million years ago, there was a coastline with a subtropical climate running through northeast London, England with mangrove palms growing at what is now the mouth of the River Thames (Fig. 5.5).[57] As well as vastly increased tropical conditions at points in the geological past (due to higher global temperatures, not just plate tectonics moving the continents around the globe), there is also the possibility that the tropics, as climatically defined, contracted significantly in size during the (geologically) recent ice ages—an idea we will return to later.

Historical rates of species formation and extinction, again

Clearly, such historical changes in climate and area have big implications for explanations that use the area occupied by the tropics as an important part of the explanation for latitudinal gradients in species richness. One possibility is that a tendency, over geological time, for higher latitudes to get colder, especially over the past few million years, may have caused increased extinctions away from the tropics. Analysis of bird data suggests that this may have been the case with increased extinction of evolutionary older groups ('clades')—potentially those better adapted to warmer climates—from higher latitudes.[58] This is consistent with the pattern we have already described for several fossil data sets, showing an increase in the steepness of the latitudinal species richness gradient in the geologically recent past—the past 2.6 million years have been characterized by significant global cooling.[59]

A clear (although not unique) prediction of the area hypothesis is that more species should originate in the tropics than at higher latitudes (including species that are now extinct). This idea is open to test using fossil and/or molecular data—although the ebb and flow of tropical climates over geological time provides a complication that few studies have really addressed. Many of these studies utilize higher taxa, rather than species data, because they are considered to be more robust against problems with incompleteness of the fossil record. This is because to record the presence a given species in the fossil record requires you to find a fossil of that species (obviously) but to record a genus or family only requires the identification of a single specimen of a single species from that group, so counting higher taxa provides a statistic that is less susceptible to the vagaries of fossilization. In addition, in collating data on the presence of fossils in rocks from various sites, higher-taxon records are less ambiguous as there is usually more agreement amongst palaeontologists in assigning a specimen to a genus than to a species. Marine taxa are often used in such studies, rather than terrestrial ones, as sediments on the bottom of shallow seas are particularly good sites for fossilization.[18]

Evidence from a range of groups, such as corals, forams, and some mammal groups, shows that the average age of genera and families decreases with decreasing palaeolatitude (i.e. the latitude in the geological past, not the current latitude of the site, which

Shells of the London Clay.

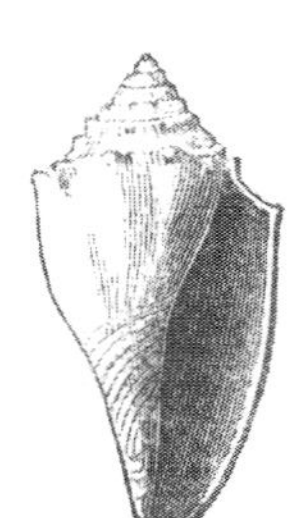

Voluta nodosa, Sow. Highgate.

Phorus extensus, Sow. Highgate.

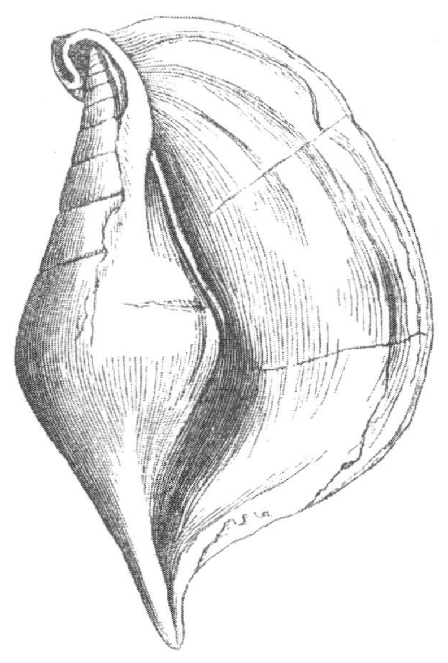

Rostellaria (Hippocrenes) ampla, Brander. $\frac{1}{3}$ of nat. size; also found in the Barton clay.

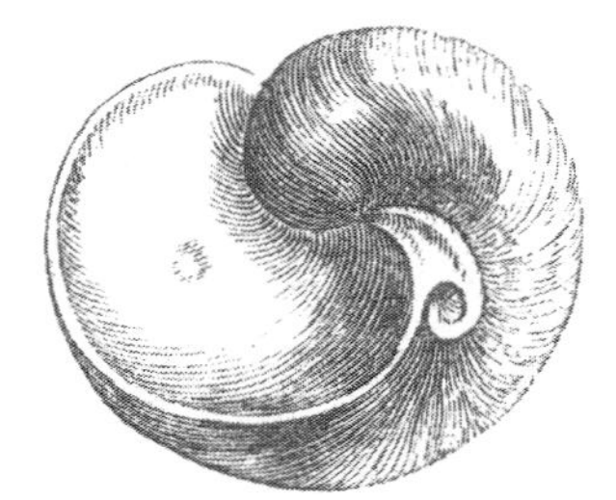

Nautilus centralis, Sow. Highgate.

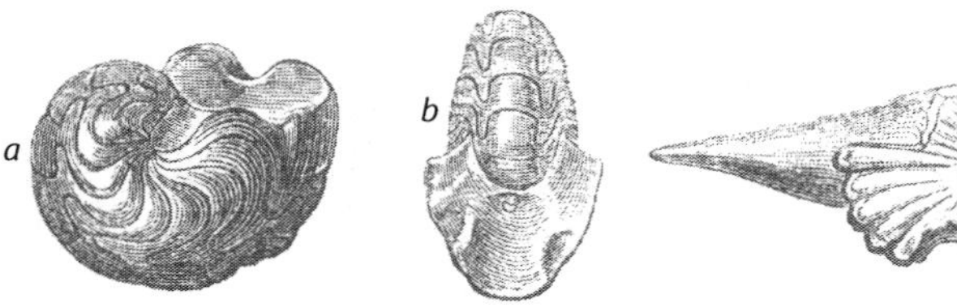

Aturia ziczac, Bronn. Syn. *Nautilus ziczac*, Sow. London clay. Sheppey.

Belosepia sepioidea, De Blainv. London clay. Sheppey.

Figure 5.5 The evidence for warm conditions around London 45 million years ago comes from a range of fossils, not just plant fossils. The climatic implications of these fossils have been discussed for well over 100 years. For example, this illustration of fossil shells in the London clay comes from a standard student textbook on geology from the second half of the nineteenth century written by the eminent geologist Charles Lyell.[100] In the text accompanying this figure, he wrote that 'Marine shells from the London clay confirm the inference derivable from the plants and reptiles in favour of a high temperature'.

may have been affected by drifting continents moving it to a latitude different from that at which the fossil organisms lived). This suggests more geologically young taxa in the tropics, hence greater rates of speciation.[51] Another good example of such a study is the work by Jablonski and colleagues[60] using marine bivalve fossils from the past 11 million years of Earth history. They found that over this time span 117 genera first appeared in the tropics while 46 first appeared outside the tropics—this difference is particularly impressive as rocks outside the tropics (especially in the northern hemisphere) tend to have been better studied by palaeontologists than tropical rocks, as most of the world's major research universities are in temperate regions. They suggested that, at least for marine bivalves, the tropics had more species because they originated there more readily, an idea which Stebbins[61] referred to as the tropics being 'cradles' of species richness. In addition, they were also probably more likely to survive there (tropics as 'museums' of species, in Stebbins'[61] terminology) and so species also spread out from the tropics into non-tropical areas.

Ecological explanations: diversity begets diversity?

There are several ways in which the tropics could act as the 'cradle' of global species richness. First, rates of evolution in the tropics could be higher, perhaps because the organisms there have shorter generation times (hence more opportunity for speciation), or the effects of higher temperature on their biochemistry may lead to higher mutation rates.[62] Another possibility is that the tropics have stronger biotic interactions, such as parasitism or predation, and this drives a greater rate of speciation.[63] In our later section on 'ecology meets evolution', we consider several specific ways in which diversity could beget more diversity in a runaway process. It is not entirely a circular argument, because similar to the tropical conservatism hypothesis discussed earlier, it stresses how initial differences can become magnified over time.

At its most basic, one might argue that the occurrence of many species of plants potentially allows more herbivores, which could then allow more carnivores. However, this 'common sense' explanation does not always match the data; for example, primate species richness in South America shows a positive correlation with plant productivity but *not* with plant species richness.[64] Clearly, both climatic effects on rates of evolution and the effects of biotic interactions could potentially operate together. As a recent review pointed out,[63] distinguishing between these two mechanisms will be challenging and we currently have little suitable data to attempt this. However, the idea that the tropics may be the 'cradle' of species richness is not just of theoretical interest; if such a view is correct then it has implications for conservation. As Jablonski and colleagues[60] pointed out, 'If the tropics are the engines of global biodiversity... then major losses of tropical taxa will have a global effect by suppressing the primary source of evolutionary novelty for all latitudes'.

Allowing for the difficulties in testing ideas that apply to geological expanses of time, there is a reasonable amount of data that is compatible with the idea of greater overall speciation rates in the tropics, as predicted by the area hypothesis and several other hypotheses. However, there are also studies that appear to run counter to some of its predictions. For example, a comparison of the DNA of closely related (sister) species of both birds and mammals suggested that taxa at higher latitudes had experienced both higher speciation and extinction rates than the tropics; such a higher turnover of species at higher latitudes is consistent with the tropics as 'museums' but not as 'cradles' of species.[65] We now shift our attention to why extinction rates might differ between temperate and tropical areas, although we note that the same processes that bring about higher extinction rates in temperate areas may also facilitate higher speciation rates in the tropics.

Can history explain Amazonian diversity?

One of the most prominent historical explanations for tropical diversity in the ecological literature of the second half of the twentieth century[66,67] has been the glacial refugia explanation for species richness in Amazonia and other areas of tropical forest. This stems from a classic paper in *Science* by Jürgen Haffer in 1969[68] on 'Speciation in Amazonian forest birds'. At the time Haffer, a keen and expert ornithologist, was working as a research geologist for the oil firm Mobil; this background perhaps explains his approach in this work—a novel mix of geology and bird biogeography. In his *Science* paper he used the distribution of Amazonian birds to identify areas of high species richness, which he then used to identify areas of the Amazon basin where tropical forest may have survived during presumed cold and dry periods of the Ice Age. His 'working model' suggested that during cold periods, when ice sheets covered much of the higher latitudes, there were dry conditions in the Amazon basin that caused the forest to contract to a series of isolated patches separated by open country. This, he argued, caused the forest birds (and other forest species) to be isolated and led to allopatric speciation causing high species richness in these areas.

During the course of the Quaternary (the most recent geological period covering the past 2.6 million years and characterized by a series of multiple 'ice age' conditions interspersed with warmer climates), these cycles of intense allopatric speciation would have occurred repeatedly. The idea—which Haffer[68] made clear was a speculative working model—could potentially explain the high species richness observed in tropical forests, compared to the lower diversity at high latitudes where ice sheets had repeatedly cleansed the landscape of most biodiversity so giving no opportunity for a build-up of species richness by repeated allopatric speciation at the same location.[69,70] Keith Bennett[69] has suggested that on a timescale of hundreds of thousands of years the climatic oscillations associated with changes in the Earth's orbit (Milankovitch cycles—see Chapter 9) mix up populations and tend to prevent allopatric speciation, with this tendency presumably being more pronounced at higher latitudes that were affected by the repeated ebb and flow of Quaternary ice sheets. Of course, you do not need both higher extinctions in temperate areas and greater rates of speciation in the tropics to generate a species diversity gradient, but it is a compelling 'double whammy'.

One obvious way to test these ideas is using data on the past distribution of forest in the Amazon catchment; however, such information was not available to Haffer in the 1960s. Today the situation is a bit better, but it is still the case that there are only a few areas of the Amazon catchment that have provided good data.[71,72] The main source of such data is from pollen preserved in lake and other sediments such as those deposited off shore by the Amazon River. Sediment cores can be collected, taken back to the laboratory, and pollen extracted and identified from different levels in the core; these can then be dated using radiocarbon and other methods. In 1996, Paul Colinvaux and colleagues[73] published a controversial paper in *Science* describing pollen data from the sediments of a lake in a part of the Amazon identified by Haffer[68] and others[74] as being within an area where the forest was likely to have been absent during full glacial conditions. These data showed that tropical rainforest had occupied the area for more than 40,000 years, although the species composition changed during colder glacial conditions. There was no evidence of a dry climate dominated by non-forest vegetation, assumed necessary for allopatric speciation. Since then, more detailed work by the same team on lake sediment cores from this area, which now covers the past 170,000 years, has confirmed these findings.[72] The obvious provisional conclusion is that Haffer's hypothesis is an example of what the nineteenth century biologist T.H. Huxley described as 'The great tragedy of Science—the slaying of a beautiful hypothesis by an ugly fact'.[75] However, we stress that this conclusion is as yet based on a very limited amount of pollen data (sediments of the correct age are rare in the Amazon), most of which has been collected by a single research team.

Although developed for the Amazon, Haffer's ideas may work better in other parts of the tropics—for example, there is better evidence for his postulated dry periods in tropical Africa.[71] Indeed early studies of African climatic history informed his original educated guesses about what may have happened in the Amazon. Colinvaux's work suggests greater climatic stability in much of the tropics, compared to glaciated higher latitudes; this should have allowed a higher number of specialist species to evolve and survive there.[69,76] So although Haffer's specific original idea may be wrong, the climatic fluctuations of the geologically recent past are likely to have played a role in the current high diversity of many tropical habitats.

The idea that between-year climatic stability is important in maintaining and/or promoting tropical diversity, is difficult to test. One reason is that climate in the geologically recent past (the Ice Ages) tends to be correlated with modern climate—so areas that are warm today had lower magnitude changes in the past. However, a recent attempt to compare flowering plant diversity with both modern climate, and the magnitude of inferred climate change since the height of the last 'ice age', found that the amount of past climate change was statistically at least as good as modern climate in predicting current diversity.[77] We will return to the important issue of stability, this time from a perspective of within year (seasonal) climatic stability in a later section.

More ecological explanations

One of the first ecological ideas that many readers of this book will have come across at high school is that of the simple food chain. Green plants fix solar energy and then are eaten by a herbivore that is later consumed by a carnivore, and some of the energy is passed up the food chain; so in one sense, lions can be described as solar-powered animals. This simple idea suggests that it might be useful to ask questions about the amount of light available for photosynthesis in the tropics compared to higher latitudes—after all, holiday brochures suggest that you should go to tropical beaches if you want lots of sunshine.

Initially, this does not appear to be a very promising idea as everywhere on Earth gets the same amount of light—half a year's worth. This is because while higher latitudes have long winter nights, they make up for it with long summer days. However, the Earth is roughly spherical which means that a fixed amount of incoming solar energy is spread over a greater amount of ground surface at higher latitudes.[78] The classic way to demonstrate this is with a globe and a flashlight in a darkened room. Shine the light beam directly at the equator and note the amount of surface of the globe that is illuminated. If you raise the flashlight vertically (keeping it horizontal) so it is now shining on the Arctic the light will be spread over more of the globe as the curve of the sphere bends away from the incoming light beam. So, low latitudes acquire more solar energy per unit area of ground than higher latitudes—this is why the tropics are warmer than high latitudes. With more energy flowing into the base of food chains, it might seem reasonable that the tropics should support more species; however, it is not immediately apparent why more energy should give rise to more species rather than just more biomass—with a few superabundant species using up all this extra energy.[79] Increased productivity in the tropics was one of the potential mechanisms for tropical diversity identified by Eric Pianka in his influential 1960s review paper[5] and was also championed by G.E. Hutchinson[80] around the same time in a highly influential paper on animal diversity published in *American Naturalist*.

This idea is often called the 'energy richness hypothesis' and has been formally described in the following way: 'Species richness varies as a function of the total number of individuals in an area. Net primary productivity (NPP) limits the number of individuals, and climate strongly affects NPP'.[63] Excluding arid areas, where water shortage limits net primary productivity (NPP) (i.e. the amount of biological production by autotrophs—such as green plants—after the effects of respiration have been subtracted), then NPP will often be strongly correlated with temperature—for the reasons outlined earlier about latitudinal gradients in solar energy. This idea looks promising, in that there is a widely described relationship between measures of productivity and species richness for many groups of organisms.[18,63] However, more detailed consideration suggests that the situation may be rather more complex than one might at first assume. One of the explicit assumptions of the energy richness hypothesis is that the density of individuals should be positively correlated with productivity—so that the density of individuals should be greatest in warm, wet places. However, analysis of several large-scale data sets (of forest trees from tropical America, North American breeding birds, and North American butterflies) show at best a weak relationship between productivity and density of individuals.[63] In addition, the causal relationships in the energy richness hypothesis suggest that NPP affects the number of individuals (I), which in turn affects the number of species (S), so that NPP → I → S. It follows from this that the correlations between NPP and I or between I and S should be stronger than those between NPP and S, because other variables not relevant to the energy richness hypothesis will be affecting both links in the chain so reducing the overall correlation between NPP and S. However, in general, correlations between NPP and S are found to be stronger than those between NPP and I; this undermines the causal relationships suggested by the energy richness hypothesis.[63] The above-mentioned tighter correlation between energy and species richness may come as something as a surprise: for example, our previous chapter on speciation dealt with the variety of ways in which new species form, but at no point did we suggest that higher temperatures, or energy input to an ecosystem, facilitated the speciation process. In this case, it is probably worth reminding oneself of the well-worn maxim repeated in a multitude of statistics textbooks that correlations do not demonstrate causality.

The role of within-year climatic stability

As well as having more solar energy per unit area, the tropics also have this energy distributed relatively evenly across the year so avoiding the cold winters of higher latitudes. The reason why this may be relevant to our question of 'Why are the tropics so diverse' can be seen by considering birds. Many high-latitude countries, such as Canada and Britain, have bird species that feed exclusively on insects during the summer, a good example being the Barn Swallow. This species feeds almost exclusively on invertebrates it catches while in flight; such food is effectively unavailable during the northern winter and the swallows migrate south to areas where this food is still plentiful.[81] So while higher-latitude sites can have a high biomass of insects in the summer (indeed there can be a greater insect biomass than in many tropical forests), they are unavailable as food for much of the year, thus restricting the diversity of specialist insectivores.[82] There is a similar situation with fruit-eating birds: fruit is available year round in many parts of the tropics allowing specialist fruit-eating birds to evolve there.[83] However, in a higher-latitude country such as Britain, there are very few fruits available from the late winter until mid-summer,[84] making it impossible to be a resident specialist fruit-eating bird. A recent analysis of seasonal variation in bird species richness in North and Central America showed that, because of migration, patterns of diversity tracked changes in temperature, precipitation, and vegetation over seasonal time sales, paralleling a similar relationship on a spatial scale.[85] Incidentally, this ability to understand temporal variation in species richness and the close relationship between temporal and spatial variation gives important support to the idea that climate variables play an important role in influencing species richness.

The bird examples we have just discussed also help to make a more general point. Clearly climate often limits species distributions, as illustrated by a multitude of gardening texts describing which plants can (and cannot) be grown in a certain climate. Since more species may be able to tolerate local conditions in the tropics year round, then this may lead to a correlation between climate and/or NPP and species richness. The exact reasons why more species should have tolerances for tropical conditions than temperate conditions are varied. One reason may be that it simply arises out of some form of fundamental physiological constraint—most species tolerate conditions in some places (warm and wet) better than others (cold and dry). However, taking an evolutionary perspective, we find ourselves coming back to a hypothesis we introduced earlier—more species may have tolerances to the tropics because of niche conservatism (many major taxa arose in the humid tropics, and what starts in the tropics tends to stay in the tropics). The relative lack of seasonality in tropical environments may also play a role, promoting increased specialization under these conditions which reduces competition and thereby allows more species to be packed in (see also Rapoport's rule later).

Nevertheless, there are problems with the tolerance idea, particularly when tolerances are viewed as some form of immutable constraint. As David Currie and colleagues[63] have pointed out, the fact that 'many species are often absent from areas whose climate they can tolerate, and to which they apparently could disperse', suggesting that tolerances do not fully dictate distributions, and cannot therefore provide a full explanation for the patterns of latitudinal species richness. Possible examples of this are the shrub *Rhododendron* along with sycamore trees in parts of Europe. Both of the aforementioned species are native to some areas of Europe but, as human introductions have demonstrated, they are able to thrive in many parts of the continent outside their natural range.[86] *Rhododendron*, in particular, appears unlikely to have been limited by dispersal as it naturally reached many of the areas, where it is now a human introduction, in previous interglacial periods.[86,87]

Ecology meets evolution

There is another potentially important idea that follows from the relative stability of tropical climates. In 1967, Dan Janzen published a paper with a characteristically alluring title 'Why mountain passes are higher in the tropics'.[88] In that paper, he pointed out that tropical mountains may form greater barriers to dispersing organisms if one thought about them as physiological rather than topographical barriers. He first assumed that one of the main ways a mountain stopped organisms from dispersing through them was because of the mountains effects on temperature—so that a particular species could not cross a mountain range because it was unable to survive at the low temperatures found high in the mountains. He then pointed out that the within-year variation in climate was greater at high latitudes, so making it more likely that a non-tropical species would be able to survive the wide range in temperatures it would experience in crossing a mountain range. It followed from this that mountains will form much more significant barriers in the tropics because of the reduced seasonal changes in climate, so leading to many tropical species having smaller range sizes and so an increased likelihood of allopatric speciation. Rather modestly, Janzen, in his original publication, made it very clear that it was 'not an attempt to explain tropical species diversity', and was more concerned with explaining the ecological observation that tropical species were often much more geographically localized than species at higher latitudes.[88] However, reduced range size and increased speciation opportunities are obviously very relevant to the evolutionary explanations for tropical diversity we have already described.

Janzen's 'mountain pass' paper has been very influential and the basic ideas behind it are obviously based on well-established climatology—such as the increased seasonality in higher latitudes. A recent review of the current status of Janzen's idea[89] concluded that while there was no direct test of the suggestion as a whole, there was a large amount of published data consistent with its main assumptions. However, this review did point out several potential problems and complications with Janzen's approach. For example, it seems likely that the idea works better for comparisons between the tropics and high-latitude sites in the northern hemisphere. This is because the large amount of ocean in the southern hemisphere reduces the extremes of climate at higher latitudes—as we have already explained in the context of hemispherical differences in the latitudinal gradient of species. Also, Janzen focused on annual variations in climate; however, the diurnal variations of climate can be quite high and this is often particularly the case in low latitudes.[89] Any mountaineer or desert traveller reading this chapter is likely to have experienced cases of sub-zero night-time temperatures followed by days where overheating is the main problem. So, organisms living in some tropical areas may be adapted to a wider range of temperatures than Janzen had assumed. Nevertheless, the basic idea that the relative lack of climatic variation in the tropics can contribute to increased geographical isolation has a reasonable amount of observational support—if no direct tests—and although Janzen suggested it in a primarily ecological context it is highly relevant to the more 'evolutionary' explanations of tropical diversity. This should come as no surprise, as the close relationship between 'ecological' and 'evolutionary' explanations for our big questions has been a recurring theme in many of our chapters (see also Chapter 11).

A related idea to Janzen's 'mountain pass' arguments is the so-called 'Rapoport's rule'. This suggests that species at higher latitudes tend to have larger range sizes and wider ecological tolerances. The rule was named after the Mexican ecologist Eduardo Rapoport, who described it in a study otherwise devoted to the biogeography of mammalian subspecies.[90,91] It is an empirical observation rather than an explanation; however, one reason why the rule may arise is because of the increased environmental variation at higher latitudes. Assuming that many species struggle to adapt to widely varying conditions, perhaps because of niche conservatism, then it follows from this 'rule' that fewer species can be fitted into the non-tropical areas.[91,92] The current consensus seems to be that while this pattern may sometimes exist at high latitudes (e.g. 40°N–50°N) there is little evidence that it is a valid generalization at lower latitudes and within the tropics.[15]

Janzen's 'mountain pass' arguments were developed in relation to climatic differences between the tropics and higher latitude; however, his fertile scientific imagination has also produced potentially important ideas about differences in the biological environments.[93] In the early 1970s, both Janzen and J.H. Connell independently suggested the same potential mechanism to help explain tropical diversity—although Connell later came to believe the idea was at best only partially correct.[94] They suggested that tropical tree species diversity could be explained by the high numbers of potential herbivores (mainly insects), and more conventional parasites, in the tropics (a 'diversity begets diversity' explanation). The suggestion was that any tropical seedling trying to grow near its parent would be overcome by herbivores/parasites that had been feeding on its parent. It follows from this that the development of dense stands of trees of the same species may be difficult in the tropics (unlike, for example, the boreal forests of Canada or northern Scandinavia), so leading to an increased diversity of tropical tree species. Some studies have produced data consistent with this hypothesis. For example, Augspurger[95] found that the seedlings of a particular wind-dispersed tree on Barro Colorado Island, in Panama, grew better if they were distant from the parent, as they were more likely to escape the effects of parasitic fungi. However, as Connell himself pointed out[94] the experimental results are rather mixed and many other studies have failed to find such a relationship between seedling success and distance from parent. Also, in many studies success was evaluated at the germination stage, whereas the true test of this idea would be to measure this as reproductive success. For trees this would take too long to document from start to finish even for the most patient and long-lived scientists, but it is now possible to estimate the genetic relatedness among mature trees different distances apart using molecular methods, so potentially allowing better tests of these ideas. Moreover, the idea that parasite (broadly defined to include organisms such as caterpillars) pressure may be important in helping to explain tropical diversity could also be relevant to animals as well as plants. For example, a comparison of pairs of closely related bird species (in each pair one was tropical and the other one was not) suggested that the tropical birds invested more in anti-parasite defence—measured by density of leukocytes in the blood and by spleen size.[96] It would be interesting to see if this intriguing pattern can be replicated in other data sets. While parasite (including herbivore) levels do not seem to provide a full explanation for tropical diversity, it is possible that they may yet turn out to have a role in speciation through mediating processes such as seed dispersal and sexual selection.

Conclusion—so why are the tropics so diverse?

It is clear from the discussion in this chapter that there are many different potential explanations for the species richness latitudinal gradient and many of them are interrelated. The phrasing of our question as 'Why are the tropics so diverse?' may be, in part, an artefact of the fact that most research ecologists (including ourselves) have tended to be based in universities in temperate latitudes. A more natural way for a tropical biologist to ask the question could be 'Why are high latitudes so species poor?'.[78] Such a person may, for example, be inclined to give particular emphasis to the effects of 'ice age' conditions reducing the probability of speciation (or increased extinction rates) at high latitudes.

So what is the explanation for tropical diversity? As with the explanation of many patterns in ecology and evolution, there is probably no single all-embracing explanation. As John Lawton[97] has written:

> Too often, ecologists seem obsessed with finding a single explanation for some process or pattern...For many phenomena, there are likely to be several contributory mechanisms, and the question is not so much about which mechanism is correct, but about the relative contributions of a plurality of mechanisms.

The null model of the mid-domain effect may be capable of explaining some, but certainly not all, of the pattern—it is most likely to work for groups such as birds that tend to have reasonably large range sizes. The effects of area, as suggested by Terborgh and Rosenzweig, could also contribute. Here geography and geometry (e.g. the large area of the tropics) potentially influence evolutionary factors such as opportunities for speciation—as do the long geological history of tropical conditions and greater climatic variation at higher latitudes including the ebb and flow of ice sheets. The most obvious ecological effect, and the factor most likely to explain patterns of species richness, is the correlation between tropical climates (both current and past) and species richness—at least for most terrestrial macroscopic organisms. However, it is currently not clear exactly how climate affects species richness because, as we have described earlier, the correlation between NPP, number of individuals, and number of species suggests that it is not mediated directly by biomass. In addition, there is the problem that current climate tends to correlate with size of past climatic changes, making it very difficult to disentangle these two effects.

As noted in our introduction, perhaps the greatest advances in the next few years will be made by attempting to understand global patterns of species richness on a finer spatial scale than simply comparing tropics and temperate areas, or the gradients of species richness with latitude. While part of the attraction of the latitudinal diversity gradient is that it appears so general, it is of little surprise that there are so many theories out there that can equally explain observations when they rest on explaining differences between only two broad geographical areas.

In reflecting on his work on Amazonian ecology, conducted in collaboration with W.D. Hamilton (see Chapters 1–3 for discussions of some of Hamilton's other insights), Peter Henderson[98] suggested that their approach was marked out by combining both evolution and ecology, including processes operating over time scales of millions of years. He points out that such an approach is unusual, with most studies focusing on a single time scale, and focusing on ecological *or* evolutionary processes. However, in reviewing the huge literature on tropical diversity during the process of writing this chapter, we are of the opinion that Henderson[10] and Hamilton were thinking along the right lines. While we are still some way from a full understanding of latitudinal diversity gradients, the correct approach must consider *both* evolutionary and ecological processes, and give considerable emphasis to processes that are difficult to study because they happen on a time scale that is much longer than the life of an individual biologist—however, long-lived. It is deeply depressing to reflect on the fact that with on-going large-scale, human-caused, loss of tropical diversity,[16] the opportunities for increasing our understanding of these phenomena may be sadly curtailed.

6

Is Nature Chaotic?

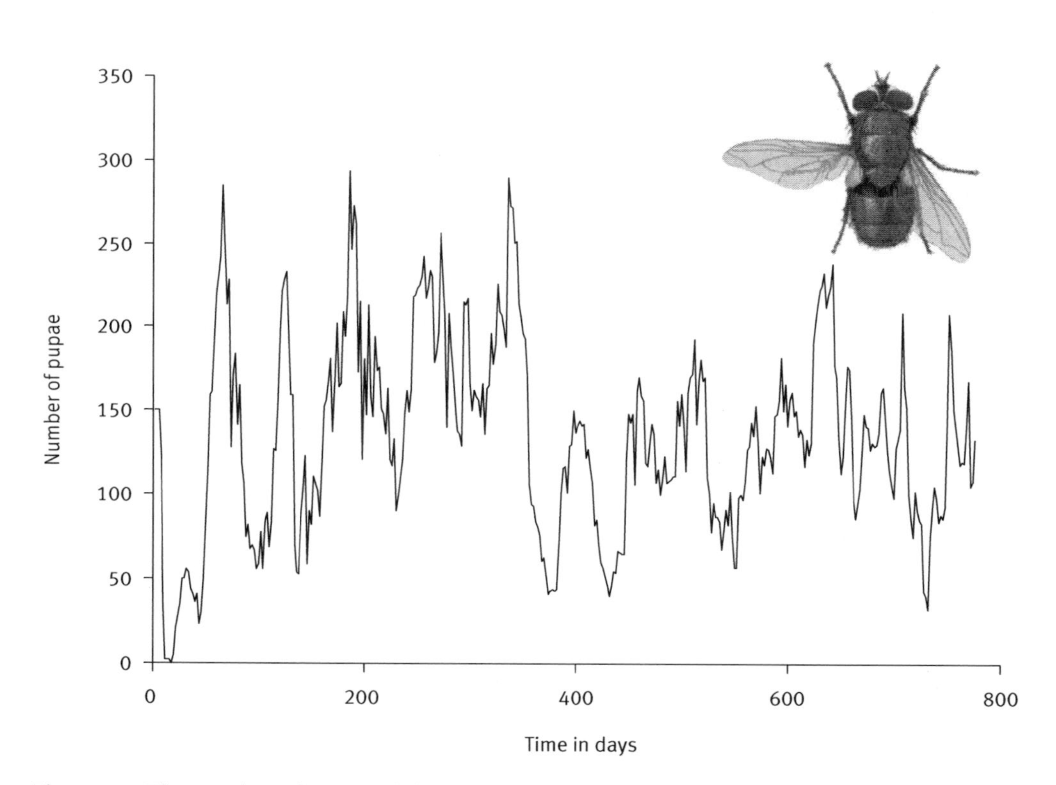

Figure 6.1 The number of pupae of the green bottle (sheep blowfly), in a laboratory population monitored every two days for two years. Data kindly made available to researchers by Robert Smith and colleagues (see http://mcs.open.ac.uk/drm48/chaos/).[1]

> *... even if it were the case that the natural laws had no longer any secret for us, we could still only know the initial situation approximately... it may happen that small differences in the initial conditions produce very great ones in the final phenomena. A small error in the former will produce an enormous error in the latter. Prediction becomes impossible...*
>
> —*Henri Poincaré (1908).*[2]

Centuries before King Harold of England famously received an arrow in the eye (AD 1066), Chinese officials in the T'ang dynasty (AD 618–907) began collecting annual reports on the abundance of migratory locusts.[3,4] The primary aim of this initiative was to make sense of the changes over time (the dynamics) of this devastating agricultural pest, and thereby predict the timing and intensity of outbreaks. Now, despite a staggering 1,300 years of faithful recording, few patterns are evident and the data look decidedly messy.[5] Irregular climatic fluctuations, particularly those involved in the drying up of grasslands on river deltas, may explain some of the variability.[4] However, one might wonder whether some of this 'messiness' was internally driven, caused by some sort of 'feedback' arising within the dynamics themselves. Many long-term data sets on population dynamics have these extremely messy qualities, ranging from the daily number of damselfish reaching maturity on the Great Barrier Reef[6] to the number of feral sheep on Scottish Islands,[7] and it is important to know where it all comes from.

The study of 'chaos' (easiest to define negatively as an absence of order, but we will get to a more formal definition later) has its roots in precisely the type of feedback processes referred to above, reflecting what mathematicians call 'non-linearities' (relationships that are not straight lines). Several mathematicians, most notably, the eminent French mathematician Henri Poincaré (1854–1912), had long noted that non-linear systems could generate some extremely unusual dynamics, such that the precise trajectory a system took was highly sensitive to the initial conditions. However, observations such as these were largely overlooked by ecologists until a new generation of researchers, notably Robert May (a physicist turned ecologist, now Lord May of Oxford), began toying with their own simple ecological models and appreciating that the behaviour of these models was not always simple.[8,9] Until ecologists were made aware of the potential effects of non-linearities in the 1970s, the prevailing view was that complex dynamics must have complex causes. One of the many benefits of the development of chaos theory is that it has led to an appreciation that sometimes extremely complicated dynamics can arise out of the simplest and most innocuous looking of mathematical models, even those without any elements of chance built in.

We begin this chapter by describing one such simple model with potentially complicated dynamics, called the 'discrete-time logistic growth model'. A version of the logistic model was introduced in 1838 by Pierre François Verhulst (and later rediscovered by Raymond Pearl[10]) in an attempt to formalize arguments he encountered in Thomas Malthus' *An Essay on the Principle of Population*[11] (an essay that also had a famous influence on Charles Darwin's ideas). This mathematical model will help define what chaos is, and how it is arrived at, before we go on to ask our ultimate question of whether natural populations fluctuate in a chaotic manner. If chaotic dynamics are a common feature of natural population fluctuations, then it has all sorts of implications for conservation biology, disease control, and many other areas of ecology; therefore, we take time to consider what a 'yes' answer would mean for ecology. We also ask some related questions, such as whether natural selection tends to produce population fluctuations that lack chaotic dynamics, and whether human intervention can make some non-chaotic populations chaotic and *vice versa*.

The fish pond

Imagine a population of fish in a pond. We census the population each year at the end of the breeding season; let the symbol x_t represent the population size of these fish in generation t, expressed as a fraction of the absolute maximum number of fish that could ever live there (this conveniently helps keep all numbers between 0 and 1). How might x_t vary over consecutive generations? At extremely low densities, each individual would have access to plenty of resources so it is likely that each individual would produce a relatively high number of offspring. In contrast at high density, individuals would be competing over resources, so that each individual would not leave as many surviving offspring. In effect, the population should 'feedback' on itself—at low population densities the *per capita* population growth rate would be relatively high, but at high population densities the *per capita* population growth rate would be relatively low. It is a good bet that something like this goes on in many populations—after all, no species on the planet goes through permanently unfettered geometric growth. We know this for sure, because (as Darwin had argued in the case of elephants[12]), were it any different, we would soon be up to our eyeballs in them.

How do we express this type of 'density-dependent' feedback mathematically? There are lots of different ways, many of which would yield qualitatively similar results, but one of the simplest is to simply let $x_{t+1} = rx_t\,(1-x_t)$ where r is a mathematical constant. Although we have largely avoided formal mathematics in this book, in this case it is worth working through the implications of this simple equation because of the insights that it provides. Here we see that when x_t is extremely small, then the index of population density in the next generation (x_{t+1}) is approximately rx_t (since $1-x_t$ is approximately equal to 1). In other words, the *per capita* population growth rate is almost r when the population size is small. Yet, when we increase x_t the feedback term $(1-x_t)$ now becomes increasingly smaller, so the *per capita* population growth rate diminishes. The mathematical function we have assumed might appear somewhat arbitrary, and probably there is not a population on Earth that actually shows precisely this dynamic, but it does the trick of introducing a feedback, and it makes sense to start with a simple rule. We also note in passing that this is a *discrete-time* version of the logistic equation, representing population size in the next generation as a function of population size in the current generation. As such, the equations used to predict the changes in population size are called 'difference equations'. However, we could let generation time tend to zero and end up with smoother, completely continuous changes in population size. Under these conditions, we would have a 'differential equation' (the way the logistic equation is often presented in ecology textbooks) and in this case the tools of calculus could be used to understand their dynamics.

Let us get back to the discrete-time logistic equation. We can see directly that the feedback involves a non-linearity when we plot x_{t+1} against x_t for a variety of values of x_t between 0 and 1 (Fig. 6.2). Thus, when $x_t = 0$ then the predicted population size in the next generation (x_{t+1}) is 0. Equally, when $x_t = 1$ then $x_{t+1} = 0$ (since $1-x_t = 0$). Hence, only intermediate values of x_t generate non-zero values for x_{t+1} and the end result is a curve that bends over on itself rather like a hairpin (a 'fold'). In fact, these particular curves are 'parabolas' (yes, the trajectory of a cannon ball we all know and love from high-school mathematics). Interestingly, increasing the value of r increases the intensity of the feedback and hence the severity of the folding (Fig. 6.2).

To see what dynamics are predicted by the model, we can start with a particular index of population density x_0 (e.g. 0.4) and simply update the equation iteratively

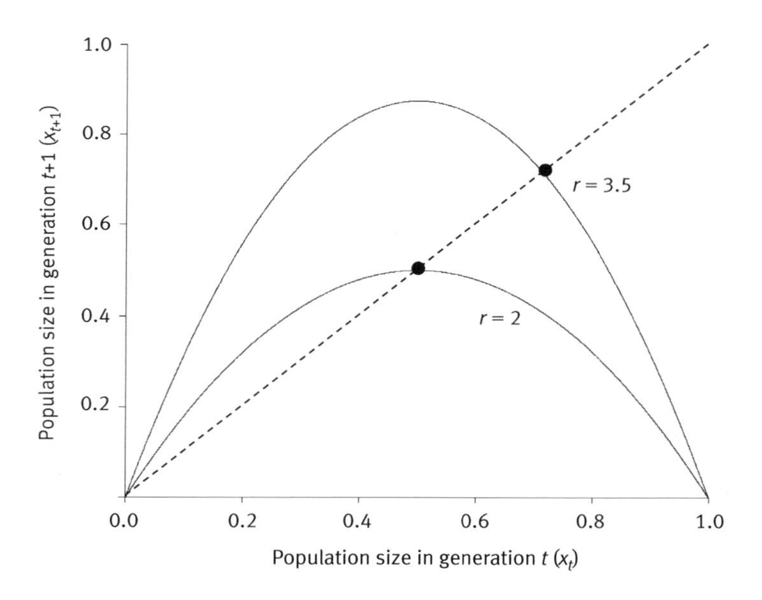

Figure 6.2 A 'map' of population density x_t against x_{t+1} as assumed by the logistic equation. Increasing the value of r increases the curvature of the relationship because changes are higher the higher the value of r. The graph also shows a line (dotted) in which x_t is plotted against x_{t+1}, allowing us to highlight where potential equilibria occur (for $r = 2$ population densities move towards this equilibria, but for $r = 3.5$ population changes are just too lumpy to allow the equilibrium to be converged upon and population densities vary around this equilibrium instead.

(i.e. in steps), calculating x_1 and placing it back into the equation to find x_2 and so on. You can try yourself—it requires no mathematics beyond arithmetic—yet had you done these simple calculations in the 1960s, *and realized their significance*, then you would have made an important scientific discovery. It turns out that the type of dynamics one predicts is only dependent on the value of r (and not the value of x in the starting generation). If r is relatively small (e.g. 2), then the fish population size always rises to a single value (the 'equilibrium') and stays there indefinitely (see Fig. 6.3a). The more mathematically minded reader might wish to confirm, by setting $x_{t+1} = x_t$, that this equilibrium is $(r - 1)/r$. Equilibria similar to these have a certain appeal and they imply a reassuring sense of stability and order. Indeed, before the 1970s, equilibrium solutions were the type of result most ecologists concentrated on when developing and exploring their models,[13] almost going out of their way to ignore complications.[14] The oversight comes in part from the absence of fast computers to help visualize the dynamics (the fastest computers in the world in 1970 were several orders of magnitude slower than a good modern desktop[15]), and it is no coincidence that the development of ideas about chaos came with the increase of computing power.

As we increase r further, then strange things happen. First, we get regular repeated cycles occurring in which the population overshoots the equilibrium then undershoots it, overshoots, then undershoots (Fig. 6.3b). This can be seen as a simple consequence of the discrete ('lumpy') nature of the change—the higher r, the higher the potential size of the changes from generation to generation, and the less fine-scale adjustment is possible (rather like adjusting temperature in a shower, in which the time delay between adjusting the handle and experiencing its effects means you can never get it just right). In Fig. 6.3a, the population compensates for being above or below the equilibrium value, so that each generation is closer to the equilibrium than the last. However, a higher r value makes the system feedbacks larger, and this higher sensitivity tends to lead to overcompensation and so the population never settles down to the equilibrium, but fluctuates around it. Note that although the system shown in Fig. 6.3b does not settle down to a single equilibrium, the dynamics are entirely predictable, so that the size of the population is always exactly as it was four generations previously. Increasing the value of r still further produces dynamics that seem to lack any sort of pattern at all (Fig. 6.3c)—it is no longer a question of consistently overshooting and undershooting but rather irregular behaviour that never quite repeats itself. Welcome to the world of chaos. There are no elements of chance whatsoever built into these dynamics (in the jargon, the model is 'deterministic', as opposed to 'stochastic')—the apparent noise is solely driven by the high degree of non-linearity in the system (it is not noise at all, but 'deterministic chaos').

Beautiful bifurcations

To see more clearly how chaos is arrived at, imagine starting at some arbitrary value ($x_0 = 0.3$ say, the exact value does not matter) and iteratively calculating the population

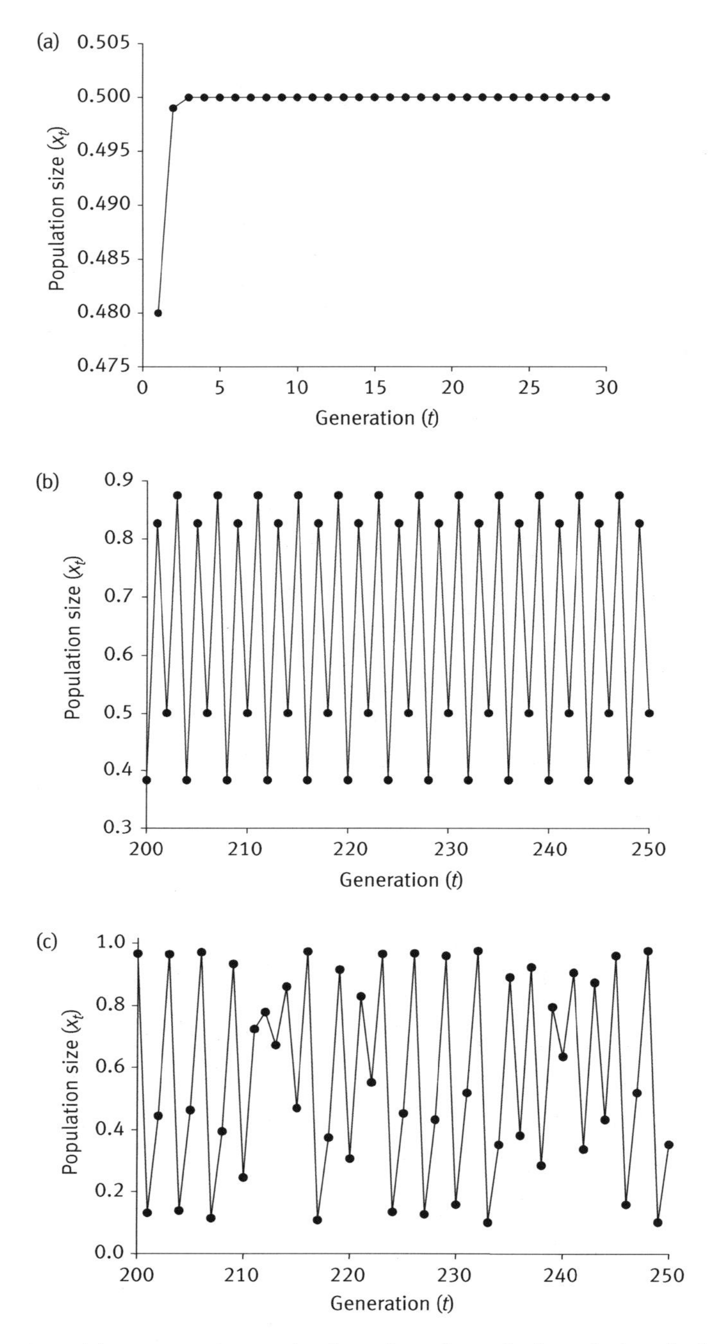

Figure 6.3 Starting with $x_0 = 0.4$, the graphs show the values of x_t iteratively calculated over multiple generations for different values of r. (a) When the parameter r is small (here $r = 2$), we observe a simple rise towards equilibrium. (b) When the parameter r is increased (here $r = 3.5$), we start to see cycles (here a four-point cycle is shown). (c) Increasing r even further (here $r = 3.9$), we start to see chaotic dynamics with no underlying pattern.

dynamics over 10,000 generations—very time-consuming to do 'by hand' but quick and easy on a computer. Now let us plot out the value of *r* on the *x*-axis, against the population size(s) calculated over the final 9,000 generations of the 10,000 generation iteration (Fig. 6.4). We ignore the first 1,000 generations because they will include the 'transient' population sizes that inevitably arise from the particular starting value (x_0). Once we have done this for one value of *r*, let us repeat the whole exercise for a slightly higher value of *r* until we have explored the full range of *r*. When *r* is low then the final 9,000 population sizes will be exactly the same as one another (the equilibrium) and so they will be represented by single point on the graph. As *r* increases, the equilibrium value changes (it increases in this case) but it is still a single value (the equilibrium) for a given value of *r* and so the 9,000 values are again represented by a single point. Nevertheless, as we increase *r* into the range at which two-point cycles arise, then the population sizes in the final 9,000 generations will fluctuate between two values and two points will start to appear on the graph (Fig. 6.4). You will see that there is a relationship between these two-point cycles and the former equilibrium, with the single line effectively

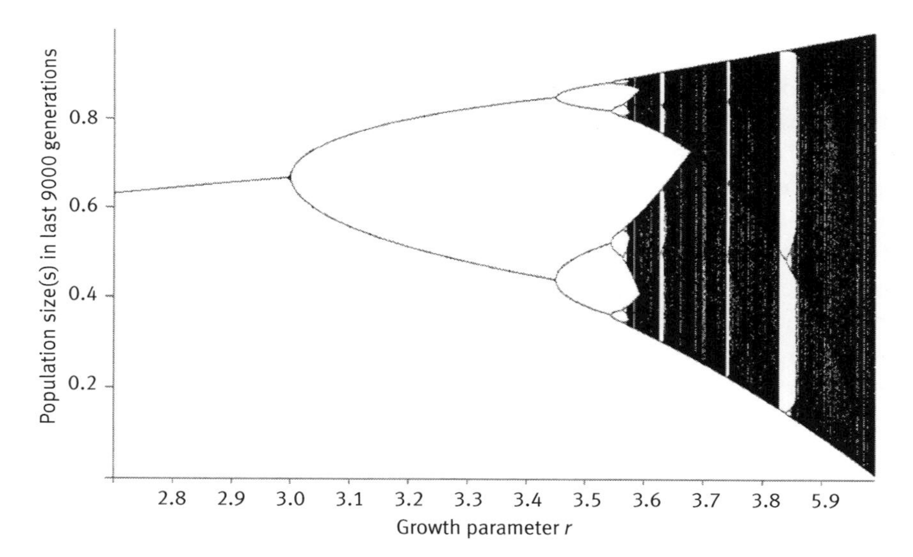

Figure 6.4 A bifurcation diagram, created here by plotting out the population sizes x_t in the final 9,000 generations of a 10,000 generation iteration of the logistic equation. We omit the early values because they may be 'transients', not typical of the steady-state dynamics. For low values of *r*, only a single value of *x* is recorded (the equilibrium) for all 9,000 generations. As *r* increases, then two values are reported (a period-2 cycle), then four values are reported (a period-4 cycle): 8, 16, 32, and so on. The range of values of *r* with a particular period cycle gets progressively smaller, allowing a point of accumulation beyond which an infinite number of points occur in the cycle and there is no pattern to the dynamics. Even within the chaotic regime, however, we can have ranges of *r* that give regular predictable cycles and these can be of odd numbers such as period 3.

'bifurcating' into two. As r increases further, we see the amplitude of the oscillations increases (another consequence of the increase in lumpiness) until another bifurcation takes place and a four-point cycle arises. After a smaller range of r then the period 4 cycle bifurcates into a period 8, then (over an increasingly smaller range) period 16, period 32, 64, 128, and so on until a 'point of accumulation' is reached where we break out to an infinite-point cycle. Beyond this point is the chaotic region, the period of the oscillation becomes infinite, and so the dynamics never repeat themselves.

The beauty of bifurcations does not end here. May and Oster[16] proposed that there was something quite predictable to the cascade of period doublings, noting that the ratio of the intervals between successive period doublings was approximately constant, and they did some mathematical work to characterize it. About the same time, Mitchell Feigenbaum took on the challenge of measuring these ratios directly. Using a (now-ancient) Hewlett-Packard HP 65 programmable calculator, he observed that the ratio of the difference between the values at which successive period-doubling bifurcations arise rapidly approached a constant as the number of period doublings increased. This constant was eventually estimated as 4.6692 (to four decimal figures). The fact that the ratios are constant is surprising, but the really surprising thing that Feigenbaum discovered (and mathematicians subsequently helped formally understand) is that a whole range of dynamical equations that likewise have a chaotic region, such as the Ricker equation ($x_{t+1} = x_t \exp[r(1-x_t)]$) used in fisheries research, and the trigonometric mapping $x_{t+1} = k \sin(-x_t)$ used in pure mathematics, all have precisely the same Feigenbaum constant of 4.6692. In other words, the 'scaling ratio' of the bifurcation does not depend on the specific equation. Indeed, Feigenbaum's constant can be used to demonstrate that a model is capable of generating chaos even if it is not directly observed.

So, chaos can be seen as dynamics with an infinite number of points (never repeating) in a cycle—they are 'aperiodic'. Yet peer into the chaotic regime past the point of accumulation and you see that for certain values of r we get regular 3-point cycles. These 3-point cycles bifurcate into 6-point, then 12-point cycles each reaching its own point of accumulation. Elsewhere we have 5-point cycles bifurcating to 10- then 20-point cycles and so on. The bifurcation diagram has what we call fractal structure (more on this later), in that if we focus on smaller and smaller ranges of r and blow them up, we would see the same complex pattern dominated by chaos but with bifurcations once again breaking out. In fact, the first scientific paper to use the word 'chaos' in this context was by Tien-Yien Li and Jim Yorke in 1975, and it highlighted the unusual occurrence of cycles with an odd number of points and explored the implications. The authors entitled their paper 'Period three implies chaos'.[17] Apparently, colleagues had suggested using a rather more sober description, but by using a catchy term the scientists (and many that followed) had an appealing banner under which to sell their work.[13] Using a colourful label for a scientific idea can be very helpful in attracting attention to it; think of 'selfish genes', 'The Red Queen', or 'Gaia'. As Stephen Jay Gould[18] argued 'phenomena without names... will probably not be recognized at all'.

Defining chaos

One of the problems with using terms with broad appeal is that it can also attract criticism for its lack of precision and scope for misunderstanding. 'Chaos' means different things to different people, so we have to be careful to use the term in a strict scientific way. Probably, the easiest definition of chaos is that it is an 'intrinsically driven' absence of order—this 'absence of order' in turn may be interpreted as dynamics that lack any underlying pattern, so that you cannot predict what is going to happen in the long term. Yet many populations have dynamics that appear to lack any form of pattern, and not all of it may be driven directly by internal feedbacks within the population itself. For example, weather may add what we might think of as extrinsic 'noise' to the underlying dynamic (think of an extremely crackly radio reception, which crackles with noise obscuring the 'signal' you are trying to listen to), and so might simple measurement 'error' (not mistakes *per se*, but chance sampling variation when attempting to estimate population size). So if chaos is 'internally driven' unpredictability, then we will need some good mathematical tools for distinguishing intrinsically driven disorder, from extrinsic 'noise'. In other words, how can we tell whether the population dynamics of fish in our pond (or antelopes on a savannah, say) are truly chaotic? As we will see, thankfully chaos has some rather different properties than a sequence of random numbers.

One popular way to ascertain whether a *mathematical model* is capable of exhibiting chaos is to examine the nature of the non-linearity involved and examine how particular parameters might affect the extent of the non-linearity, just as we have done for the discrete-time logistic equation. Once characterized in this way, one can explore the impact of the non-linearity by identifying any potential bifurcation points, and the point of accumulation beyond which chaos lies. Bifurcation diagrams are usually straightforward to generate when there is one dynamical variable of interest (such as the population size of one species), but similar techniques can be used with multiple dynamical variables (such as the population sizes of several species simultaneously). Of course, it is hard to do these types of manipulations with real observations of natural populations (although experiments using flour beetles have met with certain success—as we describe later), so other techniques must be used to look for chaos in real data.

The butterfly effect

There is one property of chaotic dynamics that we have not mentioned yet, but it is such an important and universal property of chaotic systems, that it has now become its key defining characteristic.[19,20] Mathematical models fluctuating chaotically always show extremely sensitive dependence on the initial conditions. This feature has been called the 'butterfly effect' following a 1972 talk by meteorologist, and father of modern chaos theory, Edward Lorenz,[21] who sadly died in early 2008 as were completing our book. Lorenz's original insight came in the 1960s when he recognized and documented the extreme sensitivity to initial conditions exhibited by a simple non-linear model of fluid convection in the atmosphere.[21] Thus, the story goes that if weather systems were chaotic then an almost negligible change in local wind speed in South America, such as that created by a wing flap of a butterfly, may ultimately mean the difference between having a hurricane in the northern hemisphere and not having one. Of course, this sensitivity has nothing to do with butterflies *per se*, and butterflies do not directly trigger anything—it is simply that with chaos, a small difference will always cascade to produce uncorrelated futures (not necessarily bad ones either). The nursery rhyme 'For the want of a nail, the shoe was lost; for the want of a shoe the horse was lost...' captures some of this contingency.

Naturally, the butterfly effect could be called something else, such as the 'seagull effect' (Lorenz's original metaphor). However, the butterfly neatly captures the shape of Lorenz's strange attractor (Fig. 6.5, see later for a full explanation) and, bizarrely, Ray Bradbury's 1952 short story *A Sound of Thunder* also uses a butterfly to depict the nature of extreme sensitivity. In this story, a prehistoric butterfly is crushed underfoot by a time-traveller and this perturbation to the world is sufficient to change the outcome

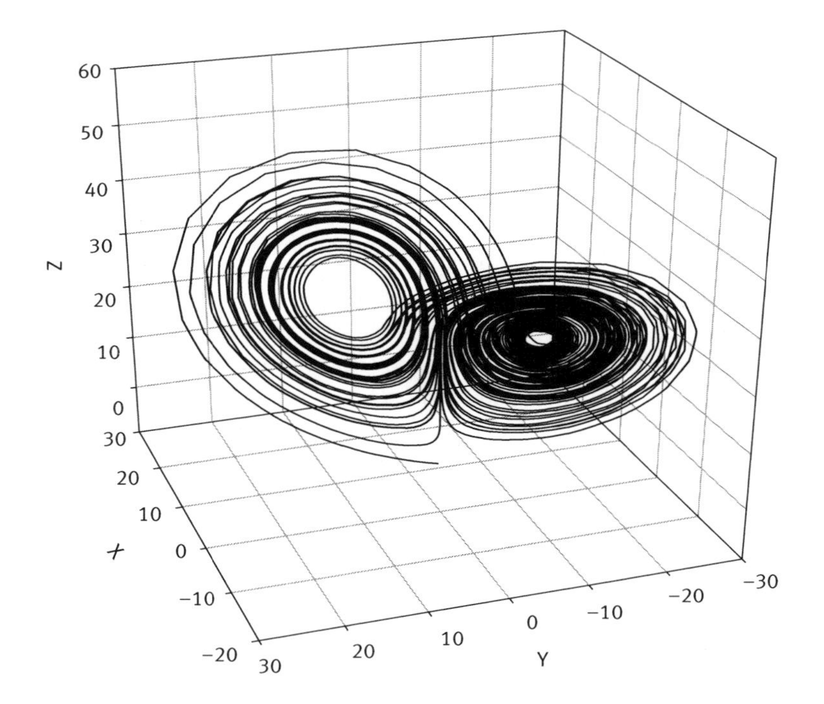

Figure 6.5 The famous Lorenz attractor, derived from a continuous-time model of fluid flow in the atmosphere involving three dynamical variables (x, y, z). Under chaotic conditions the same dynamics never repeat, so the continuous line never intersects itself.

of a presidential election many years later. Apparently, Al Gore hired a mathematician to teach him chaos theory after failing to gain the U.S. presidency in 1988.[13] In almost poetic irony, given Bradbury's short story, it appears that another butterfly—this time the infamous butterfly ballot paper in Florida—may have cost Gore the 2000 election.

Mathematicians have characterized this extreme sensitivity to initial conditions, noting that if a system has chaotic dynamics, then the difference between the trajectories of two populations that have slightly different initial conditions grows exponentially (geometrically) until this difference is essentially as large as the variation in either trajectory. At this point, the two population trajectories have no relationship to one another, although they may have started out at almost the same densities. The rate at which trajectories from similar, but not identical, starting conditions diverge from one another can be characterized by a quantity (or a series of quantities) known as a 'Lyapunov exponent(s)' (spelt in various ways) after the Russian mathematician, Aleksandr Lyapunov. A positive Lyapunov value means that the trajectories do indeed diverge exponentially from one another. In effect, due to their sensitivity to initial conditions, chaotic systems are 'noise amplifiers' while non-chaotic systems with deterministic rules tend to be 'noise mufflers'.[22]

Let us stop to think what this means. If natural populations (or the weather, or atmospheric carbon dioxide levels, or whatever dynamics we are interested in) did fluctuate chaotically, then we could give up on long-term forecasting. We cannot measure the 'start conditions' with infinite precision (imagine trying to record all aspects of the weather simultaneously across the entire globe—or even one small part of it—with complete accuracy), so even if we had the best mathematical model one could ever produce, then the difference between what we thought would happen and what will happen will diverge exponentially. This is just one 'casualty of chaos' and we will return to the full casualty list later, as it helps to demonstrate the great potential significance of chaos.

Fatal attraction

Another important way of determining whether a particular mathematical model or ecological data set exhibits chaos is to present the dynamics in a rather different way, not as population size (or whatever variable you are interested in, such as temperature) against time, but as population sizes against one another. This is most easily seen when there are two or more variables such as densities of a predator and a prey species, or densities of three competing species. Instead of plotting the number of predators, and the number of prey separately against time, we can plot the number of predators at given times against the number of prey at the same times directly, and effectively ignore time. The technical term for displaying dynamics in this way is to show the results in 'phase space'. If predators and prey quickly reach an equilibrium, then this equilibrium will appear as a single point on a graph of predators vs prey, and the dynamics will stay at that point for all the remaining time. In effect, the dynamics will appear as if predators and prey get 'sucked in' to an equilibrium point in phase space, and this equilibrium point is therefore known as an 'attractor'. Other forms of attractor are also possible. For example, if the number of predators goes up when there are plenty of prey to eat, but the number of prey goes down when there are many predators, then predators and prey might enter into regular and predictable cycles. If we plot these cycles not as predator vs time, and prey vs time, but as predators vs prey, then again we would see an attractor, but this attractor would be a regular orbit (a closed loop), with predators and prey continually circling around it.

Now let us consider what chaos would look like in phase space. Chaos almost by definition must be bounded—while lacking order, the variable(s) in question should fall in a finite range between extinction and unfettered growth. Yet at the same time, the lack of order means that the same pattern is never repeated (if it did so, then with no built-in elements of chance, the dynamics would simply have to repeat itself). Imagine, therefore, the long-term dynamics of a population as an infinitely long ball of wool. How can you get an infinitely long ball of wool into a finite space without ever crossing over (repeating) itself? The answer is by having peculiar properties of folding and self-similarity that we alluded to earlier when discussing bifurcation diagrams. In other

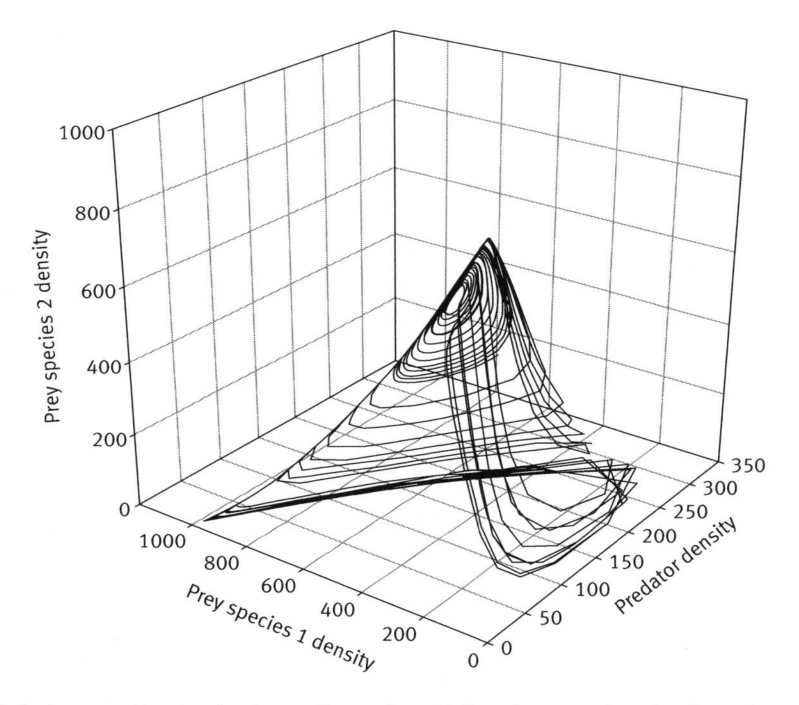

Figure 6.6 A strange attractor in three-dimensional 'phase' space, showing how the population dynamics of a predator and two prey species fluctuate together over time. These chaotic dynamics were discovered by Michael Gilpin[23] in a simple model in which predators and prey continually reproduce and interact with one another.

words, our 'attractor' has to be a strange geometrical object, and for this reason chaotic attractors are known as 'strange attractors'. Figure 6.5 (showing the relationship between dynamical variables in Edward Lorenz's metrological model[21]) and Fig. 6.6 (showing the relationship between predator and prey density in Michael E. Gilpin's three-species ecological model[23]) each depict strange attractors. They look weird, and indeed they are. Their complex beauty has not only attracted biologists, but also artists, art historians, and poets.[24] Choose points on two separate lines and you will see that the trajectories rapidly diverge from one another—the stretching and folding effectively pulls them apart—reflecting the high sensitivity to initial conditions. One quantitative measure of strangeness is a measure of their self-similarity at different scales, an attribute that is measured by their 'fractal dimension'. It is primarily for this reason that chaos is associated with the world of fractals, although we will not be exploring fractals any further in this chapter.

What chaos is, and is not

We now get to a workable definition of chaos and clear up a few misconceptions. A recent definition was proposed by Cushing and colleagues in their book *Chaos in Ecology*.[25] In their definition, which we will use ourselves, they combine elements of disorder, 'boundedness' and sensitivity to initial conditions all in one: 'a trajectory is chaotic if it is bounded in magnitude, is neither periodic nor approaches a periodic state, and is sensitive to initial conditions'. So, it is the sensitivity to initial conditions that provides a key clue to chaotic dynamics.

The first potential misconception is easily cleared up by pointing out that chaos is not only a property of mathematical models expressed in terms of difference equations. We introduced chaos through a simple difference equation, but models based on continuous changes can also exhibit chaos—indeed the two strange attractors in Figs. 6.5 and 6.6 were generated by models with continuous rates of change involving three dynamical variables. It turns out that chaos only occurs in simple differential equation systems involving three or more variables,[20] but the possibilities for chaos get richer as we increase the number of variables.[26] As mathematician Mark Kot[27] noted, 'As soon as you move to three or more species, there are hundreds of ways to get chaos'. Second, while early researchers were taken aback by the complex dynamics predicted by simple sets of equations with no elements of chance involved (so-called deterministic equations), and many investigators continue to emphasize chaos as a primarily deterministic phenomenon, work has also been done to understand the role of small random elements (noise) in these chaotic systems.[22] For example, small amounts of noise added to the dynamic can make something of a mess of bifurcation diagrams we described earlier, but the underlying bifurcations are still evident and the extreme sensitivity to initial conditions remains.[28] Despite this, depending on one's specific definitions, noise may have the potential to turn non-chaotic systems intrinsically chaotic,[29] thereby creating much *more* unpredictability than one would expect from the random elements alone.

The role of noise is currently under debate[30–32] but it is clear that noise may do much more than provide a fuzzy cloud around a deterministic skeleton. We leave further consideration of the influence of such 'stochasticities', particularly in connection with cycling populations, until later in this chapter.

The casualties of chaos

Now that we know what chaos is, we briefly ask what are its implications if it turns out that many populations do indeed exhibit chaotic dynamics. In other words, is it worth finding out whether natural populations are chaotic? We have already pointed to several potential benefits of this branch of research. In particular, if ecologists observe a fluctuating population, it is only natural to wonder whether the fluctuations are caused by external environmental events such as temperature and rainfall, or whether they are caused by internal feedbacks within the population itself. By carefully analysing the data and looking for the signatures of chaos, we can hope to find out.

In some ways, just asking the question moves the debate forward. For example, for several decades in the past century there was a heated debate over whether natural populations were regulated by internal density-dependent mechanisms (such as competition for resources) or external density-independent mechanisms (such as periods of bad weather). At first glance, one might assume (as many population biologists did) that density-dependent mechanisms would tend to produce stable dynamics, while density-independent mechanisms would tend to produce erratic fluctuations. Yet, armed with an understanding of chaos, all this is turned on its head—too strong a density-dependent feedback, and one could end up with highly erratic fluctuations. Not only does this suggest that standard tests for density dependence in time-series data are invalid for chaotic systems, but it also means that just because you see unusual fluctuations does not mean that there is no density-dependence operating. This has implications for global ecology, not just population dynamics—for example, asking questions about the potential regulation of carbon dioxide or oxygen on Earth over geological time.

Perhaps the single most important reason why it is helpful to know whether populations are chaotic relates to the sensitivity with respect to initial conditions. Spontaneous, unpredictable events are a central element of quantum theory, but the thought that natural populations could also show extreme unpredictability due to the sensitivity of their dynamics must have come as a shock to many ecologists. As already noted, if a high proportion of natural populations fluctuated chaotically due to intrinsic non-linear feedbacks, then we could hang on to very short-term forecasting, but kiss goodbye to the goal of long-term forecasting (this appears to be the case for local weather forecasting). We might be able to say statistically what the mean and likely range of population sizes were (in the same way that we can predict climate into the future, even if local weather forecasts are much more constrained), but beyond that we can simply give up. Of course most of us take long-term forecasts of any complex system, be it the weather, the economy, or a natural population, with a pinch of salt, but if they were chaotic, we know that our predictive power has built-in limitations, and we might usefully be able to quantify these limitations. For example, if global weather patterns are chaotic then we might be able to describe how the accuracy of predictions might be expected to decay over longer and longer prediction intervals.

Is mother nature a strange attractor?

Therefore, the crunch question is: do natural populations fluctuate chaotically? To answer this, we must quantitatively examine real data on the estimated size of some specified population over many generations. Early approaches to address the question involved assuming that a particular mathematical model (which was capable of exhibiting chaos under some conditions) was an accurate descriptor of the underlying dynamics. The model was then fitted to the observational data and the parameters were estimated (such as the parameter r in the discrete-time logistic). If the estimated parameter values were such that they would generate chaotic dynamics in the model, then one might be tempted to believe that the dynamics being investigated were also chaotic.

Population biologist Mike Hassell and colleagues[33] took just the above approach in 1976 when they fitted a general discrete-time population model involving three parameters (α, λ, and β) to 28 different data sets on the dynamics of insects (24 from field situations, and four from laboratory studies). It turns out that in their particular model, not one but two parameters—λ (growth rate) and β (a competition coefficient)—influence the degree of non-linearity, and that only high combinations of both λ and β would generate conditions sufficient to produce chaos. After fitting the model, the authors cautiously concluded that the vast majority of insect data sets had parameter combinations that would put the dynamics into a simple equilibrium, and only one case—the classical laboratory study of blowflies conducted by Nicholson[34]—had parameter combinations that would put the dynamics into the chaotic regime. However, as the authors pointed out, even this case may have simply arisen as a laboratory artefact—the flies were not subject to many natural mortality factors such as parasitic wasps, which may have exaggerated the non-linear qualities of the dynamics.

Of course, the entire model-fitting approach is fraught with problems,[35,36] not the least of which is that one must be extremely confident that the model you have fitted does indeed represent the underlying dynamics. Another related objection is that the dynamics of natural populations are often dependent on the influence of many other species, so fitting such a simple model is inappropriate—although one might argue (with a degree of mathematical justification) that the fitted model could be considered a representation of the outcome of all relevant species interactions.[37] Nevertheless, Hassell's approach remained an obvious and sensible way to treat the data, especially since it helped readers see the underlying chaos (still a novel concept at that time) in the mathematical model, before the model was fitted to the data.

Thomas and colleagues[38] followed suit in 1980, this time fitting a 'θ-Ricker' model capable of exhibiting chaos, to the dynamics of 27 species of fruit fly in the laboratory. They expected the functions to be highly folded because they maintained very high population growth rates by changing the flies' food regularly. Somewhat surprisingly however, when they came to analyse their data they found that the estimated parameters were not sufficient to put any of the 27 species into the chaotic regime. A similar result was obtained when Mueller and Ayala[39] examined the dynamics of 25 genetically distinct populations of the fruit fly *Drosophila melanogaster* and found no evidence that the estimated population parameters were sufficiently large to push the dynamic into the chaotic regime. Collectively these studies, on an impressively large number of species populations, were sufficient to convince many ecologists that chaos was relatively unimportant in natural populations, and therefore simply a 'plaything' for theorists. More recent studies applying much the same models, such as the Hassell model to the population dynamics of a number of species of weeds, have likewise come to the conclusion that the dynamics were not chaotic.[40]

Nevertheless it has not all been one-way traffic. In the mid-1980s, Schaffer and Kot[41] began looking at the dynamics of measles cases in New York City and Baltimore, reported monthly from 1928 to 1963. Before the widespread employment of vaccines, measles epidemics arose almost every year in large American and European cities, but major peaks were unpredictable, occurring every second or third year in New York and less frequently in Baltimore. The combination of seasonal 'forcing' (contact rates among school children are higher in the winter when schools are in session, compared to the summer) and feedbacks via gaining immunity made childhood infections very plausible candidates for chaotic dynamics. Rather than fitting a model, the authors attempted to reconstruct the attractor and test whether it had sufficient strangeness (stretching and folding) to qualify as a strange attractor. Of course with only a single dynamical variable, it is hard to produce a strange attractor directly, but thanks to a neat solution proposed by the physicist Floris Takens,[36] it is possible to plot the number of measles cases at time t against the number of cases at time $t+\tau$ and the number of cases against time $t+2\tau$ with τ a variable time difference and thereby (assuming you have enough data) build up an equivalent picture of the underlying dynamic. Although chance may have played some role in generating the measles unpredictability (e.g. some cases will go unreported, and reports will be lost), by reconstructing and analysing the attractor that the epidemics represented (followed up by an estimate of the Lyapunov exponent) the authors argued that there was a strong deterministic component to this unpredictability (chaos) in both the New York and Baltimore data sets. More recent analyses have generally supported these conclusions,[42] including an analysis based on time-series analysis.[43] However, doubts still remain,[44] most notably because underlying factors such as birth rates have changed over time, and because the amount of seasonal forcing required to generate chaos in mathematical models of measles epidemics is considerably more than actually observed.[45]

The year before (in 1984) Schaffer[46] had analysed the oscillatory dynamics of the Canadian lynx as recorded by the numbers of skins shipped yearly by the Hudson's

Bay Company over the 1800s and 1900s—these data have been a classic of ecology textbooks since the 1920s.[47,48] As the author himself has noted, these data made a somewhat less convincing case, but they were again suggestive of chaotic signal (and indeed more recent analyses provide additional support for this[49]).

By the mid-1980s, Schaffer was beginning to see sufficient signs of chaos, that he issued (with Kot) a call to arms, arguing that ecologists were ignoring the very real possibility that chaos could be an important component of ecological systems and likened chaos to 'the coals that Newcastle forgot'[50] (the implication being that this rich vein of science was under the noses of ecologists and they did not realize its potential). With titles like that, coupled with the bestseller (and all-round wonderful read) *Chaos* by James Gleick,[51] scientists were well and truly waking up to the possibility of chaos.

In 1991, Tilman and Wedin provided experimental field evidence of the signature of chaos in dynamics of the perennial grass, *Agrostis scabra*, grown at two different initial densities on 10 different soil mixtures. For progressively richer soils, the dynamics evaluated over 5 years tended to exhibit higher amplitude oscillations with the richest (highest nitrogen) soil exhibiting dynamics the authors described as chaotic. Of course, with only a 5-year data set this interpretation is at best speculative (once again model-fitting methods and parameter estimation were used), but the inherent time scale of the annual dynamics clearly poses experimental challenges. One potential source for the significant non-linearity was the accumulation of leaf litter. Thus, in high-density years the accumulation of leaf litter (dead plant material at the end of the growing season) may inhibit growth of the following year. More recent work on the dynamics of another plant species, an annual greenhouse weed *Cardamine pensylvanica* likewise found evidence of oscillatory dynamics over 15 years, but in this instance found no evidence of chaos.[52]

Chaos in small mammals?

The regular oscillations of small mammal populations such as voles and lemmings have given population ecologists plenty of data (they are often pests of forestry plantations and leave signs—such as grass clippings and bark scrapings—which can be used to estimate their densities).[53] No, lemmings do not jump off cliffs into the sea on a 'suicide drive' as Disney's 1958 documentary *White Wilderness* would have us believe (indeed, the shot of lemmings jumping was entirely contrived—not only were they pushed, but also the sequence was filmed in Alberta, Canada, which has neither lemmings nor sea). However, many populations of small mammals exhibit remarkable high-amplitude 3–5 year oscillations in population size. These dynamics appear rather different in form in different regions. In particular, in southern Fennoscandia (including the Scandinavian Peninsula, Finland, and Denmark) and central Europe, populations seem to exhibit far lower amplitude fluctuations than in northern Fennoscandia. A possible reason for this is that the density of generalist predators is low in the north, and here specialist predators, notably the weasel, drive the dynamics.[54]

It has been proposed that the shift from south to north in small mammal dynamics is not from equilibrium to an entirely regular cycle, but rather from stability to chaos. Indeed, Hanski and colleagues[55] analysed data on the population sizes of *Microtus* voles in western Finland and, on the basis of time-series analysis that revealed positive Lyapunov exponents, the authors argued that the observed dynamics in these populations were chaotic (with the exception of the most southerly population), albeit with a significant periodic component. In essence, chaos may be superimposed on top of a more regular signal. They supported their interpretation with a predator–prey model involving seasonality that readily generates the type of chaos they had revealed in the data. Nevertheless, it is fair to say that not all researchers are fully convinced, and there has been considerable debate over the issue, centring on how the Lypanunov exponent (and its likely range) is best estimated in systems involving noise.[56–58] Of course, added noise is inevitable if you are trying to evaluate vole density across a large part of Finland using indirect methods of estimation.

Chaos in the laboratory

Perhaps the most ambitious set of experiments to investigate chaos was conducted in the laboratory on a species of flour beetle *Tribolium castaneum*.[59,60] *Tribolium* is cannibalistic, with older individuals eating smaller ones, so if the population is at high density then many small larvae will get eaten by the older individuals, reducing recruitment into the next generation. After modelling *Tribolium* dynamics using an age-structured population consisting of larvae, pupae, and adults, the authors concluded that the cannibalistic feedback was capable of generating chaos, as well as unusual dynamics that never quite repeats itself but does not show the sensitivity to initial conditions ('quasi-periodicity'). Having a theoretical model to play with is a helpful way to judge when and where interesting things might happen, and understand why. More importantly, the authors combined this modelling approach with a replicated experimental study in which they artificially manipulated the adult mortality rate[59] and recruitment rates of pupae to adult stage,[60] and in each case they found good evidence of the predicted shifts in the dynamics (from stable points, to cycles and quasi-periodicity or stable points through a range of dynamical behaviours ending with chaos). One might argue that by manipulating the ecology, the authors have forced the system to match the model rather than the other way around, but this study remains convincing evidence that populations are at least capable of exhibiting chaos.

Even more recently, Becks and colleagues[61] have manipulated the dynamics of a bacteria-eating ciliate predator and two species of bacteria (rod-shaped and coccus) in a chemostat: a rearing facility ensuring approximately constant environmental conditions. By experimentally manipulating the rate of delivery of the organic food source for the bacteria to the chemostat, the authors found that they could change the underlying dynamic between equilibria, stable cycles, and chaos (based on Lyapunov exponents). Precisely why chaos was generated is unclear, but the system has parallels to

Gilpin's model[23] of a one predator and two prey system. Moreover, this is the first case we know of in which chaos has been demonstrated in a microbial system. A second related example followed in 2008, and involved culturing a functioning planktonic food web isolated from the Baltic Sea under standardized laboratory conditions.[62] Despite constant external conditions, this microscopic community, which consisted of bacteria, several phytoplankton species, herbivorous and predatory zooplankton, and detritivores, showed marked fluctuations in abundance over the 2,319-day experimental period and yet the populations still persisted intact. Moreover, the dynamics had all the hallmarks of chaos, including positive Lyapunov exponents for each species.[62] Collectively, these studies indicate that chaotic dynamics can and do arise in complex microbial communities.

The bottom line

Our review of the presence or absence of chaos in populations is not intended to be exhaustive. For example, there are scattered accounts of tests for chaos in the dynamics of bobwhite quail[63] (no evidence), water fleas[64] (no evidence), aphids[65,66] (no evidence), and moths (no evidence)[65,66] and no doubt many more. Interestingly, in a recent review of chaos in real data sets,[67] several of the data sets (including blowflies[1] and flour moths[68]) had dynamics 'on the edge of chaos'; that is, oscillations that, with a little more feedback, would have been chaotic. To this we can add a recent analysis of certain populations of Fennoscandian voles.[69] Whether this condition is common, or whether it is an artefact of the underlying statistical methodology, is currently unclear. However, it is now known[70] that noise superimposed on a regular periodic cycle can generate dynamics with no change in period but a change in amplitude—these quasi-periodic dynamics are just the sort of dynamics that give rise to zero Lyapunov exponents and dynamics at the 'edge of chaos'. So, perhaps some of these populations at the edge of chaos are simply ones that have an underlying tendency to show regular cycles in abundance, while being influenced by external noise.

Let us return to the original question we set ourselves. Ecologists have long realized that the systems that they are dealing with are non-linear, but are they sufficiently non-linear to drive chaos? In 1993, an excellent review of chaos in ecology was published using the subtitle 'is mother nature a strange attractor?'[20]—one we have borrowed for one of our own section headings. The authors knowingly avoided answering their own question directly, preferring instead to suggest that 'chaos is quite likely, but much more work is needed to obtain a fuller answer to the question'. Now 15 years later, ecologists are expressing doubts. More recent opinions have varied from 'the jury is still out'[67] to 'chaos is rare'.[71] In 1999, science journalist Carl Zimmer[72] wrote about 'life after chaos'. Our own survey leads us to conclude that there is very little good evidence for chaos in natural populations. We have to be cautious, however, because part of the problem may be that ecological population data are by their nature relatively short term and noisy, making unequivocal proof of the existence of chaos challenging at best. Perhaps this is one reason why ecologists have recently been more successful in detecting chaos in microbial systems which can be monitored for many more generations. The shortage of good data sets for multicellular organisms has occasionally led to controversy. For example, ecologists have sometimes ended up arguing over the same data sets: as Schaffer has quipped (no slur seems intended): 'novel claims conjoined with a paucity of data inevitably attract the attention of statistics, much in the manner that offal attracts flies'.[37]

We cannot rule out the possibility that mother nature is, in general, a strange attractor, but we have to say that the case is looking increasingly shaky, at least for multicellular organisms. If ecological systems are not chaotic then, given that it is a widespread property of many population models, we need to ask the reverse question posed early in the debate by Berryman and Millstein[73] in 1989—'if not, why not?'

If not, why not?

Jeff Goldblum, playing that self-confident 'chaotician' in *Jurrasic Park* who eventually met the end we could all see coming ('When you gotta go, you gotta go'), remarked before his demise 'Life will find a way'. Perhaps natural populations are not chaotic because natural selection somehow finds a way of pushing population parameters towards levels where they would not exhibit chaotic properties. Both Thomas and colleagues[38] and Berryman and Millstein[73] thought this might be the case, noting that in the chaotic region populations tend to fluctuate wildly yet spend a high proportion of their time at relatively low densities where extinctions are more likely to happen. Their argument was explicitly 'group selectionist': 'it seems reasonable that natural selection might favor parameter values that minimize the likelihood of extinction and, consequently, chaotic dynamics'.[73] However, there are several problems with this argument. First, not all chaotic dynamics suffer from a high probability of extinction—some chaotic dynamics are tightly bound well away from zero, and chaos can in some cases reduce the likelihood of species extinctions.[74] So, despite the biblical impression that chaos is all about doom and destruction, it is not necessarily the case in the ecological sense of the word. Second, it ignores the problem of individual selection for cheats that favour their own reproductive success, even if it ultimately leads to the demise of the group. Chaos is about long-term dynamical behaviour, but natural selection is driven by what genetic variants perform best right now. There are cases where group selection effectively overpowers individual selection, but we generally need rather extreme assumptions[38] (see also Chapters 1 and 2).

Perhaps natural selection on individuals, rather than groups, can favour non-chaotic dynamics. The role of natural selection in influencing population behaviour, even in short-term laboratory experiments, is now widely recognized. For example, Yoshida and colleagues[75] recently successfully produced predator–prey cycles in a laboratory microcosm involving a rotifer feeding on a green alga. However, the cycle periods were far longer than predicted, and the observed predator and prey cycles were almost exactly out of phase, which is not what one would anticipate. Only by accounting for (and testing for) the possibility of on-going natural selection in their system—in which rotifers effectively traded competitive ability for the ability to defend against predation when predation rates were high—were the authors able to reconcile their experimental and theoretical results. Thus, it seems the prey population was actively evolving at the same time it was undergoing fluctuations in density.

Alexander Nicholson's 'blowflies' represent one of the most celebrated and analysed data sets in the history of ecology[34,76] (see also Fig. 6.1). Seeking to understand how and why populations fluctuate led him to begin an intensive series of experiments in the 1950s with caged Australian sheep-blowflies. Maintained in the laboratory for several hundred days, the blowflies exhibited characteristically 'double-peaked' oscillatory dynamics. However, in some of his longer-term experiments (lasting over 700 days), the dynamics became rather irregular after about 400 days, and at the same time the period of their oscillations also dramatically halved (to a mean of approximately 38 days). Nicholson himself recognized these patterns in his data and proposed that natural selection was acting in the course of his experiment. George Oster[77] went further, proposing that natural selection had a destabilizing influence, carrying population parameters into the chaotic regime (thereby neatly explaining Hassell and colleagues[33] earlier observations—see earlier). Yet a more detailed analysis has subsequently revealed the opposite[78]—over the course of the experiment it appears as if there was a reduction in the maximum possible fecundity of adult females, moving the dynamics from unstable to more stable dynamics, tracking the regular addition of protein food supply. So, in the case of Nicholson's blowflies we may have evolution towards stable dynamics. However, we would do well to remember that selection pressures in a jar in a laboratory are likely to be quite different from those found in the wild.

Over the years, a general consensus has been building (with a few notable exceptions[79]) that there may be selection on individuals that happens to take their populations away from the realm of chaos. For example, whether a population will evolve towards stability or towards chaos appears model-dependent, but a fluctuating population with constant carrying capacity (such as that represented via a logistic) will tend to experience selection that results in the population evolving towards population stability (in effect, parents go for offspring quality rather than quantity). Likewise, a suite of general population models,[80] models of competition,[81] and those involving stage structure[82] have all been reported to involve natural selection that indirectly promotes population stability. Experimental evolution in fruit fly populations had earlier suggested little or no evolution of parameters affecting population stability,[83] but a recent study under rather different conditions did indicate that populations evolve towards stability[84] (as a consequence of individuals reducing their fecundity to develop more rapidly). Once again, we see a close interrelationship between evolution and ecology, here with natural selection generating demographic parameters that happen to facilitate population stability.

Another set of reasons why populations may fail to exhibit chaos while mathematical models readily exhibit it may have something to do with the particular type of mathematical models that have been explored. In a recent review, Scheuring[85] made the case that several biologically relevant details, such as sexual reproduction, population structure, and dispersal tend to be overlooked in simple population dynamical models, yet incorporation of these details into mathematical models generally favour dynamical stability. For example, certain population models that include sexual reproduction show a reduced propensity to exhibit chaos.[86] Similarly, when you allow small amounts of dispersal between several otherwise chaotically fluctuating populations, the resulting dynamic becomes more stable.[87–89] Of course, it is very difficult to be general, but reversing the usual statement about chaos, it seems that complicated models with realistic features can generate simple dynamics.

Conclusion

The discovery that simple non-linear relationships, common in ecological systems, could generate extremely complicated dynamics was nothing short of a revelation. The associated finding that these complicated dynamics exhibited extreme sensitivity to initial conditions carries with it implications for all of ecology. In the intervening three decades since these discoveries were made, ecologists have worked hard to find evidence for this chaos in natural populations. Chaos has been formally defined, and methods have been developed to help test for it in the short and noisy data sets that ecologists are forced to deal with. We now know that populations can indeed be manipulated to generate all the features of chaos seen in mathematical models, and there is reasonable evidence of chaos arising in certain cases, such as childhood measles and some microbial systems. So, chaos *can* occur. Nevertheless, the majority of attempts to find chaos in natural populations have either drawn a blank or remain controversial.

Early in the ecological study of chaos, Schaffer and Kot[50] likened chaos to 'the coals that Newcastle forgot'. With painful irony, their paper was published shortly after the UK national miners' strike and all of the coal pits in the Newcastle area are now closed (as from 2007, only six pits remain in operation in the entire United Kingdom). Surveying the literature here leads one to suggest that many of the richest seams (to stretch the coal metaphor) of available ecological data have now also been explored, and few have provided much return. There may be good reasons why natural populations do not exhibit chaos, but only time will tell whether chaos is indeed rare.

Given the wonderful diversity of the natural world and knowledge that many systems have the *propensity* to exhibit chaos, perhaps a better question to have asked is 'when and how often are natural systems chaotic?' rather than 'is this system chaotic?'. We have seen already that voles can exhibit very different dynamics in different populations, and both the blowfly and the lynx data are suggestive of a marked change in dynamics at some point in their history. Likewise, flour beetles can exhibit a range of different dynamics dependent on underlying experimental conditions. So, it is perhaps naive to characterize a population as 'non-chaotic' or 'chaotic', because dynamics can change according to the prevailing conditions. Human activities could yet turn non-chaotic dynamics into chaotic dynamics by increasing the degree of non-linearity involved—as has been suggested for some fisheries.[90] This may particularly be the case for insect pests, such as the migratory locust mentioned in our opening paragraph, if drastic control measures are only implemented if pest density reaches a high level.[73]

Chaos theory continues to grow and develop in a variety of scientific fields where it has found wide application. As May noted in one of his early seminal papers,[9] 'Not only in research, but also in the everyday world of politics and economics, we would all be better off if more people realized that simple non-linear systems do not necessarily possess simple dynamic properties'.

Ecologists are now much more aware of the subtle effects of non-linearities, and appreciate the wide variety of dynamical behaviours they can generate. Yet the truly surprising thing in all this is how long it took scientists to discover chaos. As James Yorke, one of the early pioneers, has recently said 'I continue to wonder, if nearly all scientists missed this pervasive phenomenon, what other obvious phenomenon might we all be missing now?'.[91]

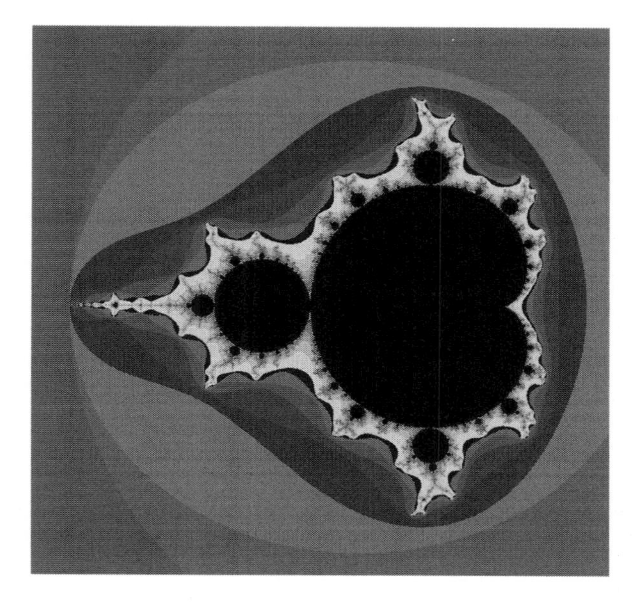

Named after the mathematician Benoit Mandelbrot, the Mandelbrot set has become one of the icons of chaos theory. This rather complex object is generated by a relatively simple set of rules and has self-similarity at different scales (hence fractal dimensions), such that zooming in one sees the same patterns at an increasingly finer scale. Due to their aesthetic appeal, fractals have long attracted the interest of graphic designers and artists. Image by TNS.

7

Why is the World Green?

Figure 7.1 The green world—secondary forest in Quebec, Canada. Photo: DMW.

And the locusts went up over all the land of Egypt... and they did eat every herb of the land, and all the fruits of the trees.

—*Exodus Chapter 10, verses 14 and 15.*[1]

Defining the question

Viewed from space by human eyes, the predominant colours of our planet are the blue of the oceans and the white of the clouds. The blue of the oceans forms the subject of another of our chapters. However, if one focuses on the land masses other colours dominate. On land the white colour still features prominently in the polar areas covered with snow and ice, but zoom in on lower latitudes and much of the land is a mix of the green of vegetation and the brown of more arid areas. Green dominates large areas of land, so unless you are reading this in a desert, during the high-latitude winter, or in a highly urban area, then green will probably feature prominently in your surrounding landscape.

One answer to the question that heads this chapter is that the climate (often rainfall) allows some parts of the land to be green with plant life, while making other areas arid and brown. However, this green of extensive plant life is still a puzzle—plants are food for a wide range of animals, so why is so much food left unused? Swarms of locusts, destroying most plants in their path (be they biblical plagues or modern day outbreaks), are the exception not the rule. But why is this so? Why are so many parts of our world green in the face of this threat from herbivores? As we will see, if herbivores are the key to our question, then what starts as a question in plant ecology ends up being a question about factors that limit the size of herbivore populations. In effect, we need to understand why herbivore populations do not increase in density to such a level that they destroy all the available plants, giving a land that is brown rather than green.

Until the middle of the twentieth century if you had put the green world question to biologists, many of them would probably have suggested that it was not in the interests of a species to consume all of its food reserves. However, such 'good of the species' arguments are now understood to suffer from a range of problems—an issue that surfaces repeatedly in Chapters 1–3 dealing with evolutionary questions. Although such arguments still appear in the commentaries to some TV nature documentaries, twenty-first century evolutionary ecologists usually demand alternative explanations.

The question that heads this chapter was famously articulated in a classic short paper published in *American Naturalist* by Nelson Hairston, Frederick Smith, and Lawrence Slobodkin (hereafter HSS) in 1960.[2] After discussing various alternatives, their tentative answer was that herbivores do not consume all of their food supplies simply because predators keep their numbers in check. They supported their arguments by pointing to exceptions where human action or 'natural events' removed predators leading to rodent plagues and insect outbreaks destroying the vegetation; as in a plague of locusts. To cut a long story short, most ecologists now suspect that their answer was wrong (or at least incomplete). However, often in science posing a good question can be as important as providing an answer. In this chapter, we explore some of the implications of HSSs' fruitful question.

On first reflection the idea of HSS that predators are crucial to answering the green world problem seems to have a lot going for it. Agricultural crops appear to provide a nice example as they suffer from several important pests, and the damage can be made even more serious through pesticides killing the predators and competitors of these pests—although it is often very difficult to prove direct causality in a totally rigorous manner.[3] Particularly interesting examples are those herbivore species that are well protected from predators and can sometimes completely exploit their plant hosts in the way HSS suggested. A good example, familiar to many naturalists living in Europe, is the caterpillar of the cinnabar moth; these are particularly relevant to our discussion as there are reasons to suppose that these caterpillars may have less trouble from predators than do many insects, although the common cuckoo will sometimes eat them. These black-and-orange-striped caterpillars are commonly seen on ragwort and a few related plant species, and will sometimes eat all the leaves and flowers of their food plant before leaving the now bare stems to crawl away in search of a new plant—often dying during their search.[4] The striking colours of the caterpillars are a warning to potential predators that these larvae contain poisonous chemicals, which include a range of alkaloids acquired from their food plants.[4,5] Certainly, the caterpillars often feed in full view of potential predators without being eaten (Fig. 7.2). As we discuss in more detail later, because the cinnabar moth caterpillars only eat a few closely related plant species, they cannot change their whole environment from green to brown; however, if all insect species behaved in a similar way the 'Green and pleasant land' of England, famously described by the poet William Blake at the start of the nineteenth century, would likely never have existed. So HSSs' original argument, that predators of herbivores may be crucial to keeping the world green, is certainly something to be taken seriously.

The earlier example can be summarized as a simple food chain, long familiar from introductory science classes at school: Plant (ragwort) → Herbivore (cinnabar moth caterpillar) → Predator (cuckoo). The predator is at the end or 'top' of the food chain, so ideas such as those suggested by HSS, which assume that predators limit herbivores and thereby allow plants to persist, are described as 'top-down' effects. The obvious alternatives are 'bottom-up' effects, where changes in the base of the food chain (the plants) have effects on organisms higher up in the chain. So if one explanation of why the world is green is based on the top-down effects of predators (and parasites) on herbivores, it will come as no surprise that the other main class of explanation makes use of bottom-up effects, with plants themselves limiting the effectiveness of herbivores. We describe both of these ideas in more detail later in the context of relevant experimental and other evidence, before going on to discuss some related ideas which do not neatly fit into this top-down/bottom-up classification. Clearly, both top-down and bottom-up effects could play a role in keeping a green world, such that if one mechanism was somehow turned off, then the other would still prevent herbivore numbers getting out of hand.

Figure 7.2 Cinnabar moth caterpillars feeding on ragwort in Lincolnshire, eastern England. Note that they are prominently positioned on the plant with no attempt to hide from potential predators. Poisonous animals with associated warning colouration often make no attempt to hide; indeed they seem adapted to display their warning colours as prominently as possible. The common cuckoo, which seems to specialize in eating poisonous caterpillars, is one of the few birds that has been recorded eating them in any numbers.[68] However, there is a problem with these types of records; are naturalists recording the behaviour of a predator which is immune to the toxin, or just a naive predator that has not yet learned to avoid such prey?[69] Photo: DMW.

Top-down? Experimental data

An obvious first step to evaluating top-down explanations is to observe what happens when the number of herbivores is altered. The most direct test of the top-down theory would involve removing all the predators and parasites of the herbivore species in question and simply seeing what happens—if the herbivores reach such high densities that they consume all the available vegetation, then we would have direct proof of the validity of the top-down hypothesis. However, such manipulations are extremely challenging, and most studies have settled for simply evaluating the impact of herbivores on vegetation in a general sense, either by removing them or by increasing their density. This can be done either by experimentally manipulating the number of herbivores, or by finding a location where natural or man-made changes have led to the required 'experimental' conditions. We will first describe three examples of such studies, and then try to draw together their main message relating to the validity of top-down explanations in answering this chapter's central question.

A common experimental approach to studying the effects of herbivores are exclosure experiments—an exclosure is a fence designed to keep animals out rather than the more familiar enclosure that keeps them in. A good example of such studies is the long-running series of experiments in the mountains of North Wales, UK. These experiments were set up in 1957 to investigate the effects of excluding sheep from patches of montane vegetation (note that small herbivores such as voles, and invertebrates such as slugs, could still access the vegetation). Initially, there was a rather complex experimental design where, by moving the fences during the year, sheep were excluded from some areas all the time but allowed to graze others at certain times of the year. This proved rather time consuming to maintain and from 1982, sheep were permanently excluded from all the exclosures.[6] The results of these experiments are typical of many similar studies[7,8] in that the vegetation inside the exclosure grew taller than the more heavily grazed surrounding vegetation and, more importantly, developed a different species composition. Several plant species, such as common heather, which were kept small and rare by the sheep grazing, developed to dominate parts of the exclosures, so even from a distance the vegetation in the exclosures appears different (Fig. 7.3). Thus, it is clear that herbivores do indeed have a measurable impact on the physical structure and species composition of vegetation, yet we still have to explain why the heavily grazed land around these exclosures was still 'green', as food for herbivores is apparently still available but not being used. Something is clearly limiting the herbivore populations outside of the exclosures and in the world in general, so that they have not ended up consuming all of the available vegetation.

This Welsh experiment used fences—rather than predators—to alter the density of herbivores, and in this sense it is rather artificial. In addition, the principal herbivore was a domestic animal. A more realistic 'experiment' has been established in eastern Venezuela by the creation of a new lake for the production of hydroelectricity. This has led to the formation of a large number of forest-covered islands, which have been the subject of detailed studies by John Terborgh and colleagues.[9]

Six of the islands investigated by Terborgh and colleagues were too small to support viable populations of medium-sized mammals such as armadillos and various primate species, which can be significant predators of invertebrate herbivores—especially leaf-cutter ants. These fascinating insects are considered to be of particular importance in forest ecosystems as they are thought to consume far more vegetation than any other major animal group (such as herbivorous mammals, or the caterpillars of butterflies and moths) in tropical South America.[10] The six small islands were found to have much lower densities of tree saplings than the larger islands, almost certainly because there were no predators to reduce the ant populations that defoliate the trees. Indeed, Terborgh and colleagues suggested that the small (predator-free) islands with higher

Figure 7.3 One of the North Wales exclosures—this one is in Cwm Idwal ('site 2' in the paper by Hill *et al.*[6]). (a) Taken in 2002—note the much shorter grass outside the exclosure where there has been sheep grazing and the dark-coloured bushes of heather which are only able to grow inside the exclosure where they are protected from sheep. As part of the management of Cwm Idwal nature reserve attempts were made to exclude all sheep from late 1998. In 2001, Britain suffered a major outbreak of the animal disease foot and mouth, which led to a decline in sheep farming in the area, making it easier to keep sheep out of Cwm Idwal (although a few still mange to get in). Slowly, the grass outside the exclosures is getting longer and it is now easier to find small heather plants growing amongst it—therefore, (a) probably shows the maximum difference in vegetation between the exclosure and the surrounding vegetation, with sheep grazing only recently stopped outside the fence but with parts of the exclosure ungrazed for 45 years. (b) Shows the same view in April 2008: note the small heather plants now visible in the grass to the right of the exclosure fence. Photos: DMW.

leafcutter ant densities were heading towards a future vegetation with very few trees that would eventually be dominated by ant-resistant lianas. They describe their results as consistent with the top-down green world hypothesis of HSS—however, they note that the absence of major predators of the herbivores is not leading to a plant-free system (a brown world) but one that is still green, albeit dominated by very different types of plants. In addition, the lianas presumably survive because of bottom-up processes, such as chemical defences, of the kind we discuss later. Moreover, the data of Terborgh and colleagues are more difficult to interpret than might initially appear. Perhaps the experimental 'treatments' differed not only in herbivores but also in a whole range of factors such as resources, and the microclimate experienced by the plants. This is a common problem in large-scale ecology; it is often almost impossible to perform experiments that only change a single factor in the environment.

Top-down? Large-scale observational data

Both of these examples come from experimental or quasi-experimental approaches, the Venezuelan study being a 'natural experiment' that cleverly makes use of islands that had been formed for other purposes. As always in science a good experimental design can make the results much easier to interpret—however experiments, by definition, modify the environment in some way so the system cannot be considered entirely natural. In addition, formal experiments often limit you to relatively small-scale studies as very large-scale modification of the environment can range between the impractical and the impossible. The alternative is to make detailed observations on relatively natural systems, preferably over many years. This can often allow studies of much larger systems than those that can be experimentally manipulated.

One good example of such an observational study is the effect of the size of the moose population that feeds on young balsam fir trees on Isle Royale in Lake Superior, in Michigan, USA, where moose are a potential top-down control on the plant life. This island is 544 km^2 in area, and both moose and wolf populations (which can prey on moose) have been studied since 1958. These long-term studies identified an apparent link between snow depth and moose population size, and also demonstrated a role for wolf predation and even global climate fluctuations in affecting the Isle Royale system.[11] At the end of the 1990s, Eric Post and colleagues published a stimulating hypothesis about the ways in which different components of the system might be interrelated in a paper in the journal *Nature*.[12] Since this story adds another level of complexity to the top-down explanations for a Green World, it is worth briefly describing.

In some winters, the island has much more snowfall than in others due to changes in the behaviour of air masses over the North Atlantic (the North Atlantic Oscillation). In these snowy winters, the wolves tend to form larger packs which makes them more able to successfully hunt bigger prey, namely moose. This causes the moose population density to decline and consequently reduces the amount of browsing on fir tree saplings. Eric Post and colleagues[12] were able to show connections between fir tree growth and moose numbers by studying the annual growth rings in the firs—trees that survived the attentions of moose showed periods of low growth when they were saplings during years with higher moose numbers. This provides another example of top-down effects on vegetation (wolves → moose → fir), at a larger scale than the other studies we have described, but we again note that it is affecting the type of vegetation (reducing the success of young trees and so potentially favouring less woody vegetation), not the presence or absence of vegetation. It also illustrates a new complication, which was the main focus of the 1999 *Nature* paper, that vegetation changes were being driven not by changes in the *number* of predators but by changes in predator *behaviour*, which in turn was caused by changes in the climate (other long-term effects of changes in predator numbers can also be seen in this system).

What are the main implications of these studies (which appear typical of many others we have not had space to describe) for explaining why parts of the world manage to remain green? They suggest that while top-down processes controlling herbivore numbers can have big effects on the type of vegetation, they do not exercise sufficient control over herbivore numbers to turn a green world into a brown one, even when the majority of predators of herbivores are removed from the system. Robert Paine[13] has summarized the position well, writing that while the world may stay 'green' in the presence of abundant herbivores 'the dynamics are substantially altered'—that is, even if herbivores cannot (for whatever reason) completely exploit their green resources, they can control the type of vegetation present (as in the earlier examples). Paine also makes what he considers a very important point, namely that it is often larger mammals rather than invertebrate herbivores that have the biggest effect on the vegetation. Many of his arguments are based on exclosure type experiments similar to the Welsh sheep example we described earlier. While much of the scientific literature on the Green World problem has continued to focus on insects, as indeed HSS did in their original paper, Paine is certainly correct that we must also think about larger (and smaller) herbivores as well.

Of course, it is almost impossible to remove all predators and parasites of herbivores from a system, so it is possible to argue that if only more top-down control was removed *then* the herbivores could eventually eat all the plants. Indeed there do seem to be some limited cases where grazing by large mammals—albeit often kept at extremely high densities by humans (and therefore hardly constituting a natural equilibrium)—can destroy most plant cover. Increased grazing in arid areas where the plants may already be struggling can sometimes have this effect. For example, historical changes in grazing around El Paso in Texas, USA, associated with the rise of cattle ranching, appear to have caused arid grassland to turn into semi-desert with lots of bare ground between isolated Mesquite and Creosote bushes.[14] However, this Texan example may be the exception rather than the rule, as most similar examples only show changes in the type of vegetation.

Despite occasional examples of apparent top-down control, the current consensus[15,16] is that in most cases top-down effects are not sufficient to solve the green world question. For one thing, we do not see experiments in which the release of herbivores from predation results in the complete exploitation of their food supply—although as noted earlier this may be possible because parasites continue to control herbivore numbers in the absence of predation. Nevertheless, as our examples demonstrate, it is clear that top-down processes can dramatically alter the nature of the vegetation cover.

Bottom-up?

If top-down processes cannot provide the full answer to our green world question what about bottom-up processes? To understand the potential for bottom-up effects, let us examine how plants affect herbivores. One might think that plants just sit there, rooted to the spot, waiting for animals to eat them. This is an easy mistake to make because, with the exception of thorns, spines, and stings, the anti-herbivore adaptations of plants are less obvious than those of animals, who can fight or run away and hide from danger. However, plants may provide much less easy pickings for herbivores than we might initially assume. In particular, plants may be difficult to eat because they contain poisonous chemicals (such as the alkaloids in ragwort) or they may be of poor nutrient quality, such as being low in nitrogen by the standards required by an animal. There are two complementary ways in which bottom-up explanations could work, namely that (i) individual herbivores will poison themselves (or cease eating) in the short term if they consume too much plant material thereby leaving the rest of the plant and (ii) there is a long-term effect of such chemicals on herbivore population size, such that plants with secondary compounds and low nutrients cannot support high herbivore densities. Not all commentators have distinguished clearly between these two interrelated phenomena, but it is clear that both may help to explain our green world. Of course, ultimately it will be the population size(s) of a herbivore species (or range of species) that determines whether their food source is exploited to destruction, not the appetites of individual organisms.

Sometimes in science it is relatively straightforward to associate a key idea with one or two important scientists, for example, the idea of natural selection with Charles Darwin and Alfred Russel Wallace, but in many other cases a theory emerges from the work of a much larger number of scientists. This is the case with the bottom-up explanation for why the world is green, which mainly developed during the 1970s from the contributions of many people including P. Feeny, D.H. Janzen, S. McNeill, and T.R.E. Southwood.[17] These concepts, which we will return to later in this chapter, built on the idea developed during the 1950s that many 'secondary plant compounds' were responsible for preventing invertebrate attack.

Secondary plant compounds are chemicals produced by plants that are clearly not used by the plant in primary metabolism (i.e. in the main energy-supplying chemical processes in a cell). Our modern understanding of the functions of these secondary compounds is generally credited to Gottfried Fraenkel,[18] although the idea can be traced back to the nineteenth century. For example, the German botanist E. Stahl wrote in 1888 that 'the great differences in the nature of chemical products... [produced by plants]...are brought nearer to our understanding, if we regard these compounds as means of protection, acquired in the struggle with the animal world' (translation from Fraenkel[18]). We now have good evidence for these ideas—for example, it has been shown experimentally that certain secondary compounds (phytoalexins) in soybean plants deter important insect pests of this crop, with varieties with higher concentrations of these chemicals being less susceptible to attack.[19]

We have already described how the cinnabar moth uses poisons taken from its food plant to protect itself. These chemicals make it difficult for most herbivores, without the cinnabar's special adaptations, to eat large quantities of ragwort—the plant is, for example, notorious for poisoning horses. To see evidence of the effect of plant chemical protection on insects, dig out a field guide to butterflies where you will see that the caterpillars of most species specialize in eating a limited number of plant species. Indeed, nineteenth-century manuals for butterfly collectors can contain extensive tables describing which plant species should be searched to find caterpillars of any given species, as collectors often wanted to catch the larvae so they could be reared to provide undamaged adults for their collections.[20] One potential explanation for this is that caterpillars have to specialize in only a few food plants because these are the only ones to which they have evolved mechanisms for dealing with potentially poisonous chemicals in the plant tissue (e.g. the cinnabar moth).

The idea that plant secondary compounds can cause dietary specialization does not just apply to insects; similar patterns are seen in some herbivorous mammals where dietary specialization allows them to deal with only a subset of all potential toxic plant chemicals. This detoxification may rely on enzymes produced by the mammals themselves or on the properties of the microorganisms that live in their guts.[21] A good example of such specialization is seen in Australian koalas which are famous for specializing in eating the leaves of *Eucalyptus* trees. Recent research by Ben Moore and colleagues has shown that koalas select certain species of eucalyptus with lower levels of particular secondary plant compounds (formulated phloroglucinol compounds) in them, and eat much less foliage if forced to feed on the better protected species.[22] Secondary plant compounds, such as nicotine, cocaine, and caffeine (all alkaloids), are also of interest to humans as many of them have effects on our nervous system. Besides alkaloids, a wide range of other defensive chemicals are used by plants including cyanogenic glycosides (produced by bracken fern amongst other species) and various toxic proteins. One of the best known of this latter group is the protein ricin from the castor oil plant, which has been linked with various assassinations and terrorist plots.[23] All of the above chemicals have probably evolved to reduce the consumption of plants by herbivores. Some of them are continually present in plants, but others are synthesized when the plant is attacked (e.g. many alkaloids).[24] Such induced defences, which are switched on when a plant is attacked, make it harder for herbivores, such as caterpillars, to defoliate whole plants so turning a green world into a brown one.

To recap, the most obvious bottom-up explanation for our green world stresses poisons, and we have good evidence that such chemicals can provide some protection from herbivores. Plant material may be green, but that does not mean it is all readily available. All the same, as the example of ragwort and the cinnabar moth showed, poisons can be circumvented by specialist predators. So, as Dan Janzen[25] has pointed out, there are really two questions tied up in the problems discussed by HSS; while chemical poisons may explain why most herbivores cannot eat most plants, they do not explain why *specialist* herbivores do not consume all their food supply.

Bottom-up; not only poisons

The second proposed bottom-up mechanism keeping the world green is slightly more subtle and suggests that plants make poor food for animals even if they are non-toxic. From the perspective of a typical animal, there is a stoichiometric problem with eating plants (stoichiometry is a chemistry term that refers to the proportion of different chemical elements). Nitrogen is a particular problem as a typical insect may contain 10 times more nitrogen in its constituent chemicals than the leaf it is eating, and if it feeds on especially nitrogen-poor plant material, such as wood, the situation is much worse.[15] Therefore, to obtain enough nitrogen an animal may need to eat very large amounts of plant material, much more than they require for energy, in order to acquire enough nitrogen to manufacture proteins and nucleic acids. There is also a connection with the previous toxic chemical explanation, because in eating this large quantity of plant material the animal may ingest a high dose of any poisonous chemicals which are present. Of course, the entire argument appears to box its proponents into a corner—if a great deal of plant material has to be eaten for an individual herbivore to make a living, then why would this lead to a greener world and not a browner world?[26] Remember that if a potential herbivore has to work harder to eat a decent meal, then it is unlikely to leave as many offspring. It therefore seems reasonable to assume that the long-term average population size of a herbivore species faced with a food shortage of palatable material would be lower than one with an abundance of palatable resources.

Plants can also produce chemicals, such as tannins, which while not directly poisonous, reduce the palatability of plants to herbivores and may also be broken down into potentially toxic compounds in the animal's gut.[23,27] In addition, plants can have tough leaves and contain structural compounds such as lignin and cellulose that are very difficult for animals to break down; plants also have lower water content than most animals.[15] Many of these compounds are probably not primarily adaptations to protect plants from herbivores; their defensive properties being pleiotropic (see Chapter 1) by-products of other selection pressures on plants. For example, lignins and cellulose, which both contribute to making plants difficult to eat, are key in providing structural support that allows plants to grow tall, and hence compete for light with other plants. The low nitrogen levels may also not have directly evolved as adaptations against herbivores, since nitrogen is often scarce in soils so it is not surprising that it is also often scarce in plant tissues as well.[28] Nevertheless, many studies have shown that the toughness of plant tissues also makes them difficult for insects to eat, and it is probably relevant that some of the earliest plant-eating insects known from the fossil record had piercing mouthparts, which did not need to chew tough plant tissues.[29]

The bottom-up ideas are very plausible explanations for why the land is green, but ideally we would like experimental tests that show that they actually work—rather like the exclosure experiments we described for evaluating the role of herbivores. The difficulty is that it is usually much easier to alter the number of herbivores, or their predators, than it is to modify the chemical make-up of plants. One of the more interesting tests of these bottom-up ideas comes from a 'natural experiment' caused by air pollution in north-west Europe. Heathlands, that is, plant communities dominated by dwarf shrubs, such as common heather, are greatly prized by European conservationists because many rare species are associated with them. Many of these heathlands have tended to lose heather during the twentieth century and purple moor grass has expanded in population size to take its place. These declines in heather abundance are usually blamed on air pollution raising the nutrient status of the heathland, although there is evidence from plant remains preserved in peat cores that such changes could also have happened before the rise of industrial air pollution.[30] Given that pollution-related nutrient levels are heavily implicated, the question is 'how do increased nutrients cause these changes'? There is evidence from Dutch heaths that bottom-up processes are involved. In experiments conducted by Brunsting and Heil,[31] fertilizer was added to heather-dominated heath. They were able to show that this raised the nutrient levels in the plants and so allowed a build-up of heather beetle populations due to this improvement in their food supply. They suggested that these beetles reduced the dominance of the heather and so allowed the purple moor grass to invade the plant community. Although not a perfect experiment—for example, it would have been helpful if they had been able to keep beetles out of a sample of nutrient-enriched sites to prove that it was not just that the fertilizer directly benefited the grass—it strongly suggests a role for bottom-up processes. In fairness to these experimenters, we should point out that it is much harder to exclude small beetles than big sheep, as in the Welsh experiments we described in the 'top-down' section. Notice also that as with the top-down experiments a brown world has not been created due to an increase in plant palatability, but only a different type of green one.

Although it has been difficult to test directly, most ecologists writing on the topic in the past couple of decades have been reasonably convinced that bottom-up changes are very important in explaining the green world. Yet these bottom-up explanations suggesting that plants are difficult to eat may seem unlikely given our own experience of eating lots of highly digestible plant material. In response, remember that many of our agricultural crops are the product of millennia of selective breeding to improve their quality for use as human food. Even so, we manage to eat very few species compared to all the plants in the world—for example, we have only domesticated around 100 species of plants out of a global total of around 200,000+ described species[32] (although the total number of plant species is rather uncertain).[33] Some of our close relatives appear to similarly struggle with most plant species; in his classic field study of the mountain gorilla, George Schaller[34] found that they only eat a 'small percentage of the total number of plant species available to them, and some of the most abundant plants are not utilized'.

More bottom-up; spatial processes and hiding places

Besides poisons and nutritionally poor ingredients there is another possibility—perhaps plants can hide from their enemies? This sounds odd, how could a plant hide when it cannot run away from its predators? We are used to the idea that many animals are camouflaged to help escape their predators and it turns out that some plants may use the same trick. For example, the stone plants of southern Africa (Fig. 7.4) give the appearance of being cryptic, each species closely matching the colours of the stones in its particular desert habitat,[35] presumably making it hard for any visually hunting herbivores to spot them. However, these plants are exceptional: the more common way that plants evade detection by herbivores is simply through their wide spatial distribution. This idea is well illustrated by a classic ecological laboratory experiment from the 1950s carried out by C.B. Huffaker.[36] Using a system with two species of mites, one a herbivore, and the other a predator of this herbivore, he showed that in simple spatial environments (oranges, which provided food for the herbivore, arranged so they were

Figure 7.4 Two living Transvaal Stone Plants in the Namibian desert (clue to spotting them; the most obvious plant is just to the right of centre of the photograph, the other is to its left). The various species of stone plants are one of the few plant groups that appear to be cryptic—matching the colour of the desert surface. Photo: TNS.

close together) the predator quickly caused the prey population to crash and hence led to its own extinction. However, when he made the laboratory environment more spatially complex (oranges, more spread out in the environment) then the predatory mites could not find all of the prey mites before they reproduced and so the two species could coexist for long periods of time. In a similar manner, an insect herbivore may struggle to find all individuals of a food plant if it is rare and scattered throughout the landscape.

Perhaps, therefore, refuges and spatial complexity in general help to keep the world green? Essentially, this suggests a view of the world as a shifting mosaic of plants getting to grow relatively unmolested in a given area, until herbivores find the resource, reproduce, and begin consuming. The herbivores then die or disperse, and plants can recover from seeds, rhizomes, or woody tissue that are protected from attack.[37] As this explanation depends on the distribution of plants, it is a bottom-up explanation (although it does not directly involve plant chemicals). Such spatial ideas have been very influential in population ecology in the past few decades[38] and scientists are just starting to think about them in the specific context of the 'green world' problem.[39]

Explicit consideration of the challenges of finding host plants also raises other interesting issues. During the 1960s and 1970s, several scientists (especially D.H. Janzen, P. Feeny, D.F. Rhodes, and R.G. Cates) realized that some species of plants would be more easily found by herbivores, and that this has potentially important implications for their anti-herbivore adaptations. The idea is known as 'apparency'[15] and proposes that large long-lived plants such as trees may be more easily found by herbivores than small short-lived plants such as annuals. It suggests that the more apparent plants (e.g. most trees) should invest in chemicals that reduce digestibility—such as tannins, lignin, and so on which will be discussed in more detail later—while less apparent plants should use the kinds of toxins which we have already described. The logic behind this is that apparent plants are likely to be found by specialist herbivores which will have evolved ways of neutralizing a plant's toxic secondary compounds, so the plants are better off using other means of defence. There are problems with this idea; for example, trees produce many of their indigestible compounds for structural reasons so their presence may have little to do with anti-herbivore adaptations. In addition, it has been very difficult to come up with a convincing way of scoring a plant's apparency—which makes testing the idea difficult.[15] Because of such problems, the idea has tended to fall out of favour with many plant ecologists. However, in 1992 Peter Grubb[40] considered the apparency idea in the context of plant spines, rather than chemicals, and suggested that despite the problems we have just outlined he considered it a useful idea, albeit a bit too simplistic. So, although the idea of apparency has not been very fashionable with plant ecologists in recent years it is probably capturing something of importance for bottom-up explanations of the green world. Collectively these spatial processes, coupled with phenomena such as apparency, provide a rather mixed explanation for why specialist herbivores do not destroy all plant life—herbivores simply cannot find all available plants, and when they do, the plant is not entirely palatable.

What about microbes?

The arguments so far in this chapter may lead one to think that there are effectively two main types of explanation (top-down and bottom-up) to what R.M.M. Crawford[23] has called 'one of the marvels of nature' namely, that plants 'while providing the original source of food for all animal and microbial life are not themselves consumed to the point where they are no longer able to support their predators'. In making our case, we have largely considered insects and mammals as herbivores but we have been forced to ignore microbes as there is little in the way of relevant experiments to discuss.

Despite this lack of study, microbes may play a role in keeping the world green and some fascinating relationships are starting to be discovered. For example, one recent study showed that antibiotic-producing bacteria in southern beech trees apparently provide protection from fungal attack.[41] Likewise, it has been known for some time that some species of fungi that live within grasses can produce alkaloids that can help protect both fungi and grass from attack.[42] So, microbes may feature in bottom-up processes. Micro-parasites are also probably very important in many top-down processes, by potentially controlling the population size of many herbivores. Of course, it is a two-way street because microbes can facilitate as well as prevent defoliation—for example, gut-living microbes are also important in allowing many animals to detoxify plant secondary compounds[21] and can also directly destroy plant material. The role of microbial processes in each of these contexts is under studied and we may find out a lot more over the next few decades.

So bottom-up is the current favoured explanation, but there are some problems

If the top-down mechanisms do not provide a full explanation for why the world is green, then what about bottom-up explanations? Do these provide an adequate answer? They certainly look more promising than most top-down explanations. As we have previously outlined, the bottom-up explanations fall into two main categories. First, either plants are poor sources of the nutrients needed by animals, or plants make poor food because they contain poisons and other defences. Second, spatial processes may play a role in reducing the accessibility of plants to herbivores.

There are a wide range of ways in which plants can provide poor ingredients by the standards of what is required to sustain most animals—for example, we have already mentioned that plants are low in nitrogen compared to what is required by animals. However, there may be a small complication in the specific argument about low nitrogen reducing the density of herbivores if we consider its role in a geological context. This is because there are good reasons for thinking that the average nitrogen of plants is higher now than it has been in the geological past.[43] For example, nitrogen concentrations are known to be higher in modern flowering plants than in ferns or cycads—which have a much longer geological history than flowering plants. In addition, the nitrogen concentration in modern leaves is related to carbon dioxide levels in the atmosphere, with high concentrations of atmospheric carbon dioxide being correlated with low leaf nitrogen.

It is well established that carbon dioxide has declined during geological history—although we are currently causing a potentially extremely important rise in its level (see Chapter 9). For example, there was a particularly large drop in atmospheric carbon dioxide during the Permian and Carboniferous periods (approximately 250–350 million years ago). This long-term decline is partly because of the effects of land plants on the weathering of various types of minerals, which leads to the removal of carbon dioxide from the atmosphere[44,45] (see Chapter 10), and suggests that forests of the past had leaves of rather lower nitrogen content. As well as these changes in plant nitrogen, early forests were dominated by huge ferns and club mosses which were much richer in lignin than modern trees.[46]

Collectively, the above observations cause us to be a little more cautious in suggesting a strong role for nitrogen limitation (and other bottom-up effects) today—as herbivores have apparently coped with even lower levels in the past. For example, in detailed studies of fossil leaves from rocks around 200 million years old in South Africa (which matches another 'peak' in carbon dioxide levels[44]), Andrew Scott and colleagues[47] found that some of the plant species had up to 50% of their leaves showing signs of insect damage. In addition, the largest land herbivores known were various species of dinosaur, and they apparently had no problem feeding on vegetation—although their large size may have helped these animals process poor quality plant food by allowing them to retain food in the gut for longer, so giving more time for the gut microbes to do their work.

One apparent prediction of the bottom-up approach is that plants with more diverse or unusual secondary plant compounds may be expected to have fewer problems with herbivores. Clive Jones and John Lawton[48] compared the biochemical make-up of British plants in the carrot family and could not find any evidence that biochemically diverse or unusual plant species supported a less (or more) species-rich insect assemblage. Unfortunately, one could also make the reverse argument and say that only plants with lots of different herbivore species need to evolve complex chemical defences. So data such as Jones and Lawton's, while of interest, cannot be used to test the validity of bottom-up argument.

Top-down and bottom-up; the story so far

Where does this leave us when it comes to explaining why the world is green? Similar to many questions in ecology there is no single clear answer. The current consensus (which we broadly support) is that bottom-up processes are the most important in maintaining a green world. So, herbivore population densities tend to be limited by their food supply, not their predators. Indeed, taking an evolutionary perspective, members of a plant species with few defences and low capacity to reproduce would probably go rapidly extinct, so almost by definition we will be left with populations of plant species that can persist despite the ravages of herbivores, even without the help of predators to keep herbivore numbers low. More generally, it seems likely that in many cases herbivorous relationships would evolve before predators begin to attack the herbivores, and yet the plants have somehow survived this onslaught.

Bottom-up processes involving secondary plant compounds probably explain why all herbivores do not consume all plant species—many herbivores being required to specialize, or alternatively eat small amounts of many different plants to avoid poisoning. Nevertheless, it is quite possible that top-down processes can play a role in controlling herbivore numbers in some contexts, where predators happen to be at particularly high local densities. The dual role of these processes is nicely illustrated by recent work in tropical forests in Panama, where predatory control of herbivores appeared most important in clearings that allowed lots of plant growth, but bottom-up processes dominated in deep shade where plants grew slowly (and had lower nitrogen levels) and food for herbivores was in shorter supply.[49]

In the late 1970s, Lawton and McNeill[17] neatly encapsulated this diversity of explanations for the green world by describing herbivorous insects as caught between the devil (of top-down processes) and the deep blue sea (of bottom-up processes). They extended this metaphor to the dilemmas facing scientists trying to make sense of this area of ecology, describing them as also caught between 'the devil of oversimplification on the one hand and a deep blue sea of endless unrelated facts on the other'. This tension between unrealistic simplification and a bewildering array of detail often faces ecologists. Yet in the green world case there are even more complications. Recently, several ecologists have been arguing that to fully understand this problem we have to add yet another factor to the top-down and bottom-up processes we have so far been describing—namely fire.

A sideways look; is it all just top-down and bottom-up?

There are several approaches to explaining 'Why the world is green?' that do not neatly fit into the classic top-down/bottom-up classification; fire is one of these that has received increasing attention in recent years. Plant ecologists usually define the effects of fire as a type of disturbance, that is, a process 'associated with the partial or total destruction of the plant biomass'.[50] In addition to fire, disturbance includes the effects of herbivores but also things such as storm damage, trampling, and flood damage.

As William Bond and colleagues have pointed out,[51,52] herbivory and fire have a lot in common that differentiates them from other types of disturbance; indeed in an attempt to draw ecologists' attention to the importance of fire they have described it as effectively a 'global herbivore' because it consumes large amounts of plant material in many parts of the world. Fire has the potential to be a particularly successful 'herbivore' as it is unconstrained by plant poisons, woody tissue, or low nitrogen; indeed plants with all these anti-herbivore mechanisms will burn well if reasonably dry. As such, fire is a herbivore substitute that is unaffected by all the bottom-up (and top-down) processes described in this chapter (although some vegetation may be more combustible than others).

In the past, fire has been given relatively little prominence in ecology textbooks, unless they were on particularly fire prone systems such as mediterranean type vegetations. Indeed, it is interesting to note that the plant ecologists who have recently been stressing the importance of fire are mainly based in South Africa or California—both areas rich in 'mediterranean' vegetation. One reason for ecologists underplaying the importance of fire in the past may be the assumption that it is a largely modern phenomenon associated with human activity. While it is true that our actions have greatly increased fire frequency in many parts of the world, while decreasing it in others, geological evidence from charcoal preserved in rocks shows that fire (caused by lighting strikes and volcanic activity, among other processes) has been a regular occurrence as long as there has been widespread terrestrial vegetation—that is, for at least 420 million years.[53] As such, fire is clearly potentially important for any explanations of why the world is green because of the additional challenges it poses—it seems to have the potential to turn everything brown or black and yet has not done so.

Since fire does not neatly fit into the classic division of top-down and bottom-up processes, Bond and Keeley[52] have suggested that both fire and herbivores are better considered together and described as 'consumer control' processes acting on the green world of vegetation, rather than focusing on the traditional top-down/bottom-up dichotomy. However, fire could also interact with traditional top-down mechanisms by killing herbivores and their predators. Plants may recover more quickly after fires than do herbivores, for example, recovering from seeds and rhizomes protected in the soil, and for a period of time they would be able to grow in conditions with few herbivores—which cannot return in any numbers until the land is green again to provide the required food supplies.

It is obviously difficult to do large-scale experiments with fire, but an alternative is to run computer models that attempt to predict global vegetation patterns based on climate—while excluding fire from the models. These can then be compared with what we see in the real fire prone world. When this is done the results suggest that vast areas of what are currently grassland and savannah in Africa and South America have the potential to support forest in a fire-free world.[54] Another (not mutually exclusive) explanation for the existence of these grasslands is that large herbivores may also be involved in preventing forests from developing. There is currently intense interest amongst scientists who try to model the relationships between global climate and life in trying to include both the effects of fires and large herbivores (such as elephants which are well known to affect savannah trees) in their models. Such models, when available, are likely to give us a better understanding of the importance of fire consumer control and mammal-driven top-down processes in the next decade.

However, it is important to note that once again fire is apparently not making a green world brown (or black), but affecting the type of green vegetation we see growing at a particular place. As such, it is behaving more similar to the sheep of North Wales or the leafcutter ants of Venezuela described earlier. There are several reasons why fire does not destroy the green world. First and foremost, spatial heterogeneity can prevent it spreading just as we have seen in the case of herbivores. Second, paralleling another bottom-up control, some types of vegetation will not burn even when dry and most plant life will not burn when wet. Indeed the oxygen levels required for wet plant material to easily burn (something over 30%, the current atmospheric level being 21%) are thought to set a limit to oxygen levels in the geological past—as global vegetation was never destroyed in a great conflagration.[45] So, similar to herbivores, fire usually only influences the type of vegetation, not its presence or absence.

Population biology and the sideways perspective

Consideration of population ecology also contributes additional explanations at various levels, or at very least new ways of looking at the same phenomena. For example, the long-known reduction in biomass along food chains because of loss of energy at each stage[55] can be thought of as reflecting constraints driven by bottom-up processes. The time taken to find plants, and the defences of plants even when they are discovered, all contribute to this reduction in efficiency of energy transfer between levels, and may help explain why herbivore populations are not high enough to completely exploit their resource.

Other factors influencing herbivore population size may also be relevant. One such factor comes under the general umbrella of 'density-independent processes'—so-called, because the density of the organisms involved does not feed back into the process causing changes in its population size. It may be that herbivore density is kept in check by these density-independent mechanisms, thereby indirectly controlling the extent of herbivory. A classic example of such a density-independent process occurs where populations of herbivores can be greatly affected by changes in their environment, such as rainfall or temperature (it is a density-independent process because the likelihood of a cold winter is not usually affected by a species population size). In an influential series of studies by Davidson and Andrewartha[56] in the 1930s and 1940s, these researchers showed that population sizes of apple blossom thrips, in Australia, were largely determined by year-to-year climatic variation. Likewise, in Britain combinations of mild winters and hot summers can lead to population explosions of some species of ladybird beetles ('ladybugs' in North America)—such outbreaks occasionally hit the TV news as large numbers of starving beetles can sometimes bite people, chasing swimmers, and sunbathers from beaches and picnic areas.[57] As important predators on aphids, these ladybirds are potentially important players in top-down explanations, providing one reason why herbivore densities do not get out of control. H.G. Andrewartha was particularly impressed by examples such as these (as was another of his colleagues L.C. Birch); and when he came to draw together his ideas on animal ecology in the early 1960s for an introductory textbook, he gave major emphasis to the role of weather in influencing animal population size.[58]

So, clearly one possibility is that herbivores do not achieve the numbers required to turn a green world brown because environmental (often climatic) constraints prevent herbivores from gaining sufficiently high densities to cause complete exploitation. In addition, as illustrated by ladybirds, these climate effects can potentially interact with top-down processes, most notably control of herbivores by predators. Insectivorous birds may provide another example; in reviewing the effects of birds as predators of insects Şekercioğlu[59] suggested that their ability to control insect population sizes may be limited in temperate latitudes by the vicissitudes of annual climatic changes (Fig. 7.5) but that they may be more important in the tropics—although as he pointed out, we are short of good tropical studies from which to generalize.

An additional sideways population mechanism is competition between herbivores, both within the same species and between species, which again may limit their population sizes to levels that cannot destroy the green world. For example, analysis of 40 years worth of data on blue wildebeest populations on the Serengeti, in East Africa, showed that levels of rainfall in the dry season was a key factor in their mortality rates, but this effect only operated when wildebeest numbers were high and so competition for limited food in dry conditions was also high.[60] Similarly in the Isle Royale moose population, there was a large crash in numbers during late winter/early spring 1996 apparently due to a combination of high moose numbers (due to a reduction in the wolf population)

Figure 7.5 Deciduous woodland in winter; this example is Tattershall Carrs, a woodland nature reserve in Lincolnshire, England. Many of the insect-eating bird species found in such woodland are migratory—only present in summer. This photograph was taken only a half an hour walk from the location of Fig. 7.2, but at a very different time of year—highlighting the potential importance of seasonal changes for plants, herbivores, and predators in the temperate regions of the Earth. The distinctive growth form of these trees is because the traditional management practice of 'coppicing' is implemented here, that is, the repeated cutting back of the trees to ground level to produce a supply of thin wooden poles. Photo: DMW.

increasing competition and extreme winter weather.[61] So the population effects of both environmental fluctuations and competition between herbivores could in theory be important 'sideways' processes as well as the effects of fire described earlier.

Why is the soil brown?

As well as its effects on type of vegetation, fire is also globally important because it can release large quantities of carbon stored in plants back into the atmosphere; this obviously has potential climatic implications.[62] However, carbon is not just stored in growing vegetation; large quantities are also locked up in soils. Indeed globally more carbon is stored in soils than in either vegetation or the atmosphere.[63] Computer modelling suggests that one of the major uncertainties affecting the future levels of carbon dioxide in the atmosphere is the uncertain effects of increased temperatures on the rate at which organisms break down soil organic matter so releasing carbon dioxide back to the atmosphere.[64] As Steven Allison[65] has pointed out, the amount of organic matter in soils—which often gives them their brown colour—is a problem very similar to the green world ideas of HSS. This organic matter is potential food for all sorts of soil organisms, especially microbes, so why is so much left unused—that is, 'why is the soil brown?' The logic of HSS can also be applied to this question. One possibility is that the soil remains brown because of top-down processes, with predators controlling the number of decomposer organisms feeding on the organic matter. Allison suggests that this is unlikely if we are thinking of conventional predators; however, we know very little about the possible role of parasites (such as viruses infecting bacteria) in the soil and these could turn out to provide important top-down controls. A more likely explanation relies on bottom-up processes, suggesting that much of the organic matter in soils is in a form that is difficult for organisms to use. This links directly with the main question of this chapter, as much of the organic matter in soils comes from plants and so has the same low nutritive content that we have described as posing a problem for herbivores. Soil also shows a strong spatial structure and in addition soil organisms are also affected by the 'sideways' population processes of density-independent environmental effects and density-dependent competition. So all the explanations described earlier for the green world are potentially involved in maintaining the brown world of soils.

So why is the world green?

Our question has been described as 'one of the most basic yet astonishingly complex questions in ecological research'.[66] The apparently simple question raised by HSS illustrates the complexity of the real ecological world with both bottom-up and top-down control of herbivores contributing to the explanation for the persistence of vegetation, along with other processes such as fire and several aspects of population ecology which do not neatly fit into this simple classification. Indeed, one of the most influential (and best) of the university level ecology textbooks[67] just lists all these possibilities, suggests they all contribute, and makes no attempt to distinguish between them in their relative importance.

Yet the studies we have discussed in this chapter suggest that there is an imbalance in the relative importance of these processes, with top-down explanations only appearing convincing in a few special cases such as the artificial 'overgrazing' of arid areas by livestock maintained at high density. In trying to answer this apparently simple question, scientists have also produced important data on how herbivores and fire contribute to determining the type of 'green' found at a particular location; indeed many of the research papers that describe themselves as investigating the green world problem of HSS are really studying the factors that determine the type of green (not why green rather than brown).

Having pointed out these complexities and the multiple processes involved in a full explanation, it appears that bottom-up processes are probably the most widely applicable explanation for why herbivores do not destroy all vegetation and so they provide an important part of the answer to the green world problem. One reason limiting the efficiency of herbivores is not just the defences of plants but also their spatial distribution. Bottom-up processes may help explain 'Why is the soil brown?', although studies on this question are in their infancy. Gary Polis[16] summarized this bottom-up view in a widely cited review paper in 1999; he wrote:

> The implication is clear: even in a world full of green energy, many/most herbivores cannot obtain enough requisite resources to grow, survive, or reproduce at high rates. Nutritional shortages regulate herbivore numbers, often limit their effects on plant biomass, and form one important reason why much of the world is green.

8

Why is the Sea Blue?

Figure 8.1 Sunset over the Irish Sea. Photo: DMW.

I would peer up the sheer cliff, searching in vain for other divers. Seeing none, I turned outward to the void... I sank through rays of light like a particle in eternity.

—*Peter Matthiessen, Blue Meridian.*[1]

Defining the question

One answer to this chapter's question is straightforward and based on high-school physics. The early SCUBA divers quickly discovered that if they took underwater colour photographs, even if they were only a few metres down, their pictures had a strong blue cast to them. However, if they illuminated their subjects with a flash, then a more colourful world emerged in their pictures—especially if they were photographing the rich diversity of highly coloured fish that can be found in some parts of the tropics.[2] The reason for the blueness is that as sunlight passes through water the colours of the spectrum are absorbed at different rates, with the long wavelengths (e.g. red) absorbed first and the higher-energy shorter wavelengths (e.g. blue) penetrating deeper into the depths. It follows that underwater available light is predominantly blue and that any light reflected from within the water body is more likely to be from the bluer end of the spectrum of visible light. So, light coming from the sea to our eyes is mainly blue because these wavelengths are least absorbed; indeed oceanographers who have studied some of the cleanest waters describe them as looking 'violet blue'.[3]

As biologists we are interested in a more ecological answer to the question, 'Why is the sea blue'? The physics explanation only works if seawater is reasonably clear, and it is this clarity that biologists need to explain. Consider our opening quotation, which comes from Peter Matthiessen's book describing early attempts to film the great white shark in its natural habitat. It raises an interesting ecological question—why can a SCUBA diver or snorkeler see where they are going in the ocean? Put another way, why is the sea blue rather than green?

The upper layer of the ocean with enough light for photosynthesis is called the euphotic zone (defined as extending down to the point where only 1% of photosynthetically usable light is present compared with surface light levels); this is often only a few tens of metres deep, but in extremely clear water near Easter Island in the Pacific it has recently been found to extend down to 170 m depth.[3] The lower limit of the euphotic zone clearly explains why most of the volume of the oceans is not green—it is simply too dark for photosynthesis. However, the rays of light Matthiessen poetically describes penetrating the upper levels of the ocean provide all the energy needed for photosynthesis, so why are these upper levels not thick with plants, or at least photosynthetic microorganisms?

The question we ask in this chapter also formed a chapter in Paul Colinvaux's book *Why big fierce animals are rare*?[4] Indeed, as we explained in our preface, the approach taken in his book provided one of the inspirations for the book you are now reading. Colinvaux wrote: 'The sea is blue, this is a very odd thing because the sea is also wet and spread out under the sun. It ought to be green with plants, as is the land'.[4] Now, over a quarter of a century later researchers are able to give a much more comprehensive answer to this question that has involved several remarkable discoveries, most especially with respect to the role of photosynthetic microbes.

Like most of the topics we have chosen to discuss, we believe the question, 'why is the sea blue' is about as big as you can get. The oceans make up 71% of the surface of our planet,[5] so the lack of a 'green' ocean makes a huge difference to the amount of solar energy utilized by life on Earth. We will start by considering the relatively simple question of why large plants do not survive in numbers in the oceans, and then go on to consider the more complex questions relating to photosynthetic microorganisms. This latter question is the area where our scientific understanding has most dramatically changed over the past few decades. For example, the commonest marine photosynthetic microbes on the planet were completely unknown to science when Colinvaux was writing his book in the late 1970s, and since the mid-1980s our understanding of the way ocean nutrients affect plankton numbers has changed radically.

Where are the large 'plants'?

In many temperate seas where there are large tidal ranges the most obvious marine 'plants' are the seaweeds exposed at low tide. These macroalgae are composed of members of three phyla (informally referred to as the 'green', 'red', and 'brown' algae after their most obvious pigments), which have traditionally been classified as true plants and, as such, have usually been studied in botany departments—in the days when biology in most universities was normally split into departments of botany and zoology. However, the taxonomic affiliation of these macroalgae remains somewhat unclear. More recently, these organisms have tended to be placed in the kingdom protista (sometimes called protoctista) along with the protozoa and many other, often single-celled, eukaryotes[6]; a classification that has been followed by many university-level biology textbooks. However, many areas of biological classification are in a state of flux with the avalanche of new data from molecular biology, and now some scientists would like to put these seaweeds back into the plant kingdom (plantae).[7,8] These revisions are starting to be reflected by some more general texts, for example, the 2008 edition of one widely used biology textbook puts the 'green algae' back into the plantae, but keeps the 'red' and 'brown' algae in the protist kingdom.[9] However we choose to classify seaweeds, ecologically they have much in common with traditional plants in being large photosynthetic organisms, which usually spend much of their life attached to the substrate.

When conditions are suitable, there can be a considerable biomass of these seaweeds. For example, furbelows, a kelp that is the largest European seaweed, has fronds in excess of 4 m in length—amazingly, each individual frond usually survives only for less than a year, which illustrates that under the right conditions marine 'plants' can grow remarkably quickly.[10] Seaweeds are limited to shallow water by their need for light and they also appear to be limited by other physical aspects of their environment—this is clearly seen by their scarcity (in extreme cases, total absence) from coasts that are highly exposed to wave action. High waves can make it impossible for them to establish the anchorage on the rocks that they need to prevent themselves from being killed by being washed up on beaches.[10]

We note in passing (since it is relevant to another of our 'questions') that seaweeds can sometimes completely cover the rocky bed of marine shallows, and this raises the additional question of how do they survive predation. Seaweeds are potential food to a whole range of marine organisms and so the 'green world' question (discussed in our previous chapter) also applies to this system. The answer in this case appears to be a similar mix of bottom-up, top-down, and 'sideways' processes as that used to explain the persistence of vegetation on land. For example, there are experiments, similar to the exclosure ones we described in the green world chapter, that appear to show that reducing grazing can modify the type of macroalgae growing on the seabed, and bottom-up chemical defences are also common in these algae.[11] In an analysis of 54 published field experiments, Burkepile and Hay[12] suggested that in tropical seas top-down explanations dominated but that the situation was more complex in cooler waters where the amount of nutrients in the environment appeared important—with predators (top-down) dominating only in low-nutrient waters.

One of the most striking features of marine botany is the near absence of flowering plants, which are easily the most diverse group of plants on land. Why should the shallow seas of the coastal edge be dominated by macroalgae rather than flowering plants? Indeed, globally there are only around 30 true plant species recorded from the oceans, and these tend to be mainly found around estuaries.[13] One suggestion is that pollination is more difficult in water than it is in air partly because insects, which are key to much pollination on land, are largely absent from the oceans.[13] However, many of the marine macroalgae reproduce using processes similar to wind pollination, with gametes dispersed through water in a manner similar to pollen of land plants being dispersed through wind, so if this is the explanation then the relevant differences between seaweed fertilization and plant pollination must be quite subtle.[14] Whatever the explanation, with the exception of a few sea grass beds, flowering plants do not dominate any shallow marine habitats. As far as our chapter's question is concerned, it is clear that neither sea grasses nor macroalgae attached to the substrate can make the sea green because they can only survive in very shallow water around coasts, and so are absent from most of the oceans.

What about floating 'plants'?

As the lack of light prevents 'plants' growing on most of the seabed, then an obvious question is, 'what about floating plants?'. In freshwater systems we are used to seeing plants whose leaves float on the water surface. Some of these, such as various species of water lily, are rooted in the sediment on the bottom of the pond; clearly this approach would not work in much of the ocean where the bottom can be several kilometres down.

However, many freshwater plants live by floating on the surface without any connection to the sediments below—examples include water hyacinths and duckweed[15] (Fig. 8.2). Why is the ocean's surface not covered by such surface-floating plants making use of the abundant solar energy and water supply?

Many marine ecologists have tended to take the dominance of microscopic plankton in the ocean as a given and not really asked questions about floating 'plants'—after all, they hardly exist in the ocean so why think about them? However, as Paul Colinvaux has pointed out several times over the past few decades[4,16] there are potentially big advantages to being a large (at least non-microscopic) plant (e.g. in storing lots of nutrients) which makes their absence from the sea surface rather strange. In addition, if large plants covered the surface of the sea, such as duckweed on a pond, then there would be little light available for use by photosynthetic plankton and the ecology of the Earth's oceans would be very different from what we observe. The reason why this floating way

Figure 8.2 Common duckweed covering the surface of a freshwater pond in northwest England. Very little light gets through this layer to be available for use by photosynthetic microbes in the water below. Wind and wave action prevent duckweed from covering the surface of larger lakes. Duckweed species are an example of a floating plant that many readers will be familiar with, as they have a global distribution in freshwater systems, excluding the polar or very arid areas.[63] Photo: DMW.

of life will not work in the sea is the same as why, in freshwater, it only tends to be a successful strategy on ponds and small lakes—you do not see large lakes covered in duckweed. On a large water body, wind and the associated water movements lead to things floating on the water tending to get washed up on the beach; we also see this when we look at plastic and other rubbish washed up on the strand line of a beach on the edge of the ocean (Fig. 8.3). With no roots anchoring them to the bottom and no way to swim back into position (unlike many 'floating' jellyfish), any plant that tried to float on the sea surface would be moved away from where it was growing by currents and wind. Besides the sheer physical wear and tear that this movement inevitably involves and the challenges of being stranded on a beach, if the movement is extensive it may take the plants into environments with radically different temperatures and chemical conditions that they are poorly adapted to cope with.

A graphic illustration of the potential for ocean currents to displace floating plants can be seen by considering the fate of nearly 38,800 yellow plastic toys that fell overboard from a container ship in the Pacific in January 1992. Some of them floated south

Figure 8.3 Plastic rubbish washed up on a beach on the west coast of Scotland, illustrating the fate of things which float on the surface of the Atlantic Ocean. This would be fatal to any floating plant and explains why the ocean's surface is not covered with large plants or macroalgae. Photo: DMW.

through the tropics, landing months later on the shores of Japan, Indonesia, Australia, and South America. However, others headed north and by the end of the year were off Alaska. Eventually, some even managed to get through the Bering Strait, and enter the North Atlantic—subsequently beginning to wash up on the beaches of eastern United States and even Scotland.[17]

These general challenges facing floating plants are also nicely illustrated by one of the few exceptions to the observation that floating 'plants' do not exist in the ocean. The Sargasso Sea, in the North Atlantic (between 20° to 35° N and 30° to 70° W), is effectively a giant whirlpool (technically called a 'gyre') that allows floating material to maintain its position in its centre, so the floating macroalgae called Sargasso weed can live in large patches on the water surface. This sea became infamous in the days of sailing ships when vessels could easily become becalmed and marooned in the gyre. Early travellers' tales of the Sargasso Sea overemphasized the thickness of floating plants for dramatic effect and it has been suggested that they gave poetic inspiration to Coleridge whose fictitious ancient mariner proclaimed: 'The very deep did rot: O Christ!/That ever this should be!/Yea, slimy things did crawl with legs/Upon the slimy sea'. In reality even in the Sargasso Sea floating plants fail to cover the whole surface, often being arranged into lines by wind action; there is much open water and the water surface is certainly not slimy.[18,19] However, given Colinvaux's point about larger 'plants' being able to store nutrients, it is interesting to note that the gyre that keeps the Sargasso weed in place also creates very 'pure' nutrient-poor water.[3]

So life as a floating 'plant' is not possible except in exceptional places, most famously in the Sargasso Sea, although some rafts of the Sargasso weed can be found further away, such as in the Gulf of Mexico.[19] With large plants and macroalgae unable to live in or on most of the ocean, this allows marine photosynthesis to be dominated by microorganisms, but even these can seldom turn the sea green.

The phytoplankton

If we are to explain why the sea is not green then we need to understand the ecology of these photosynthetic microbes, referred to as the phytoplankton. As they are microscopic, these organisms are much less familiar to most people, including biologists, than the seaweeds and sea grasses we have discussed earlier. Therefore, before describing the current ideas about what limits their numbers in the ocean we briefly review the diversity of marine phytoplankton. This is not simply for academic interest—these organisms are extremely important in the ecology of the Earth, being responsible for approximately half of all global primary production (i.e. the production of biomass from inorganic energy sources).[20,21]

Until the late twentieth century, studies of phytoplankton concentrated on relatively large eukaryotic organisms. For example, marine diatom studies were started during the great scientific voyages of the nineteenth century by biologists such as J.D. Hooker.[22] By the mid-twentieth century, it was recognized that there were also vast numbers of smaller photosynthetic microbes in the sea, which were potentially very important but difficult to study with the technology then available.[23] More recently, it has become apparent that extremely small photosynthetic plankton are very important in the ecology of the oceans. These are often referred to as the picophytoplankton (the prefix 'pico' referring to their very small size, formally denoting 10^{-12} m), and they are defined as photosynthetic microbes that will pass through a 2 μm diameter sieve.[24] In contrast, the microplankton, such as diatoms, which were the subject of much of the nineteenth century study, are between 20 and 200 μm in length (a micron, μm, is millionth of a metre).

Some of the most important members of the picophytoplankton are prokaryotes, especially the cyanobacteria. One of the most important of these is *Prochlorococcus*; although not described by science until 1988, this genus dominates primary production in the tropical and subtropical oceans. Indeed, despite its recent discovery, it is probably a good candidate for the title of the commonest organism on Earth.[25,26] It has a diameter of about 0.6 μm and the smallest genome of any photosynthetic organism yet studied.[27] The biodiversity of seawater, when assessed using modern (molecular) methods, has come as a surprise to many biologists, and each new survey seems to increase the estimated microbial diversity of the oceans.[28] For example, one litre of seawater can contain an estimated 20,000 species of microbes (see Chapter 4 for the difficulties of defining microbial species) most of which will only be represented by a few cells.[22]

The methods of molecular biology have the potential to revolutionize our knowledge of marine microbes. For example, Craig Venter, one of the pioneers of large-scale genome sequencing, has coordinated a large team working on water samples he collected while sailing his yacht, *Sorcerer II*, around the world. Early results from this work have suggested large planktonic diversity in the Atlantic and into the tropical Pacific.[29] In many ways, Venter's study is a 'molecular' version of the great nineteenth century voyages of scientific exploration, and indeed he writes that the British Challenger Expedition (1872–1876) was one of his inspirations for this study.

The key observation, relevant to this chapter's question, is that despite this great diversity of phytoplankton, it is rare for them to occur in densities that colour the seawater. When this does happen it tends to be in parts of the oceans where nutrients are unusually plentiful. For example, 'blooms' of phytoplankton covering several hundred square kilometres can occur off the south coast of Newfoundland, Canada, in late spring as light levels increase, allowing plankton to utilize the flood of nutrients that currents bring up from the deep ocean into the euphotic zone where photosynthesis is possible.[20] This suggests that a lack of nutrients is involved in maintaining a 'blue' sea (as do formal experiments we will describe later). Of course, this then raises the follow-on question, 'Why are nutrients relatively scarce in most of the oceans?'.

The scarcity of nutrients in the oceans: the iron story

Seawater is not pure water; as everyone who has swum in it knows it has a salty taste. These salts comprise a range of chemicals that are dissolved in the sea, the most common being chloride and sodium, followed by sulphate, magnesium, and calcium. In addition, there is a long list of less common 'salts', which include many of the key nutrients for plankton, such as phosphate, nitrate, and iron. The main source of marine salts is the weathering of rocks on land and the subsequent addition of chemicals released from these rocks to the sea via rivers or wind-blown dust, along with some nutrients from hydrothermal vents.[30] All of these chemicals are at low concentrations in seawater and this has made them difficult to study in the past. Indeed, it was not until after the First World War that accurate methods for measuring the phosphate and nitrogen components (nitrite, nitrate, ammonium) were devised, mainly by William Atkins, Hildebrand Harvey, and Leslie Cooper, working in Plymouth on the south coast of England. They were able to show that these nutrients were in such short supply that they limited the production of phytoplankton, which in turn limited the food available for the small animals of the zooplankton. Indeed, they likened the seas of southern England to 'a closely grazed pasture'.[22,23]

An obvious way to attempt to prove that an ecological system is limited by nutrients is to experimentally add more and record the results. We will describe such experiments for iron (a key limiting nutrient in approximately one-third of the surface area of the world's oceans[31]) and postpone the discussion of other important nutrients (especially nitrate and phosphate) until the next section.

Iron (chemical symbol Fe) is widely used by organisms in a variety of enzyme systems and since iron is the fourth most abundant element in the Earth's crust one might guess that access to it by microbes is straightforward. The problem for the plankton is that under oxidizing conditions iron is very insoluble above pH 4 (seawater usually has a pH of around 8), and in the open ocean, away from the continental shelves where rivers can wash in new supplies, biologically available iron is very scarce. The main input of iron to these oceanic waters is from dust blown from arid areas such as the Sahara desert,[32] although in the Southern Ocean upwelling deep water also appears to be an important source.[33] These facts have interesting implications for the marine ecology of the past. For example, during glacial periods there was increased aridity in many areas and therefore iron-rich dust was entering the ocean at a greater rate. This may have caused an increased removal of carbon dioxide from the atmosphere by increased plankton growth, so reducing the 'greenhouse effect' and making the glacial climate even colder.[33] This has led to ambitious proposals of modifying the amount of iron in the modern ocean as a way of combating human-caused global warming—we will briefly discuss this idea later after we have summarized the relevant experiments.

Interestingly, the current problems plankton experience with access to iron may not have existed in the distant past. Early in the history of life, iron may have been less limiting for marine plankton as the ocean was less oxygen-rich, since photosynthesis had not yet greatly increased the oxygen concentration of the atmosphere. Consequently, any iron would have been in an oxidation state, which was more available to life (because this less-oxidized iron is more soluble in water), potentially allowing greater biomass of photosynthetic bacteria in these early oceans.[34] Atmospheric oxygen levels are thought to have been exceedingly low until the 'Great Oxidation Event' approximately 2–2.4 billion years ago (a billion being one thousand million), although photosynthetic microbes probably evolved rather earlier than this, but did not immediately lead to an oxygen-rich atmosphere and ocean.[35,36] Indeed, the deep ocean does not appear to have become oxygen-rich until around 551 million years ago (based on the molybdenum chemistry of the rocks of the time).[37]

Our understanding of the role of iron in ocean ecology is relatively recent. At the beginning of the 1990s ship-based experiments, where iron was added to marine water samples in onboard laboratories, suggested that this element was often important in limiting the growth of phytoplankton. However, such small-scale experiments are difficult to interpret—are the results an artefact of the artificial conditions in the research ship's laboratory or do they represent what really happens in the open ocean? Owing to this, larger-scale ('mesoscale', bigger than laboratory scale but smaller than whole-ocean scale) iron-enrichment experiments were started, where iron was added to patches of ocean and the effects monitored. As of early 2007 there had been 12 such experiments carried out around the world, with a range of 350–2,820 kg of iron added to the water depending on the details of the particular experiment.[31] All of these experiments led to increased plankton primary production, with diatoms often being the organism that responded the most to this treatment. One problem with these experiments is that they are of relatively short duration. Recently, Stéphane Blain and colleagues[38] attempted to address this by studying plankton at a natural, and so more long-lived, upwelling of iron-rich water near the island of Kerguelen in the Southern Ocean. They estimated that the amount of carbon being exported to deep water (effectively a measure of plankton productivity) was at least 10 times greater than that seen in the short-term mesoscale experiments, probably in part due to the more steady input of iron from deep water (rather than being added as a large experimental 'pulse'). So, plankton productivity was certainly benefiting from the presence of iron.

The simplest way to view these experiments is to suggest that in much of the ocean phytoplankton production is simply limited by iron. However, this may often be an over-simplification as different limiting factors can interact with each other, and not all species react the same way. For example, as iron is needed for the construction of some of the proteins involved in photosynthesis, there is an interaction between light levels and iron requirements. In the late 1990s, William Suda and Susan Huntsman[39] showed that in culture (i.e. growing the microbes in the laboratory) levels of iron that were not limiting at high light levels were limiting when there was less light and hence a need for the plankton to produce more photosynthetic machinery to capture the available photons. In addition, the picophytoplankton outperform larger phytoplankton in low iron conditions because iron uptake varies with cell surface area, so that smaller cells can get proportionately more iron per unit time from seawater. In addition, their absolute requirement for iron per cell is much lower—giving a system that can work at much lower ambient iron levels.

As we have mentioned in the context of ice age dust, on a large enough scale these increases in plankton production could have global implications by possibly reducing the amount of carbon dioxide in the atmosphere. Some marine plankton can have other effects on the Earth's climate too, for example, by increasing the production of sulphur-rich chemicals (dimethyl sulphoniopropionate and its break-down product dimethyl sulphide), which are important chemicals in cloud formation.[40]

Because of the role of plankton in removing atmospheric carbon dioxide, some people have suggested adding iron to the oceans to try and combat 'global warming'; however, there are currently large uncertainties over the likely effects of such large-scale human interventions. For example, it is currently not known whether most of the extra plankton are recycled after their death in the upper layers of the ocean—so releasing their carbon back to the atmosphere—or if enough sink into the depths to effectively remove their carbon from the atmosphere. In addition, the rate of loss of iron from the upper waters is poorly known.[31,33] Because of these, and other uncertainties, few of the scientists who have been involved in these experiments are currently enthusiastic about this as an approach to addressing 'global warming'.[41]

As well as the uncertainties listed earlier, there is the problem that iron fertilization does not benefit all marine organisms. During an exceptionally long hot summer in the southern hemisphere in 1997, there were extensive fires affecting large areas of Indonesia.[42] The dust and ash this created added large amounts of iron to the surrounding seas and caused dramatic increases in phytoplankton, which led to extensive mortality in coral and fish due to oxygen shortages when the plankton decomposed.[43] This dramatically illustrates the links between terrestrial vegetation (and its effects on dust levels and fire frequency—described in the previous chapter), and the limitation of phytoplankton growth by nutrient shortages, which gives us a planet dominated by a blue sea.

Plankton, nutrients, and Alfred Redfield

Phytoplankton require iron in relatively small amounts; however, some other nutrients are required in larger quantities and these are often referred to as macronutrients—important examples being nitrogen and phosphorus. Both of these nutrients are supplied to the ocean from the land—in addition, nitrogen can also be 'fixed' from molecular nitrogen in the atmosphere or in water by nitrogen-fixing microbes (such as the cyanobacterium *Trichodesmium*). Just as iron is common, but often biologically unavailable, there is a similar situation with nitrogen. This element is abundant in the atmosphere and seawater, but as a stable molecule with two atoms of nitrogen joined by a triple bond which takes a lot of energy to break—a problem only a few organisms have managed to crack, namely some prokaryotes and human chemists.[44] Marine biological nitrogen fixation is currently a source of great research interest as it has become apparent that we have greatly underestimated its magnitude. Typical 'textbook' global estimates have in the past been in the order of 10–20 Tg N/year (a tera g [Tg] is 10^{12} g); however, currently many estimates are around 100–200 Tg N/year; the large uncertainty in this figure is a measure of the uncertain and fast-changing nature of the science of marine nitrogen fixation.[21] In addition, human air pollution, from burning fossil fuels and agriculture, is now adding large amounts of biologically available nitrogen to the open ocean.[45]

How can the levels of nitrogen, and other chemicals, in seawater be explained? During the period from the 1930s to the 1960s, Alfred Redfield pointed out, in a series of papers, that there appeared to be a strong link between the chemistry of seawater and that of plankton.[46] Redfield was a biologist with a long and varied career; in over six decades of research he published on subjects as diverse as the effects of hormones on the pigments of toads[47] to the development of salt marshes.[48] His most famous observation was that the ratio of nitrogen to phosphorus in plankton is approximately the same (16:1) as the ratio of nitrate to phosphate in seawater; this result has now been confirmed on many occasions around the world (although the ratio is nearer to 15:1 in the deep ocean) and these ratios have been extended to include other elements (so that the extended Redfield ratios are 110 C:250 H:75 O:16 N:1 P:0.01 Fe—although there is some variation and different authors give slightly different values).[21,30,49] Probably, the most obvious idea that would have occurred to most biologists in the 1930s would have been to suggest that various chemical and/or geological processes determined the ratio of nitrogen to phosphorus in seawater and the plankton evolved to utilize these chemicals in these readily available proportions without having to continually fight diffusion by pumping them across their cell membranes. However, Redfield made the radical suggestion that the real explanation was the other way round and that the plankton determined the ocean chemistry.

Redfield pointed out that the elements he observed in these fixed ratios in seawater were all involved in biology, hence he suggested that biology controlled the water chemistry rather than the converse.[46] He also suggested that if phosphate increased in seawater, then nitrogen fixation would allow plankton numbers to increase to a point where they had used up all the additional phosphate. So, a change in the ratio of nitrogen to phosphorus causes changes in the plankton, which effectively use up the nutrient that is in excess—leading to the nutrient levels in the world's oceans being driven by the production and decomposition of organic matter. Redfield's basic ideas appear correct and the logic behind them has been shown to work in recent computer models of ocean chemistry[50,51]; indeed Redfield's suggestion that life drives water chemistry is now the standard textbook account,[30] although the details of the mechanisms are still under study. It is now clear that the Redfield ratios in plankton are really a global average; while the classic Redfield ratio of N:P is 16:1, modelling of phytoplankton growth and physiology (along with field and laboratory measurements) suggests that it can vary from about 8:1 to 45:1.[52] This variation is tied into the ecology of the plankton—species that can survive in resource-poor conditions have high N:P ratios while those adapted to rapid growth in good conditions have low N:P ratios. The explanation is that most resource (light or nutrients) acquisition machinery in these cells requires much more N than P.[20,52] The fact that the Redfield ratios are an 'emergent' average of the plankton community, rather than obligate for all plankton, suggests that if the make-up of the ocean's plankton community changes then so will the chemistry of seawater; so ocean chemistry could have been different in the past or could change in the future as our actions affect the ocean's plankton.

The earlier discussion potentially explains why some macronutrients are found in highly predictable ratios in the ocean—although a lot of the details are not yet understood. However, to understand why the sea is blue rather than green with plankton, we also need to understand something even more basic—why these macronutrients tend to be in short supply in most of the oceans. The answer may be found in both the ecology of plankton and the physics of the oceans. First, we describe the role of biology and then we describe the physical explanations.

The biological pump—removing nutrients from the surface ocean

While much dead plankton and other biological materials such as the faeces of zooplankton are broken down and recycled in the upper layers of the ocean, some of them fall into much deeper water where they can become entombed in ocean sediments—and ultimately sedimentary rocks. The life in the euphotic zone is effectively pumping nutrients into deeper water and so keeping its nutrient levels low; this balances the input of nutrients from the land.[53] The extent of this so-called biological pump became apparent during the 1970s due to a large U.S. study, Geochemical Ocean Section Study (GEOSECS), which used a wide range of what were then 'state-of-the-art' analytical methods to study nutrient movement in the ocean. These methods included making use of radioactive tritium and carbon that had entered the ocean from atomic bomb testing, allowing the fate of carbon from plankton remains to be tracked throughout the body of the ocean.[22] The biological pump is also responsible for removing iron from the surface waters and imprinting the Redfield ratios, produced by plankton in the euphotic zone, on the whole ocean, especially deep water—which makes up much of the volume of the Earth's oceans.

An interesting question is what would happen if the biological pump was switched off. In theory, ocean circulation would mix the waters and cause a relatively uniform distribution of nutrients within a few thousand years, increasing nutrient levels in the euphotic zone. This sounds like something that is impossible to experiment on, outside the confines of a computer simulation, but nature may have performed the experiment for us on several occasions over geological time—the most well studied being just over 65 million years ago. This is the time of the extinction of the dinosaurs, and many other groups of organisms, probably due to the impact of a large extraterrestrial object (see Chapter 10). Studies of the chemistry of marine sediments from this time (especially the ratios of different isotopes of carbon) strongly suggest that the biological pump was switched off—presumably because of widespread death of plankton.[54,55] This produced a very strange ocean compared to what we are used to, which has been dubbed the 'Strangelove Ocean' after *Dr Strangelove*, the 1964 Stanley Kubrick movie about nuclear war.[30] Isotope studies suggest that it took more than 3 million years for the biological pump to recover. This seems rather surprising as plankton populations can grow quickly; therefore one might expect a rapid recovery. One possible explanation is that following the mass extinction event a marine plankton flora developed that was dominated by picophytoplankton; these sink more slowly than larger cells and so weaken the biological pump. In addition, they are too small for larger zooplankton to feed on (their main predators are protozoa). The relatively large faecal pellets produced by zooplankton are another way in which biology 'pumps' nutrients out of the euphotic zone so this idea is at least plausible.[55] However, if the Strangelove Ocean was really as devoid of life as some scientists suggest, then it may have been even bluer than our current one.

Thermal stratification—the role of physics

Plankton and the biological pump are not the only reasons why the upper layers of the oceans are nutrient poor. Just as in lakes, thermal stratification is important in the oceans. In both lakes and oceans, warmer water tends to sit on top of colder water (unless the water temperature falls below 4°C) and this stratification of the water body can be very stable unless disturbed by wind or currents—in the oceans salinity is an additional complication as saltier water is more dense.[30] This stratification reduces mixing between deep and surface waters and means that nutrients in deeper water are trapped there and find it difficult to get back to the warmer surface layers where light is available for photosynthetic plankton to thrive.

Physics is also important in the two-dimensional world of the ocean surface, as well as the three-dimensional one of the effect of thermal stratification with depth. Look at a map of surface ocean currents and you will see large circular circulation patterns known as gyres. The circular motion of these currents tends to isolate water within them and so they can become very nutrient poor[30]—such as the Sargasso Sea we described earlier. As sea surface temperatures increase with global warming, most models suggest that thermal stratification will increase and these nutrient-poor areas of the ocean will increase in size. This appears to be happening; remotely sensed data collected by NASA between 1998 and 2007 shows increasing areas of very nutrient-poor waters associated with such gyres—indeed the increase is greater than expected given the temperature increase over this period.[56] Such changes are potentially worrying as they suggest the possibility that the biological pump will be less able to remove carbon dioxide from the atmosphere in a warming world because fewer plankton will be able to survive in the surface ocean.

Thermal stratification could be a mechanism for amplifying warm conditions on Earth. As we described in our chapter on tropical diversity (Chapter 5) around 100 million years ago the Earth was significantly warmer than today, with tropical vegetation at high latitudes. Lee Kump and David Pollard[57] have suggested that higher temperatures produce much lower plankton numbers (because of nutrient shortages due to enhanced thermal stratification) which reduces the amount of plankton-produced dimethyl sulphide (DMS)—which, as we described earlier, is important in cloud formation. Less DMS leads to fewer clouds and so increased solar energy reaching the Earth surface. So, reduced planktonic activity could lead to increased warming. This mechanism is unlikely to be as important in modern 'global warming', as today many of the atmospheric particles involved in cloud formation come from human pollution, not marine microbes.

So why is the sea blue?

So the sea is blue, and not green, for a variety of reasons. Much of the world's seawater is too dark for photosynthesis. In the euphotic zone, where there is enough light to support a green sea, other factors come into play. Large 'rooted' plants and macroalgae can only survive in the very small area of the ocean that is shallow enough for them to have access to light while being attached to the seabed, and large plants cannot survive by a floating way of life in most of the ocean, because they will be washed up by wind and currents. This leaves the question of why phytoplankton do not make most of the ocean green? The answer to this is that most seawater is too low in the nutrients needed for their growth; one of the reasons for this impoverishment is the action of the plankton themselves through the biological pump. Physical processes, such as thermal stratification, are also important in maintaining low nutrients in the euphotic zone and may change with alterations to the Earth's climate.

Finally, it is worth reflecting on the role of plankton in determining ocean chemistry, as this is really rather important. Ecology textbooks have tended to describe the physics and chemistry of the environment as the background to which organisms evolve—the 'ecological theatre and the evolutionary play' in G.E. Hutchinson's memorable phrase.[58] Back in the 1930s Alfred Redfield realized that, at least for ocean chemistry, this was not the case and that the organisms themselves were largely responsible for creating the chemical environment in which they lived. Readers familiar with James Lovelock's 'Gaia hypothesis' will notice Lovelock and Redfield have much in common. In its most recent formulation Gaia is defined as the idea that 'organisms and their material environment evolve as a single-coupled system, from which emerges the sustained self-regulation of climate and chemistry at a habitable state for whatever is the current biota'.[59] Indeed, Lovelock has named Redfield as a forerunner of his Gaia theory.[60] These two-way interactions between life and the physics and chemistry of the Earth are now at the forefront of several areas of ecology and evolutionary biology, such as the ideas of 'ecological engineering'[61] and 'niche construction'.[62] We suspect that they will form some of the big growth areas for the future, especially because of their relevance in understanding all the changes we are now making to our planet.

9

When did We Start to Change Things?

Figure 9.1 The winding wheel at Astley Moss Colliery, northwest England. One of the many coal mines that developed in Britain during the nineteenth and early twentieth centuries; it is now a museum. Photo: DMW.

> *We are now so abusing the Earth that it may rise and move back to the hot state it was in fifty-five million years ago, and if it does most of us, and our descendants will die.*
> —*The Revenge of Gaia. James Lovelock*[1]

As we wrote the first draft of this chapter (during early summer 2007), the potential dangers of 'global warming' had moved up the news agenda to a point where most major politicians were starting to take the problem seriously. Our opening quotation comes from a book published in early 2006, which seemed to coincide with the growth of this wider concern with global warming. Lovelock was not alone in trying to raise awareness of the problem; around the same time another book on climate change by the zoologist and palaeontologist Tim Flannery[2] also attracted global attention to this issue, as did the lecture tours (and Oscar-winning film) of Al Gore—the former US presidential candidate and campaigner on the dangers of climate change.[3] Indeed, in his role as a climate campaigner Gore won a share in the 2007 Nobel Peace Prize. It is possible that future historians will see the period 2005–2007 as the start of a crucial wider engagement with these problems.

Things *may* not be as bad as James Lovelock suggests—in his book he deliberately emphasized the most worrying scenarios coming from computer models, and other evidence, in an attempt to draw attention to the critical nature of the problem. However, all these worst case scenarios were drawn from within the range of results that most climate scientists believed could plausibly happen—not extreme cases with little current evidence to support them. That one of the major environmental scientists of the second half of the twentieth century could write such prose as science—rather than science fiction—is clearly a case for concern about future climate change. It also raises another important question, relating to the history of human influence on our planet: when in our history did we start to have major environmental impacts on Earth as a whole? This is clearly an important issue from a historical perspective, but the answers may also have implications for some of our attempts to rectify the damage.

Our discussion of this question comes with various caveats. Many of the arguments we consider in this chapter are still the subject of academic disagreement. A major reason for this is that the changes we describe happened in the past, and therefore have to be reconstructed from incomplete evidence preserved in the archaeological and geological record. However, there are ways to test such historical reconstructions. Much of the evidence can be checked against knowledge of current ecology, and several of the ideas can have their consistency investigated by mathematical and computer models. Thus, while any interpretation is likely to be tentative, recent scientific developments allow these questions to be approached in a more rigorous (often a more quantitative) manner than has been the case in the past.

Are humans unique in their impact?

In an attempt to place human effects on the Earth into a wider context, it is worth first briefly considering the environmental impact of other organisms. To survive, all organisms must be taking in energy from their environment and releasing waste products (pollution in the human context) back into their environment—this simple idea is central to much of ecology. So, humans are clearly not unique in utilizing their local environment for resources and as a dumping ground for their waste products. All organisms do this—beavers modify watercourses; elephants can reduce tree cover (especially when they are at high density); bog mosses in the genus *Sphagnum* make their environment both wetter and more acidic. In short, all species have local effects on their environment.

If humans are not that special in changing their local environment, then the next obvious question is: does our ability to affect a whole region (such as North America) or even the whole globe make us special? Again the answer is 'no' on both counts; however, humans are unusual as we are large animals with global effects; most organisms with planet-wide impacts have previously been microbes or plants. For example, as described in Chapter 8, many marine phytoplankton release chemicals that are quickly broken down into dimethyl sulphide (DMS for short), and once in the atmosphere DMS has global effects. First, these chemicals complete an important link in the sulphur cycle: sulphur is lost from rocks and soils into rivers and so eventually into the ocean. On a geological timescale this could lead to soils becoming depleted in sulphur, a biologically important chemical element (it is found in the amino acids cysteine and methionine). However, DMS from plankton blows over the land and returns sulphur to the soils via rainfall.[4,5] Second, DMS plays an important role in cloud formation, so marine phytoplankton have effects on the global climate.[5,6]

A geological example of the global effects of organisms is the suggested role for microbially produced methane in the early history of the Earth. Methane is a greenhouse gas (as will be discussed later in this chapter) and the presence of a large quantity of methane in the early atmosphere may have been crucial in keeping the Earth warm enough for liquid water and therefore life[7] (see Chapter 10). In addition, as pointed out by Lovelock,[8] a methane smog could have been important in shielding life from ultraviolet rays before the formation of the ozone layer—which performs this role on the modern Earth.[7]

So, humans are not unique in either changing their environment through resource use and their waste products, or in having global effects on the composition of the atmosphere. However, our effects are now so large that they are giving real cause for concern about the future. In this context, it is interesting (and important) to ask when did we first start to have really big effects on our regional environment, and when did these effects become global?

'Pleistocene overkill'—early human impact in the Americas?

A good candidate for one of our first major regional environmental effects is what has been referred to as 'Pleistocene overkill' (the Pleistocene is the geological epoch covering the period between 2.6 million years ago and approximately 11,000 years ago). This phenomenon has been studied in most detail in the Americas, so we will describe the evidence from there in some detail before discussing the rest of the world. Between 10,000 and 11,000 years ago, a dramatic series of extinctions were seen in North and South America, including several species of elephants, giant ground sloths, and large carnivores.[9–11] This coincides with a warming climate associated with the ending of the last glaciation ('ice age').

It is important to note that the above dates are largely based on radiocarbon dating of bone, and other biological remains, and that differences in the proportion of ^{14}C in the atmosphere over time make radiocarbon years depart from true calendar years. The time around the end of the last glacial period (which is the period of interest for these American extinctions) is a particular problem because of the release of carbon dioxide from the ocean. In this context, a radiocarbon date of 11,000 years is roughly the equivalent of a calendar date of 13,000 years.[9] In this chapter, we could have converted the radiocarbon dates to 'true dates' using published correction factors, but this could give future readers problems as new improved correction factors are published (making our corrected dates no longer up-to-date). To avoid confusion, all the dates given in this section of the chapter are in 'radiocarbon years' and so only approximate true calendar years. If you are reading this chapter some years after its publication, you should be able to look up the most recent correction factors and calculate the current best estimate of calendar years based on these radiocarbon dates.

There are two main explanations for extinctions of North and South American species, and these theories have been debated for well over 100 years. The first is that the extinctions were brought about by climate change and associated changes to vegetation. The second theory is based on human hunting—as the first widely accepted evidence of humans in the Americas also comes from this time.[9,12–14] Clearly, it is also possible that these two processes acted together.

The idea that humans triggered the American extinctions has a long history, and was suggested by several of the greatest names in biology and geology during the late nineteenth century, such as Charles Lyell, Richard Owen, and Alfred Russel Wallace.[15] Paul Martin has worked on this topic in more recent decades, arguing that human hunters may have caused rapid extinctions through hunting once they arrived in America; hence, the name 'Pleistocene overkill'.[9] However, some archaeologists remain highly sceptical.[14]

In addition to hunting and climate change, which are covered in more detail later, several other theories have been suggested. Disease introduced by the arriving humans along with their domesticated dogs is one possibility. However, it has been argued that disease seldom leads directly to extinctions because as an animal gets rarer it becomes increasingly difficult for the disease to spread,[16] and it would be unusual for a disease to affect such a wide range of different species.[17] Nevertheless, there are exceptions. The dramatic extinctions seen within the Hawaiian lowland bird species community during the twentieth century are thought to have arisen from human introduction of mosquitoes capable of carrying avian malaria, which 'reached epizootic proportions' sometime after the 1920s.[18]

Another, recently suggested theory for these Pleistocene extinctions is that a large extraterrestrial object exploded on or over North America at this time, representing a smaller-scale version of the type of event that many think caused the eradication of a variety of species, from dinosaurs to plankton, a little over 65 million years ago (the Cretaceous-Tertiary or 'K-T' extinction event—see Chapter 10).[19] Although this latter research presents strong evidence for some sort of extraterrestrial impact in North America, we are sceptical about its importance in explaining the Pleistocene extinctions and will return to this possibility once we have described the nature of the extinctions in more detail.

An obvious question for an ecologist to ask about the American extinctions is, 'Are there any patterns to which species became extinct and which did not?' For a long time people had noted that many of the extinct species were larger mammals—the 'megafauna'. The size distribution of these extinct mammals has recently been quantified in detail by Kathleen Lyons and colleagues.[20] A key aspect of their study is that they did not just look at the sizes of the extinct animals, but also at those that survived. All mammals over about 600 kg adult weight in North and South America became extinct—while some smaller species were also lost, the bias towards large species was highly statistically significant. As the authors pointed out, such a size bias in extinctions is very unusual in the geological record. If human hunters were indeed responsible, it is easy to understand the selection of larger animals—after all, a dead elephant provides much more meat than a dead rabbit. Yet large mammals can also be particularly vulnerable to extinction because of aspects of their population ecology, such as their typical population size and capacity to reproduce. In general, the maximum rate of population increase declines with size, and this makes it difficult for large animals to recover from decreases in their population size, whatever their cause.[11]

Stronger evidence for the role of human hunting comes from work on the extinction of large sloth species in the Americas.[21] Sloths suffered a major extinction event with only two (small body size) genera surviving, while 24 genera became extinct during the geologically recent past in the Americas. This study showed that while the sloths became extinct in continental America between 11,000 and 10,500 years ago, on West Indian islands extinctions did not happen until around 4,400 years ago, around the time people first arrived on these islands. This strongly suggests a role for hunting and is less compatible with the climate change idea, which would predict all the extinctions to correspond with the end of the last glacial period—especially for island populations that are likely to be more vulnerable to extinction because of small population sizes (although their climates may have been subtly different).

It is also not obvious why a warming climate—and its associated vegetation changes—should selectively kill off larger species, many of which were *not* adapted to cold conditions and should have been benefiting from the changing climate. In addition, a very important and long-standing problem with the climate hypothesis has been the question of why these extinctions only happened at the end of the last glacial period when there have been repeated switches between glacial and interglacial conditions during the Pleistocene.[10,22] Indeed, for these reasons, the human hunting cause of these American extinctions is becoming textbook orthodoxy. For example, it is the preferred explanation in several widely read popular science books[23–25] and the current edition of 'Raven and Johnson'—one of the most widely used undergraduate biology texts—also backs human hunting in their chapter on Conservation Biology.[26]

The size bias in extinction and survival is also important in the evaluation of the recent ideas about extraterrestrial impacts, as it is not obvious why it should have preferentially affected large species of terrestrial animals. The extinction of the dinosaurs in the K-T event was associated with the loss of many smaller species, plants, and microbes as well as animals[27]—although of course this postulated impact was much larger than the one being suggested for North America at the end of the Pleistocene.[19] Astronomical data on the frequency of objects hitting the Earth's atmosphere suggests that there should have been numerous such smaller impacts over archaeological time; indeed the surprising thing is the rarity of evidence for impacts in the geological and archaeological record.[28] This makes us sceptical that such an event had a large role to play in the American extinctions, as the late Pleistocene extinctions are the only known extinctions with this strange size bias. In addition, the proposed impact event (possibly an air-burst by a comet) is described by the scientists working on this idea[19] as likely to have affected continental North America—its relevance to South American extinctions happening around the same time is not clear. Despite this growing consensus for human hunting, there are still several problems we need to consider. In particular, if we want to claim these end-Pleistocene extinctions were something as important as one of the earliest substantial human effects on the environment at a continental scale, then we clearly need to base this on strong evidence and rule out alternatives.

The main arguments in favour of a climatic explanation, other than just noticing the coincidence of the times of climate change and the extinctions, have been based on scepticism that humans could have killed enough animals to cause these extinctions. As the palaeontologist Tony Stuart wrote over 20 years ago, 'It's difficult to imagine how a few hunters, with what is to us a primitive technology, could have exterminated numerous species of large mammals'.[22] In addition, there are only a few known sites in America with evidence of humans killing mammoths (there were two mammoth species with wide distributions in America: the woolly mammoth and the columbian mammoth—Fig. 9.2) and many of the extinct large mammal species (such as the giant ground sloths) have no known archaeological 'kill sites'.[14,29] However, this lack of kill sites may be more of a problem with the archaeological record than a sign that people rarely killed mammoths—or other large mammals. Jared Diamond has illustrated the point by comparing these putative archaeological extinctions with the recent extinctions and population reductions of tigers on Indonesian Islands, where we know from historical records what really happened. He pointed out that although we know humans were responsible, the future archaeological record will, at best, be sparse; 'future archaeologists will find virtually no direct evidence of human causation. Hunted animals represented only a fraction of total tiger deaths, and there will be few or no butchered carcasses'.[10]

Figure 9.2 Columbian mammoths' skulls embedded in volcanic ash deposits found at Tocuila, in the Basin of Mexico. These mammoths died in a volcanic mud flow. The modelling studies we discuss suggest that only a small proportion of megafauna needs to be killed by humans for hunting to make the difference between survival and extinction. Even if humans are responsible for these North American extinctions, most mammoths and other megafauna will have died from non-human causes. Photo: Silvia Gonzalez.

Could small groups of people with 'stone age' technology exterminate large species such as mammoths? As many people have pointed out, animals that have not seen humans before are often unafraid of us and therefore easy to kill. Charles Darwin describes such a situation when he visited the Galápagos in 1835, writing; 'a gun here is almost superfluous; for with the muzzle of one I pushed a hawk off the branch of a tree'.[30] This makes it sound more plausible that the first humans in America could have a dramatic effect on their prey; however, there are other less anecdotal approaches to the problem.

One approach is to attempt an experiment. Around 11,000 radiocarbon years ago in America, there was a relatively short-lived but characteristic stone tool technology in the archaeological record referred to as the Clovis culture after the site from which it was first described (traditionally a slightly earlier start of 11,500 years is quoted for this but is based on some problematic dates[31]). The archaeologist George Frison[32] has described these Clovis artefacts as demonstrating 'an accomplished stone tool technology and hunting ability'. Using replicas of these tools as spear points, experimental archaeologists have shown that they are able to pierce the hide of modern elephants, both recently dead and mortally wounded by guns, during controlled management in an African game park. These experiments rather bravely included some animals that, although seriously wounded, were still standing.[29,33] The best way to kill an elephant with such a spear appears to be to aim at the lungs, as the skull is too thick for anything other than a bullet to penetrate and the heart is well protected by the ribs which can damage the stone spear heads requiring time-consuming repairs. In his report of these experiments, Frison noted that if anyone wanted to try similar experiments on wild uninjured elephants then the experimentalists should be 'younger, physically fit persons' as 'hunting of this nature requires the agility and strength of individuals in their physical prime'[33]; although the ethics of such experiments are open to question.

Another approach is to try to model the hunting scenario using mathematics in an attempt to see if hunting could plausibly lead to extinctions. John Alroy[34] has produced a computer model of human and large mammal populations for North America at the end of the last glaciation. One problem with constructing such a model is that we have little idea as to the success rate for Clovis people hunting large animals. To address this question, Alroy ran his model repeatedly using a wide range of different success rates, in the majority of the runs of the model the humans caused extinctions on a timescale of around 1,000 years (which appears to match the usual interpretations of the archaeological record, although the most recent attempt to re-date the Clovis culture had it flourishing for only a few hundred years[31]). As Alroy pointed out, such an event is effectively instantaneous in the geological record but happens too slowly for people to be aware of the big effects they (in concert with their ancestors) are having on their prey.

Another challenge with implicating humans in these American extinctions is uncertainty over the arrival dates of humans in the 'New World'.[14] The most widely accepted view has been that humans arrived around 11,500 years ago—these are the Clovis people, and this view is often called 'Clovis first'. However, there are some putative sites, especially the South American site of Monte Verde in Chile, from at least a few thousand years earlier which many archaeologists and palaeoecologists now accept.[12,14,35] Possibly these people arrived by moving down the American coast, using boats, rather than overland—which would explain the shortage of evidence from terrestrial archaeological sites. The oldest directly dated human bones (dated using radiocarbon techniques and quoted here in 'radiocarbon' years) are from around 11,500 years—with several examples from around 10,000 years.[36] Recently, human faeces, radiocarbon dated to 12,300 years ago, have been described from a cave in Oregon, USA.[37] Conclusive identification of these faeces as human relies on DNA evidence, and there is always a potential contamination problem with human DNA in an archaeological context—however, in this case considerable effort has been put into attempting to rule out contamination as an explanation. The crucial aspect of this study is that a date of 12,300 is pre-Clovis.

In fact, there is a long history of claims of much older archaeological sites in the Americas, which include Californian excavations during the 1960s by the renowned palaeoanthropologist Louis Leakey[38]; although so far none of these have convinced the majority of archaeologists. For example, one of the most high-profile claims of recent years has been for humans in central Mexico over 40,000 years ago—based on footprints preserved in volcanic ash.[39] The dating of this site is complex and the nature of the ash means that the footprints are not sharp enough to convince the more sceptical commentators—although the photographs and plaster casts of these prints do look intriguingly human.

Clearly, the possibility of substantially pre-Clovis people in the Americas has implications for the 'overkill' explanation of the late Pleistocene extinctions. In our view, one of the key problems with accepting very early human occupancy of the Americas is the lack of well-dated, unambiguously human-made stone tools.[32] Stone tools are usually much easier to find than human bones, as a single person may make many tools during their lifetime and they survive in the archaeological record more easily than do bones. As such, unambiguously dated tools should have been found even if the remains of the people themselves prove elusive. The most obvious conclusion is that 'Clovis first' is basically correct, but with other human cultures probably in the Americas a few thousand years before Clovis. In this scenario, it is difficult to be sure if the Clovis tools spread through much of the Americas with an 'invading' population or if the Clovis technology spread through an already existing human population, which may have arrived a few thousand years earlier.[31] Differentiating between movements of technology along with the people who used it or diffusion of technologies through existing populations is a long-standing problem in archaeology.

What of the claims for much older humans? If there were humans in the Americas 20,000 or more years ago it would seem they used few if any stone tools—as these should have been found by now in at least one or two well-dated contexts. Perhaps they only used wooden digging sticks and spears, along with stone tools so crude they are difficult to identify as being of human manufacture. At first sight this appears very unlikely as unambiguous stone tools have been made for over 2 million years in Africa; however, other groups of people are known to have abandoned apparently very useful technology. For example, Charles Darwin,[30] and others, described with surprise the nineteenth century native inhabitants of the southern tip of South America, who had abandoned the use of almost all clothing although they lived in a cold (and sometimes snowy) climate. If there were stone-tool-free inhabitants in the Americas, then they are unlikely to have been big game hunters. While we can envisage killing large animals with wooden spears (some preserved in waterlogged conditions from the European archaeological record look lethal), it is difficult to envisage butchering a mammoth with wooden tools, and even wooden spears would be hard to sharpen without stone tools. So, even if these early Americans existed they do not undermine the hunting explanation; it is the arrival of the technologically more sophisticated Clovis-type technology that did the damage.

The other main problem that some scientists have had with the overkill theory is the fact that some large mammals, such as the bison, survived. There are several points to make here. First, however many large species humans exterminated something would survive, so there would always be a question 'Why did this mammal survive?',[9] whether the ultimate source of their compatriots' extinctions was climate change or hunting. As Martin[9] has pointed out, it is at least interesting that many of the larger survivors were from species that were not just restricted to America but were also found in Eurasia—so their ancestors are likely to have been exposed to human hunting. Second, while people often assume that extinction or survival must have a definite reason, in fact, chance can also play a large role. For example, computer simulations (using stochastic population models) of Pleistocene large carnivore populations in ice age Europe suggest that survival in isolated populations (such as being trapped in Italy during the height of the last glaciation) probably involved an element of chance; in repeated runs of these models with identical starting conditions sometimes the population survived and sometimes it did not.[40] These models were based on simple population ecology without any complications from hunting, disease outbreaks, or other environmental catastrophes. So, in summary, it is possible that in some cases there is no answer to the question why specific species survive, other than good luck.

The longer a species survived in the Americas after the arrival of Clovis technology, the more likely it was to learn to be fearful of humans and so its chance of continued survival would increase. There is some evidence for this from observations on brown bear predation on moose in areas of both North America and Scandinavia that have been recently recolonized by bears as a result of nature conservation schemes. Here moose were less vigilant than normal and bears found it easier to catch them, although in a single generation the moose became more wary.[41] However, unlike some of the extinct American species whose ancestors had never met human predators, these moose had only been living in bear-free environments for between 50 and 130 years (only a few generations) and so may have some form of genetically inherited predisposition for rapidly learning about bears.

Allowing for the difficulty of reconstructing things that happened 13,000 calendar years ago, we think that there is a good case for implicating human hunting in the extinction of these large American species. This does not mean that the changing climate had no effect—but these species had survived many such climate changes in the past and both cold- and warm-adapted animals become extinct at the same time. What was different this time was the presence of people with what appears to be an advanced stone age big game hunting technology.[32] Indeed Alroy's[34] model suggests that they could have caused all the extinctions without any contribution from the changing climate, although Koch and Barnosky[42] plausibly argue that the changing climate coinciding with human impact may have exacerbated the rate of extinction. This does not necessarily mean that humans directly hunted all the species to extinction. For example, it is likely that some of the carnivores, such as sabre-toothed cats and the American lion, became extinct because humans had reduced numbers of their prey. In addition, as pointed out by Norman Owen Smith,[11] changes in the numbers of large mammals such as elephants in modern Africa can have big impact on the vegetation. So the loss of large grazing and browsing mammals will have affected the habitat for many other species. These extinctions could also have affected American plant species by removing some key seed dispersers,[43] although interestingly there is evidence of only one plant extinction in the Americas at this time.[44] The potential large-scale ecological effects of the large mammal extinctions have led Paul Martin and others[9,45] to controversially suggest introducing elephants and other species back into American parks to try to fill the ecological roles of the extinct mammals. In this case, an understanding of our past effects is informing cutting-edge ideas in conservation biology.

What about the rest of the world?

If the human hunting explanation is correct, then one might expect it to apply to places other than America; what then, are the global patterns? Africa is a particularly interesting case as it has suffered no obvious pulses of large mammal extinctions over the past few million years,[9] although those extinctions that did take place appear to have preferentially affected larger species.[20] Africa was the continent where early humans evolved (our genus *Homo* is probably somewhat over 2 million years old) and also where our own species evolved (*Homo sapiens* appeared around 200,000 years ago).[25] The suggestion is that as large mammals in Africa lived alongside early humans, then they slowly evolved defence mechanisms against us as our hunting skills developed.[9] This contrasts with the Americas and Australia (discussed later) where the first humans the animals met appear to have already been sophisticated hunters.

In Eurasia (which also has a long history of human presence) the Pleistocene extinctions were more spread out over time but as with the Americas there is again the complication that many of them were associated with periods of climate change, and in Southeast Asia there is the additional complication that climate-driven sea-level change created many islands and peninsulas.[46] Currently, the evidence from Southeast Asia is too patchy to say anything definitive about the involvement of humans in extinctions in this area[46]; however, much more evidence is available for Europe and Russia. For example, based on an extensive data set of radiocarbon dates, Tony Stuart and colleagues[47] have shown that both the giant deer (sometimes called the Irish elk) and the woolly mammoth contracted their range from the end of the last glaciation, with the last giant deer dated to 6,900 radiocarbon years (approximately 7,700 calendar years ago) and the last woolly mammoth to 3,700 radiocarbon years (approximately 4,000 calendar years, on Wrangel Island in the Russian Arctic). However, as with the American extinctions, most of the populations of these two species became extinct between 10,000 and 11,000 radiocarbon years ago. It is interesting that the last surviving woolly mammoth populations were on small islands around the Bering Strait, with radiocarbon dates of 7,908 for the last mammoths from St Paul Island, as well as the more recent dates from Wrangel Island.[48] Owing to the restricted population sizes, small islands would not be likely places for these persistent populations, unless they provided refuge from humans.

The possible links between climate and human hunting in mammoth extinction are illustrated in a study by David Nogués-Bravo and colleagues which utilized a mix of climate and population ecology models.[49] They showed that climate change between 42,000 and 6,000 years ago potentially reduced mammoth habitat by 90%—with the last extensive areas of good habitat being in Arctic Siberia. Their models also predicted a similar range reduction for mammoths 126,000 years ago (in the previous interglacial), which did not lead to extinction. However, very low levels of human hunting could have caused extinction when the mammoths had already been restricted by climate. So in these models climate is important but humans deliver the coup de grâce—without humans the mammoths may well have hung on until the start of the next glaciation and then once again expanded their range as they had done in the past.

Islands, as well as being the last refuges of mammoths, also provide many uncontroversial examples of extinction driven by the arrival of humans—the classic example being the dodo (Fig. 9.3). Extensive work by David Steadman and his collaborators[50] has shown widespread bird extinctions on Pacific islands when humans arrive—the mechanisms probably include habitat modification and the introduction of non-native species such as rats, as well as direct hunting. The timings of these extinctions differ from island to island depending on the history of human colonization, a clear indication that in these cases climate change is seldom involved. On occasion the arrival of humans on islands may have had surprisingly large-scale effects. For example, it has been suggested that decreases in the numbers of the migratory sooty shearwater observed in archaeological deposits in the western United States between AD 1000 and AD 1600 may have been due to the effects of Maori hunting of these birds on their breeding grounds in New Zealand, although (as so often in these cases) it is difficult to definitively rule out effects of climate change—in this case potentially affecting marine productivity.[51]

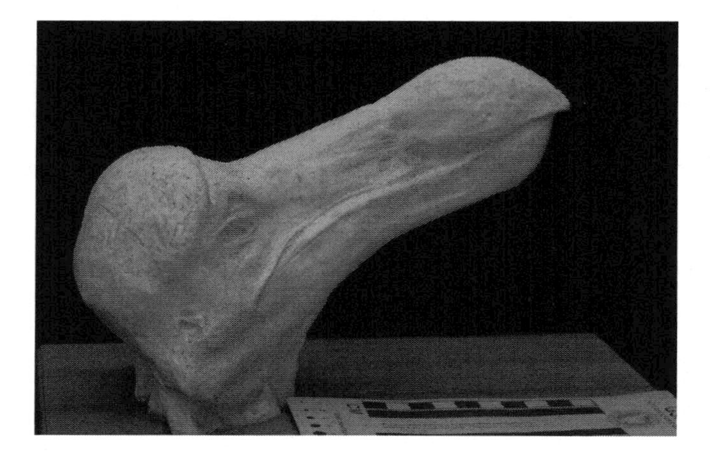

Figure 9.3 The dodo of Mauritius, which became extinct in the late seventeenth century, is an icon of conservation biology. The extinction of the dodo is usually attributed to a mixture of hunting and introduced animals (rats, cats, pigs, monkeys) after the discovery of the island in 1598. Surprisingly for such a famous bird we are not even sure what it looked like, as very little soft tissue is preserved in museums around the world and contemporary drawings are often contradictory. There was an entire stuffed specimen in the Museum at the University of Oxford; however, its condition deteriorated and despite it being the only complete specimen in the world it was destroyed on 8 January 1755.[117] The head and feet survived and are still in Oxford. The photograph shows a plaster cast of the head from a mould made before its partial dissection in the 1840s. Photo: DMW.

Extinctions in Australasia

The Pleistocene animals of Australia also experienced a wave of extinction, which preferentially affected the largest species—including marsupial 'lions', several large kangaroos, and a huge lizard, which was estimated to be 10 times the weight of the surviving Komodo dragon of Indonesia.[9,20] Interestingly, the more recent extinctions of a wide variety of species in Australia, associated with the arrival of Europeans, were not size selective[20]—as is typical of most extinction events in the geological record. Presumably, these recent extinctions were driven more by habitat change associated with European-style agriculture rather than hunting for food. It has been suspected for decades that the earlier Australian extinctions pre-dated those in the Americas and so could not be the product of the same global climate changes that some use to explain the American extinctions. It has also been apparent that humans arrived in Australia earlier than the Clovis people in the Americas, which is surprising given that even in times of low sea levels during the last glaciation they would have had to cross the sea. Determining the exact timing of arrival of the first Australians and the large animal extinctions has been difficult as they happened at or before the time limit for accurate radiocarbon dating (around 40,000 years, i.e. seven half-lives of ^{14}C).[9,52]

The earliest reasonably uncontroversial human remains from Australia are dated to around 50,000–46,000 years ago—mainly by using 'optically stimulated luminescence' (OSL) which identifies when quartz grains from archaeological sites were last exposed to light (along with other methods such as uranium series and electron spin resonance dating). Some ages of just over 60,000 years have also been published in the past but these are much more controversial and it is currently questionable if human artefacts recovered from such early contexts can be reliably associated with the dated material—as the stone tools cannot be directly dated.[53,54] In the past 10 years, OSL and other, non-radiocarbon, dating methods have greatly improved our knowledge of the timings of these Australian extinctions. One of the most detailed studies is on the extinction of the ostrich-sized flightless bird *Genyornis newtoni*, with more than 700 dates, using a variety of techniques, on egg shell remains from across Australia.[55] This work showed that it became extinct about 50,000 years ago, coincident with the earliest widely accepted dates for people in Australia. Given the large sample size and the range of independent dating methods applied, this study provides good evidence for at least one Australian extinction at the time of the apparent arrival of humans.

In 2001, Richard Roberts, Tim Flannery, and colleagues published an extensive list of OSL and uranium series dates for the extinction of giant marsupials and reptiles in Australia.[56] They showed that these extinctions centred on 46,400 years ago (with statistical 95% confidence limits between 51,200 and 39,800 years ago). This is after a period of very arid climate. With the climate 'improving' at the time, the effects of humans—through hunting and possibly human-caused fires affecting the vegetation—seem the most likely explanation for most of these extinctions. This would make the Australian extinction the first major regional environmental effect of humans for which we currently have reasonable evidence.

One of the reasons that people have looked to environmental modification by fire as well as straightforward hunting in the Australian extinctions is that the stone tool technology of these early Australians was not as sophisticated as the Clovis people, and there is also currently a lack of archaeological kill sites[42] (but see Diamond's arguments about modern tiger extinctions described earlier). However, the coincidence of the new dates for the extinctions with the earliest evidence of humans strongly suggests, but cannot formally prove (remember the old statistics maxim that correlations can never prove causality) human involvement. As ecologists we are particularly struck by the fact that the extinctions show the same size bias as the better-studied American extinctions—we think this strongly suggests human involvement.

Early regional environmental modification—fire

As we discussed earlier, the extinctions of large animals are likely to have had cascading effects on the vegetation and thus the habitat for many other species. The other plausible way that pre-agricultural people could have had widespread effects on their environment is through their use of fire, as has been suggested for Australia. Once early humans were using fire then, either by accident or design, they are likely to have affected the frequency of fires—at least in parts of the world with vegetation that will easily burn. As described in Chapter 7, fires can lead to changes in the type of vegetation growing in an area. Therefore, an important question in the history of our effects on the environment is: when did we start using fire?

The problem in writing anything at all definite about this topic is that it is almost impossible to differentiate the remains of a human-caused fire from a natural fire. Even if there is evidence of burning associated with archaeological remains, unless there is a completely unambiguous hearth, it is difficult to be sure if the fire was of human origin. There is relatively good evidence for use of fire in Europe around 250,000 years ago by members of our genus *Homo*, which has been widely accepted by many archaeologists,[57] and reasonable circumstantial evidence for earlier fire use—for example, around 400,000 years ago in both southern England[58] and in an Israeli cave.[59] There are also many claims for even earlier fire use in the literature, for example, older books[60] accept fire use in China over 0.5 million years ago and Brain[61] has claimed fire use in South Africa over 1 million years ago. The eminent evolutionary biologist Ernst Mayr[62] even speculated that fire use may have started well over 2 million years ago as a way for relatively defenceless 'naked apes' to defend themselves against large African carnivores; although he did note that 'the date when fire was tamed is particularly uncertain'. Over 10 years ago Pitts and Roberts[63] discussing the Boxgrove archaeological site in southern England (approximately 0.5 million years old) wrote: 'It is still unclear if hominids at this date were regularly, systematically using fire anywhere in the world'. This uncertainty remains today; while it is likely that fire was our first big effect on the local and presumably regional environment (predating the Australian extinctions) it is currently impossible to prove this.

Early regional environmental modification—agriculture

Arguably, the most important development in the recent (last 13,000 calendar years) history of our species is the invention of agriculture.[64] This has had huge effects on the environment; for example, conservation biology is full of examples of the negative impact of modern agriculture on biodiversity. The idea of agriculture appears to have been independently invented in somewhere between five and nine locations around the world. The earliest known agriculture is from the eastern end of the Mediterranean, although it also appears not much later in parts of China.[64] Since this chapter's question focuses on the *earliest* environmental effects of humans, we briefly describe the origin and environmental effects of agriculture in the so-called fertile crescent—in modern-day Jordan, Israel, Syria, Turkey, and Iraq—and its subsequent spread across Europe. Using the distribution of wild plant species ancestral to modern crops as a guide, Simcha Lev-Yadum and colleagues[65] have suggested that south-eastern Turkey and northern Syria may have been the location of the very earliest agriculture in this area.

Pre-agricultural people living by harvesting and hunting wild food in the fertile crescent had a long history of collecting the seeds of wild grasses; however, by 9000–8000 BC there is good evidence of domesticated plants whose seeds show subtle changes from the wild type (dates in calendar years—for the rest of this chapter all dates are given in calendar years even if, as here, they derive from radiocarbon dates. The 'calibration' of more recent radiocarbon dates is more certain as it is possible to use data on radiocarbon from tree rings of known age). After around 8,700 BC agriculture started to spread to central Turkey, Cyprus, and Crete,[66] and then spread across Europe arriving in southern France by around 5,800 BC, northern France a few hundred years later, and Britain sometime before 4,000 BC.[67] However, this spread was not uniform with populations in some areas, such as some parts of the Balkans, being slow to take up agriculture.[68] The extent to which early farmers themselves moved across Europe, or just the idea of farming moved, is a long-standing problem for archaeologists; most likely it was a combination of both processes.[67]

The factors that gave rise to the start of agriculture are still the subject of debate; the earliest agriculture is associated with major climate changes at the end of the last glaciation—however, while this is 'suspicious' it does not necessarily mean that climate change triggered agriculture. In addition, it is not clear to what extent rising human population sizes may have driven agriculture or if agriculture allowed population sizes to grow;[69,70] a classic 'chicken and egg' problem. Certainly, agriculture allowed the possibility of large permanent urban settlements, which are impossible for most pre-agriculturalists who have to move after depleting local wild food resources.[70]

A key archaeological site is Abu Hureyra in modern Syria. Around 9,500 BC this was a small, apparently permanent, settlement whose inhabitants subsisted on collecting wild plant food along with the of hunting gazelles and other animals. These food resources must have been unusually rich to support non-nomadic hunter-gatherers. Indeed, an obvious speculation is that those conditions favouring non-nomadic hunter-gatherers also favour the invention of agriculture as people will remain in the same area and so can harvest crops grown from spilled seeds of wild grasses and other sources. At Abu Hureyra, agriculture started to supplement the collection of wild food about 9,000 BC and by 6,000 BC most food came from agriculture and the settlement had grown to cover 16 ha.[71] Major towns and cities, of which Abu Hureyra was a forerunner, cannot be provisioned solely by hunting and gathering wild plants. Without large agriculture settlements, our current big population size is impossible; so the rise of agriculture is crucial to our growing impact on the global environment.

While the first agriculturalists must have had an impact on their local environment, there is a lag of several thousand years before we start to see evidence of regional impacts—for example, changes in regional vegetation as reconstructed from pollen grains preserved in sediments.[68,72,73] The conventional explanation for this is that these early agriculturalists of the eastern Mediterranean were present in numbers that were too small to have a regional impact. However, Neil Roberts[73] has suggested another interesting possibility. In a European context, we often see the effect of early agriculture as a clearing of woodland to create open vegetation more suitable for agriculture. One of the reasons that any possible early regional effects of agriculture are hard to find around the fertile crescent is that trees were very slow to spread in this area after the end of the last glaciation; this could be natural but, as Roberts points out, it could also be an early environmental effect of people modifying the landscape—possibly with the help of fire—and so keeping the landscape open. In which case, this lack of evidence for regional environmental effects of the first agriculture is actually due to human-caused modifications of the landscape making its detection difficult. Nevertheless, there are also lags identified as agriculture spread to more wooded areas in Greece and the Balkans;[68,72] here, the low population explanation currently appears the most likely. Certainly by the Bronze Age in the eastern Mediterranean (around 3,000 BC in places such as Crete), agriculture was starting to have widespread regional environmental effects.[72,73]

The kinds of environmental effects associated with the expansion of agriculture across Europe can be illustrated by parts of the coastal plain of Cumbria, in northwest England, around 3,000 BC. The vegetation history has been reconstructed from excavations which one of us (DMW) was involved in. Crops present included both emmer wheat and flax; this agriculture was associated both with clearance of the forest and with evidence of soil erosion into a nearby lake—reducing the diversity of diatoms living in the water.[74] Therefore, these early farmers were altering both terrestrial and freshwater habitats.

When did we start to affect the whole planet?—the Anthropocene

All the changes we have described so far in this chapter range between the local and continental scale. However, in the context of current concerns about human actions causing global climate change, it is instructive to ask when did we start to have global effects on the Earth? Most environmental scientists would suggest that our global effects started a few hundred years ago. Paul Crutzen—who shared the Nobel Prize in chemistry for work on the destruction of atmospheric ozone—has called this period of global human influence the Anthropocene and suggests that it started around 1784 with the invention of the steam engine.[75,76]

Crutzen's reason for focusing on the time around the invention of the steam engine as the start of the Anthropocene was because the late eighteenth century was marked by a major increase in the use of fossil fuels. As he pointed out, this event was marked by an increase in both carbon dioxide and methane in the atmosphere—as reconstructed from ancient air recovered from ice cores from Greenland and the Antarctic.[76] This rise in carbon dioxide can also be reconstructed from more biological evidence. For example, in the 1980s, Woodward[77] showed—using dried plant material that had been collected over several hundred years and stored in museum collections—that the density of stomata (fine pores used for gaseous exchange) fell over this period as plants found it easier to acquire carbon dioxide for photosynthesis from the atmosphere. A longer record of this type has been constructed using olive leaves from archaeological sites (including the tomb of the Egyptian pharaoh Tutankhamen) along with more recent museum material. The number of stomata in olive leaves fell from approximately 790 per mm^2 in 1370 BC to 530 per mm^2 in AD 1991, with the largest drop in the past couple of hundred years.[78] Therefore, Crutzen has good reason to point at the past few hundred years as a time when the industrial revolution was starting to affect the global atmosphere.

A recurring difficulty in this chapter is that it is seldom straightforward to point at a particular date in human history and say, with confidence, that this is where a particular thing started. This is certainly the case with the idea of the Anthropocene. Britain was one of the first countries to undergo industrialization, based on burning fossil fuels, in the late eighteenth and early nineteenth centuries—however, coal had been quarried and burned in Britain on a smaller scale since Roman times (nearly 2,000 years ago).[79] Indeed, dating the 'invention' of the steam engine to 1784—as Crutzen does—is also an oversimplification. In fact, this is the date of major improvements to steam engine design by James Watt, not the date of its invention or even its first industrial use. For over a hundred years before this date, steam engines had been used in Britain to help pump water from mines.[80] Indeed, the earliest known steam engine was described by Hero of Alexandria around AD 60 (admittedly, this was only a toy model, genuinely useful steam engines being beyond Greek technology at this time[81]). Therefore, there is an element of judgement in exactly where you define the start of Crutzen's Anthropocene—however, it is undeniably the case that during the eighteenth century some countries were starting to make increasing use of coal rather than 'renewables' (such as water and wood) as a source of power, so this century seems a good candidate for when we started to have planet-wide effects on atmosphere and climate. Recently, however, William Ruddiman[82] has complicated the idea of the Anthropocene through his interesting and controversial arguments that the lower boundary of the Anthropocene should be placed thousands of years earlier.

An early start to the Anthropocene?

The large-scale burning of fossil fuels is not the only way to affect global climate; land use change can also be very important. Clearing forests releases the carbon locked up inside the plants and land use changes can potentially release large amounts of carbon from organic matter in the soils. In addition, altering vegetation changes the land's albedo (the amount of solar energy reflected back into space) which affects climate, as does the extent of wetlands as this is correlated with the amount of the greenhouse gas methane released into the atmosphere. As all of these things are affected by agriculture, it raises the possibility that as agriculture spread around the world it may have started to affect the atmosphere and climate on a global scale, long before the industrial revolution. While it sounds plausible that land use change associated with agriculture may have had global effects, what is needed to really establish this is a rigorous—preferably quantitative—approach to the question. Recent increases in our knowledge of past atmospheric chemistry (especially from gases trapped in ice cores), modern computer models of climate, and our increasing knowledge of environmental archaeology start to make this a realistic prospect, and William Ruddiman has recently attempted such a quantitative approach in a series of provocative and controversial publications.[82–84]

Most of us were taught at school that a very powerful approach to science is the use of replicated experiments—in which treatments (usually involving the change of one variable) are repeatedly compared with 'controls', that is, other experiments where (typically) no changes had been made to the system. Ideally, then we would rerun the past 13,000 years of Earth history multiple times, in some cases with, and in others without the development of agriculture and study the average effect on the atmosphere and climate. Clearly, this is impossible—indeed the inability to follow this experimental approach is a major problem whenever we try to gain a scientific understanding of past events. Ruddiman has attempted to address these problems by using previous interglacials as a control for our current post-glacial period and by running experiments with computerized climate models.

The recent geological history of the Earth has been marked by repeated swings between cold and warmer conditions—the glacial/interglacial cycles. The number of such swings depends on your exact definition of 'glacial' and 'interglacial'; however, it is now clear that there were at least 40–50 such transitions over the course of the geologically recent past[83] (about 2.5 million years). This geological period is usually referred to as the Quaternary (although the correct name for this time period is currently a matter of controversy amongst geologists[85]). The Quaternary is best considered as starting 2.6 million years ago, associated with evidence for the occurrence of glacial ('ice age') conditions, and running up to the present day—although some scientists argue for a start date of 1.8 million years ago. The most recent part of the Quaternary, since the end of the last glacial and marked by the rise of agriculture, is called the Holocene which started around 11,600 calendar years ago (technically this is what geologists refer to as a series or epoch; the Pleistocene is the other series which collectively comprise the Quaternary).[85]

Before we continue with examining the effects of land use changes, it is necessary to give a little more background on glacial events. The idea that there had been 'ice ages' in the past gained currency during the first half on the nineteenth century, based on observations of the effects of former ice sheets and glaciers etched on the landscape of Europe. The obvious question was what causes these repeated glaciations? In 1842, Joseph-Alphonse Adhémer suggested that the cause might be astronomical, with ice ages driven by changes in the Earth's orbit affecting the amount of solar energy arriving on Earth.[86] These ideas were further developed by James Croll during the 1860s. At this time, Croll was working as a janitor in a Scottish college—an injury had forced him to give up his work as a carpenter.[86] He must have been far and away the most mathematically gifted janitor in Scotland, if not the world, and by 1870 he had published several papers on his work and convinced the eminent and influential Quaternary geologist James Geikie that the astronomical explanation was correct. Geikie championed Croll's ideas in his seminal book *The Great Ice Age*.[87] These ideas were further developed by Milutin Milankovich at Belgrade University in the early twentieth century, a task that involved spending years on detailed calculations which could today be done in at most a few days with a computer. Indeed the changes in aspects of the Earth's orbit that are thought to drive glacial/interglacial cycles are now often referred to as 'Milankovich cycles' in his honour[86] (see also Chapter 5 on the role of Milankovich cycles in mediating speciation rates). By the mid-1970s, these ideas were put on a much firmer foundation as it was shown that the predictions from astronomical cycles broadly matched changes in temperature reconstructed from ocean sediment cores.[88]

The repeated glacial/interglacial cycles provide one potential way to sidestep the impossibility of replicated experiments on the global effects of agriculture. In 2003, Ruddiman[82] pointed out that the trends for carbon dioxide (CO_2) in the Holocene were different from those reconstructed for the previous three interglacials, in that previously CO_2 levels had fallen throughout the interglacial while in the Holocene they had started to rise over the past 8,000 years. Previously, working with one of his undergraduate students, he had pointed out that the methane (CH_4) trends for the Holocene also appeared to behave oddly, unexpectedly starting to rise around 5,000 years ago.[89] CO_2 and CH_4 are key greenhouse gases and Ruddiman even suggested that their rise had prevented the onset of the next glaciation.[82,90] In brief, his suggestion is that the increase in CO_2 has come from land clearance for agriculture while the CH_4 has come from the expansion of wetlands associated with rice cultivation—we will discuss these mechanisms in more detail later.

One problem with comparing current glaciations with past glaciations is that not all glacial/interglacial cycles are the same: there are subtle differences in the variations in the Earth's orbit which makes the details of the climate different from cycle to cycle. A criticism of Ruddiman's approach has been that in using the previous three interglacials as a 'control' for the Holocene he had not been comparing like with like. An earlier interglacial—known as marine isotope stage (MIS) 11—is supposed to be the closest analogue to the Holocene in various orbital parameters.[91] This interglacial lasted much longer than the current length of the Holocene, thus apparently undermining Ruddiman's suggestions that human changes in the levels of greenhouse gases over the past 8,000 years had prevented the onset of glaciation in parts of Canada. However, data from pollen grains preserved in sediments from a very long series of cores of lake mud from France show that MIS 11 was not the best match for the Holocene for European vegetation patterns,[92] so illustrating the difficulty in identifying the best interglacial to use as a control. Despite this, as Ruddiman[93] and others[92] have pointed out MIS 11 provides some support for Ruddiman's ideas because it also fails to show the increases in greenhouse gases during the interglacial that appear anomalous in the Holocene.

Ruddiman[82,83] outlines a range of mechanisms by which humans could have caused an early start to the Anthropocene. He suggests that the most plausible explanation for the rise in CH_4 is an expansion of rice growing in irrigated fields. Rice was domesticated in China between 10,000 and 9,000 years ago[94] but growing rice in flooded fields appears to be a later phenomenon becoming common from around 5,000 years ago.[83] The low-oxygen conditions in wetland sediments—including the artificial wetlands of flooded rice fields—provide good habitats for methane-producing bacteria. Indeed, the predicted increase in rice cultivation over the next 40 years is seen as a potential problem in the context of global warming.[95] Rice cultivation looks a particularly promising source of this methane as there is evidence from comparing Arctic and Antarctic ice cores that the main source of this extra methane was in the tropics—if, for example, it had come from the extensive high-latitude peatlands of the northern hemisphere, then it should have been more prominent in Greenland ice cores compared to ones from the Antarctic.[83] In addition to rice cultivation, Ruddiman suggests that increased numbers of livestock—which harbour methane-producing microbes in the low oxygen parts of their guts—and burning of forests will also have tended to increase atmospheric methane levels. His main mechanism for increases in CO_2 is the destruction of forests as agriculture expands. For example, as agriculture spread across Europe the amount of tree cover started to decrease quite dramatically[96] (Fig. 9.4).

Ruddiman's calculations suggested that human-derived changes before the industrial revolution have added around 40 ppm (parts per million) of CO_2 to our atmosphere. However, the magnitude of this change appears problematic to many other climate modellers. On a millennial timescale, around 85% of any atmospheric CO_2 increase is absorbed by the oceans—this means that for an atmospheric increase of 40 ppm around 566 ppm CO_2 needs to be emitted through landscape and vegetation changes. This seems implausible based on ice core records and computer climate models, which suggest that the human impact was more likely to be on the scale of 4–6 ppm CO_2 (after equilibrium with the ocean had been obtained).[97] Ruddiman[83] now accepts that his initial estimates of the amount of CO_2 added to the atmosphere may have been too high. He now views the role of humans as a two-stage process with a more modest direct effect of humans on CO_2 level causing a warming, which affects factors such as sea ice preventing 'natural' drops in CO_2 levels that have happened in previous interglacials. Initial attempts to model the effects of land use changes associated with the rise of

Figure 9.4 Langdale, in the Lake District National Park in northwest England. The 'wild' looking scenery is deceptive. The treeless mountain slopes are not natural; before forest clearance by early prehistoric farmers much of this landscape was forested.[118] During the early Neolithic, approximately 4000–3000 BC in Britain, the Pike O'Stickle (the highest summit seen in the photograph) was a source of stone for making axes which were transported all over Britain. Photo: DMW.

agriculture over the past 6,000 years and their effects on the carbon cycle also produce values lower than Ruddiman's initial estimates.[98]

Ruddiman's basic idea appears plausible—although for the reasons given earlier, the effects of humans may be less than he originally estimated, and it is currently an open question if they were large enough to have had a substantial impact on the Earth's climate before the industrial revolution. However, there are a number of potential issues that have not yet been addressed in a quantitative manner which are highly relevant to Ruddiman's ideas. First, as several people have recently pointed out[99,100] the role of carbon in soils needs to be considered. In general, agriculture reduces the amount of organic carbon stored in soils[101]; this suggests that Ruddiman's estimates of the amount of CO_2 produced by agricultural change may be an *underestimate*. That said, in some cases, clearance of forest can increase the amount of carbon stored in soils, as in parts of the uplands of Britain where forest clearance led to the formation of peat 'soils', which are almost 100% organic matter.[102] A further complication is that increased amounts of organic carbon in soils can lead to a rise in microbially produced nitrous oxide, which is itself a greenhouse gas.[103] However, on balance, incorporating the effects of agriculture on soils probably makes Ruddiman's hypothesis more likely—although better quantitative models addressing the question are needed before firm conclusions can be drawn.

Certainly, soils have been important in the context of more recent agricultural change. In a study considering only the period between 1860 and 1980, Houghton and colleagues[104] estimated that recent land use changes had led to a significant addition of CO_2 to the atmosphere from soils—for example, they estimated that in 1980 as much carbon was lost from soils worldwide as from the burning of forest trees.

At least two other problems require consideration for a full test of Ruddiman's hypothesis, that humans made an early start to altering the atmosphere as a by-product of their actions. The first is the effects of land use change on albedo. Clearing forest adds CO_2 to the atmosphere and is a central mechanism in Ruddiman's hypothesis. However, in many cases, tree cover absorbs more solar radiation than grassland and so increases the Earth's temperature—for example, Gibbard and colleagues[105] calculated that albedo changes caused by the replacement of the world's vegetation by trees would lead to a global warming of 1.3°C, while replacement by grasslands would give a cooling of 0.4°C. Because of the large amounts of water they transpire, tropical forests tend to lead to a net cooling, but at higher latitudes this transpiration does not offset the albedo effect in this way. These albedo effects have implications for Ruddiman's suggested feedbacks between forest clearance and sea ice. Clearing forest will release CO_2 and so cause warming (and ice melting) but unless it was tropical forest that was cleared then Ruddiman's current calculations probably overestimate the warming effects as he does not factor in vegetation albedo.

The final complication we will mention applies to methane—a potent 'greenhouse' gas. In 2006 Keppler and colleagues[106] caused considerable surprise with a paper in the prestigious journal *Nature* suggesting that plants produce large amounts of methane—the main 'textbook' source of methane is from microbes in oxygen-free conditions (such as the rice fields and animal guts described earlier). The great surprise was that this new methane source appeared to be produced in the presence of oxygen. If this is correct then it provides an added complication to Ruddiman's hypothesis as forest clearance could reduce methane production; yet, as Ruddiman points out, it shows a rise during the Holocene. It is currently unclear how the work of Keppler *et al.*[106] should be viewed; however, the first published independent test of their ideas has failed to find plant-produced methane,[107] so such methane may not be an important complication in discussions about early starts to the Anthropocene.

In summary, it seems likely that for thousands of years before the industrial revolution humans have been having an effect on the concentration of greenhouse gases in the atmosphere, but the magnitude of this effect (e.g. 4 ppm or 40 ppm increase in CO_2) is currently unclear. As agriculture started to spread around the world, the impact of humans started to become global but it has accelerated tremendously in the past few hundred years. Ruddiman has also suggested that some of the 'wiggles' in the CO_2 curve can also be explained through the effects of disease on human population size in the past. We now consider this and the related question of early modification of tropical forests—including some which have, until very recently, been considered relatively pristine ecosystems by many conservationists.

The role of disease and the importance of population size

Ruddiman[82,83] also drew attention to the fact that the reconstructed CO_2 curve for the past 2,000 years looked surprisingly messy, and is interrupted by several declines in atmospheric CO_2—the largest of these being AD 1300–1400 and AD 1500–1750. There are few obvious natural explanations for these declines. Nevertheless both periods of decline are marked by an absence of 'sunspots' (described in historical astronomical records) which would have been associated with a reduced solar output that may have had effects on the climate and so indirectly on greenhouse gases, although this is controversial.[83] There is, however, a good candidate for an ecological explanation for these CO_2 minima—namely, significant reductions in human population size caused by disease.[81,82] The basic idea is that if large numbers of people die, many agricultural fields are abandoned and quickly revert to forest—thereby, taking up CO_2. This is well established for the recent historical past in North America, where during the nineteenth and early twentieth centuries large numbers of people abandoned the land (for economic rather than disease-driven reasons) and moved to the cities. In this case, large amounts of farmland reverted to forest often on a timescale of decades[108] (Fig. 9.5).

Figure 9.5 Regenerated forest in southern Ontario, Canada. This used to be the site of a farmstead now abandoned like so many farms in nineteenth and early twentieth centuries eastern North America. At this site all the forest was cleared, just after 1900, and a homestead built approximately where the photographer was standing. The open vegetation in the foreground is maintained by periodic flooding from beaver activity and possibly by some grazing by deer. The woodland has regenerated since the farmland was abandoned and is probably 50–60 years old. Photo: DMW.

In the mid-fourteenth century Europe was hit by a great plague pandemic—often referred to as the 'black death'. The pandemic is thought to have arisen in China and caused something like 30–60% human mortality in Europe.[109,110] Ruddiman's[82] assumption is that this must have led to major abandonment of agricultural land in Europe (and other less well-documented areas); certainly, if a mortality of around 50% led to the reforestation of a similar percentage of agricultural land, then this could account for the reduction in CO_2.[82] However, can we assume that a decrease in population automatically causes a return to forest? For example, if the number of people available to work the land decreases, another possibility is a change from labour-intensive arable agriculture to pastoral agriculture, which has less need for large numbers of people.[111] Such a change would not allow forest to return and have at most a modest effect on the amount of carbon locked up in the system—perhaps through some increase in soil organic matter. An additional complication is that there is some evidence for a decline in arable agriculture in parts of western Europe *before* the arrival of the black death—presumably due to a combination of economic and climate changes around this time.[110] We need independent evidence of what happened to the landscape and vegetation at this time. Some information can be gleaned from historical sources such as contemporary legal documents describing the inheritance of land, but a major potential source of information comes from pollen grains preserved in sediments such as lake muds or peats.

Pollen data can be informative, but presents several challenges with respect to understanding changes in agriculture over the last millennium. First, there is a relative shortage of studies of well-dated sites with fourteenth century pollen.[110] Second, an even greater problem is that there is no consensus on how to turn information on numbers of pollen grains preserved in sediments into an estimate of plant biomass. The currently available pollen data for the fourteenth century in Europe show a mixed picture (as do the historical documents)—some sites showing evidence of returning scrub and forest while others seem to show a continuation of open agricultural landscapes. A recent review[110] suggested that 'agricultural decline and reforestation varied geographically, and in some areas did not occur at all, making the estimate of a 25–45% increase in forest cover [from Ruddiman[82]] appear unrealistic'. However, they point out that these conclusions are currently based on rather limited evidence.

The other decline in CO_2 levels identified by Ruddiman was from AD 1500 to AD 1750 and roughly corresponded with a period called the 'Little Ice Age'. This reached its peak in Europe around 1650, with average temperatures approximately 1°C colder than present, harsh winters and cool summers.[112] For example, between 1564 and 1813 the River Thames in London, England, froze on at least 20 occasions (11 of these were during the seventeenth century).[113] As with the changes in the fourteenth century, a minimum in sunspot numbers is recorded in the fourteenth century and may provide part of the explanation for the cooler climate and lower CO_2.[112] However, there is growing evidence that ecological changes in the Americas may also have a role. The arrival of Europeans in the New World at the end of the fifteenth century is now known to have caused a massive collapse in the size of the native American populations many of whom were practicing agriculture.[12] Diseases, such as smallpox, accidentally introduced by the Europeans probably played the main role in this population crash, and it is now thought that over 90% of the native Americans may have died.[70,112] This had major environmental effects; for example, in the Darien Region of Panama, fossil pollen shows that what now appears to be pristine tropical forest had a 4,000 year history of human disturbance and agriculture, which was brought to an end by the arrival of the Spanish—with the current forest being only 350 years old.[114]

It is now appreciated that even large areas of the Amazon supported agriculture before European 'conquest'; indeed evidence of former human modification of apparently pristine tropical forest is being identified by archaeologists in Africa and Southeast Asia as well.[115] This is a discovery with interesting implications for conservation biology and the practicality of regenerating tropical forest habitats. A recent review of the evidence for the effects of a population crash of native Americans on world climate, authored by a team of anthropologists and climatologists, concluded that forest regrowth after the collapse of American agriculture could have contributed (along with other factors such as reduced solar activity) to triggering the Little Ice Age. In particular, they drew attention to the New World bamboos, which will grow very rapidly when tropical agriculture is abandoned—potentially taking large amounts of CO_2 out of the atmosphere over a short time period. They concluded their paper by writing that 'This relationship between a historic event and the climate suggests that human behaviour had a significant effect on climate centuries before the known human consequences of industrialization on weather patterns'.[112]

Conclusion: our environmental effects have a long history

It seems likely that our first major effect on our local environment was through the use of fire; however, it is currently impossible to attach a reliable date to this. The first well-dated regional environmental effect of humans evident in the archaeological record is the likely human involvement with the Australian extinctions approximately 50,000 years ago. The much better documented extinctions in the Americas around 13,000 calendar years ago were almost certainly mainly due to human hunting and related habitat modifications, affecting animal populations that may already have been stressed by climate change. During most of the Holocene agriculture has been the major mechanism through which we have affected local and regional environments. The ideas of Ruddiman suggest that agriculture may have started to affect global atmosphere, and possibly climate, thousands of years ago—although the magnitude of this effect is currently very uncertain. With the start of the industrial revolution in the eighteenth century our global effects started to increase, leading to the current concerns about our potential to cause major climate change. Population size is crucial to our environmental effects; the more people the larger the effect—as seen by the possibility that disease-driven population crashes caused reductions in atmospheric CO_2. Most organisms have larger environmental effects at high numbers (or high biomass);

humans differ from other organisms because they also show the effects of technological change. For example, the rise of agriculture and, later, industries driven by fossil fuels have played key roles in our effects on the global environment.

An understanding of past changes can inform modern ecology; for example, a recent discussion of the ecology of tropical savannahs[116] makes no reference to the fact that while Africa still has most of its large grazing and browsing mammals, the savannahs of South America and Australia are now missing such species—although they were there in the past and must have affected the evolution of the plants. We have made such dramatic changes to the ecology of most of our planet that it is often impossible to define the 'natural' state without reference to archaeology and environmental history. Only with an understanding of these past changes can we fully understand the ecology of the present and future.

10

How will the Biosphere End?

Figure 10.1 A dried-up basin in the Namib desert, Namibia. Photo courtesy of Oliver Sherratt.

Temperatures had risen throughout the equatorial and sub tropical zones... In the developed countries people remained alive in places where communal power supplies were still operative through the help of air-conditioning equipment. Yet few such people had escaped the deadly radiation which now poured through the atmosphere.

—*Fred Hoyle and Geoffrey Hoyle, The Inferno, 1973.*[1]

This fictional description of the destruction of much of life on Earth comes from a novel by the astronomer Fred Hoyle, co-authored with his son Geoffrey. In the story, the formation of a quasar in the centre of our galaxy leads to the destruction of all life on Earth, except at a few fortuitously sheltered locations. Quasars—first described in 1963—are colossally energetic astronomical objects with extremely high output of radio waves.[2] The novel built on some of Fred Hoyle's own scientific interests because in the early 1960s, along with the astrophysicist W.A. Fowler, he had predicted that the collapse of a super-massive object could form a distinctive radio source—just before the discovery of the real thing. Although Hoyle and Fowler had the theoretical head start in explaining quasars, being busy with other work they failed to follow up on this advantage, and the current best explanation for these objects is largely due to Donald Lyndon-Bell and Martin Rees. Building on the ideas of Hoyle and Fowler, they argued that a quasar is formed by a rotating super-massive black hole, fed by a disk of in-falling matter, with jets of matter flying away from the system along its axis of rotation.[3]

Like the Hoyles' novel, this chapter focuses on ways the biosphere could end; a fitting question for the close of a book on the ecology and evolution of Earth-based life. However, any answer to a question set in the far future can necessarily be only speculative and, of course, nobody will be around to put the theory to its ultimate test. This raises a philosophical problem namely, has such a question a place in science, or should it be left to science fiction writers? We believe that such questions count as science, not least because it would be good to know the answer (especially if something could be done to postpone the end), but also because in attempting to answer the question, we can extend our understanding of processes that are currently operating. Indeed, J.B.S. Haldane, one of the greatest scientists of the past century, wrote an early essay on much the same topic we consider here.[4] As to the role of such speculation in science, we can do no better than quote from John Maynard Smith and Eörs Szathmáry[5] who wrote in their book *The Major Transitions in Evolution*, which covered similarly speculative discussions in their case of the distant evolutionary past—'these are matters on which we must speculate. Why else would we study evolution?'

Speculation should not be seen as pure guesswork however; as we discuss in this chapter, our question can be approached through testable quantitative models, not just plausible-sounding qualitative arguments. This means that the 'internal consistency' and underlying assumptions of many of the ideas can be evaluated, and theories refined through a combination of critical analysis and more data. Ultimately, nobody will be around to validate the prediction, but we hope to convince you that it remains a question worth asking.

Defining the biosphere

Before addressing how the biosphere will end, we must briefly explain our use of the term 'biosphere'. This was first coined by the Austrian geologist Eduard Suess in the 1870s for the totality, and physical location of, all life on Earth. So in this sense, the biosphere of today starts in the deepest rocks that contain living microbes and runs out high in the atmosphere at the point where it becomes difficult to find any microbes. This usage of the term was popularized in Europe in the first half of the twentieth century by Pierre Teilhard de Chardin[6]—who was an intriguing mix of Catholic priest, mystical philosopher, and vertebrate palaeontologist. However, 'biosphere' has also been used in another sense, meaning life and the whole collection of life support systems (such as biogeochemical cycles and feedbacks between life and climate), an approach that came from the Russian tradition in ecology in the early twentieth century, and is particularly associated with Vladimir Vernadsky.[6,7] Both definitions of biosphere are used in different modern ecology texts, and in the context of our chapter either makes sense when one asks 'How will the biosphere end?'. James Lovelock's well-known concept of 'Gaia'[8,9] is based on a similar view of the biosphere to Vernadsky's, albeit with a much stronger emphasis on the regulatory aspects of the feedbacks between life and the abiotic environment than appears to be present in Vernadsky's original concept. Indeed, one of the key papers we discuss in this chapter is deeply rooted in this view of the biosphere, being co-authored by Lovelock.

Destruction from space—danger from supernovae?

Let us continue exploring the possibility that some form of astronomical event might cause a cataclysmic end to the biosphere. Quasars are extremely rare, and the course of events described in the Hoyles' novel is therefore correspondingly unlikely to happen to Earth. A commoner astronomical event that potentially could affect life on our planet is a supernova. This is the collapse and catastrophic explosion of a large star (at least six times more massive than our sun) in which for a few weeks the star's power output can exceed that of a whole galaxy.[2,10]

During the 1970s, several scientists attempted to explain the demise of the dinosaurs, and many other organisms, at the end of the Cretaceous (65.5 million years ago) by the effects of a nearby supernova. The mechanism underlying the extinction-by-supernova idea is that cosmic rays generated by these exploding stars can generate nitrogen oxides in the Earth's atmosphere, which react with ozone (O_3) to produce 'normal' molecular oxygen (O_2).[11,12] As is now well known from widespread concern about modern ozone depletion and polar ozone holes, the reduction in ozone high in the Earth's atmosphere can allow dangerous levels of ultraviolet light through to the Earth's surface.[13] This can in turn lead to increased levels of cancer and genetic mutations, potentially causing high rates of mortality.

In the context of the effects of depleted ozone, it is interesting to note that there is now evidence for unusually high frequencies of deformed pollen grains and spores from the end of the Permian (251 million years ago),[14] arguably the largest extinction event over the past 542 million years of Earth history, and one of the few suggested 'mass extinctions' for which we have really watertight evidence in the geological record (the 'end Cretaceous' extinction is the other reasonably well-documented case).[15] However, the scientists studying these deformed pollen grains prefer explanations based on volcanic activity, rather than astronomy, to explain changes in ozone levels and hence mutation rates.[14,16] More importantly for this chapter's question, recent attempts to model the potential effects of a supernova on the Earth's ozone layer, taking full advantage of the speed of modern computers, suggest that the danger is not as great as early studies suggested. For example, work by Gehrels and colleagues[12] suggests that a supernova 26 light years from Earth (extremely close by astronomical standards) would only double the amount of biologically active UV reaching the ground. Therefore, the vast majority of supernovae would be too far away to cause even limited damage. Even if this were not the case, then we should still not worry unduly about elevated UV in the context of ending the entire biosphere. While increased UV levels are a problem for larger terrestrial organisms, such as ourselves, this is not the case for many species protected by several metres of water. In addition, many microorganisms are surprisingly difficult to kill, even with high levels of UV, and this is especially the case if they are protected by a thin film of organic matter—as would be the case with most soil microbes.[17]

More destruction from space—asteroids and comets

If quasars and supernovae do not seem to be likely methods for ending all life on Earth, are there other astronomical events that may be more likely to destroy the biosphere? Around 1980, interest in extraterrestrial explanations for extinctions switched from supernovae to the effects of collisions between the Earth and comets or asteroids.[18] For the past 25 years, this idea has been the front-runner for the explanation of the end Cretaceous ('K-T') extinction mentioned earlier, coinciding with the disappearance of the dinosaurs among other notable groups. There is now a wide range of evidence consistent with an impact at approximately this time, and even a possible impact site located in the Yucatan in Mexico.[19,20] Clearly, this event did not destroy the whole biosphere, but could more catastrophic impacts have happened in the remote geological past—or indeed happen in the future?

The surface of the moon is a good place to start when thinking about these questions, as it is covered in impact craters in a way the Earth is not. Indeed the maria (dark 'seas' on the moon's surface) are enormous craters filled with solidified lava created by massive past impact events.[21] Our planet should have suffered even more impacts than the moon, because it is larger (and so a bigger target) and has a stronger gravity[22]—indeed it has been estimated that for every object that hit the Moon around 20 objects should have hit the Earth.[23] However, the more active surface of the Earth, with both more erosion and rock recycling by plate tectonics, has removed much of the evidence for impacts in the distant past.

Early in the history of our solar system, impacts from extraterrestrial objects were much more common due to the greater abundance of asteroids and comets—the population sizes of these objects have been reduced over time as they crash into things. Indeed, the period around 4.4 to 3.8 billion years ago (here we are defining billion as 1,000 million, as is now standard in science) is often referred to as the 'heavy bombardment', because of the high frequency of such collisions.[24] Some of these early collisions seem highly likely to have been capable of destroying any life that might have started on the early Earth. For example, about 100 million years after the formation of the solar system (thought to have formed just over 4.5 billion years ago) there is widely accepted evidence that a Mars-sized object collided with the Earth, and that debris from this collision eventually formed our Moon. This collision is predicted to have caused temperatures that would have melted the rocks forming the proto-Earth[25]—a situation presumably incompatible with life.

Lesser collisions could still have the potential to end the biosphere, either in the past or in the future. For example, Norman Sleep and colleagues[22] calculated that a collision with a 440 km diameter rock (the size of one of the larger asteroids) would release enough energy to evaporate all of the Earth's oceans. Such events could certainly have happened during the heavy bombardment—but could similar collisions happen in the future, so ending our biosphere? Eros, a slightly smaller asteroid than those modelled by Sleep and colleagues, currently has an orbit that on occasion crosses the Earth's trajectory. Calculations suggest that there is approximately a 50% chance of a collision between Eros and Earth sometime in the next 100 million years.[26] Although this would not produce enough energy to vaporize the oceans in the way described earlier, it is estimated that the collision would release an order of magnitude more energy than the impact that is suspected of ending the Cretaceous.[26] The effect on the biosphere would certainly be severe.

Clearly, there is an element of chance in the Earth avoiding a potentially biosphere-destroying event, such as a collision with a large asteroid. Such events could certainly lead to the extinction of humans, elephants, oak trees, and the like—however, it is much more difficult to kill all microbial life on Earth.[27] Even for the very largest collisions there is the possibility of microbes surviving inside rock material ejected from the surface of the Earth during the collision. Such microbe-rich rocks could return to Earth, attracted by its gravity, after conditions had improved and the microbes could potentially restart life on our planet[24]—the biosphere equivalent of rebooting a crashed computer. We will return to the difficulty in destroying the Earth's microbial biosphere later in this chapter.

More astronomy—on red giants and the faint young sun

Stellar astronomy sits at the heart of ecology. This may seem a strange assertion for two biologists to make; however, almost all energy in our ecosystems ultimately comes from the Sun through photosynthesis. In addition, without the Sun the Earth would be impossibly cold for life, and without photosynthesis there would be almost no atmospheric oxygen and so much less biodiversity on Earth—all plants, fungi, and almost all animals require oxygen to survive.[28] Over millions of years the Sun's behaviour can change, and this has huge implications for thinking about the long-term ecology of our planet.

The ultimate fate of life on Earth has been clear since the first half of the twentieth century, when the basic life cycle of stars was worked out. Typically, stars are powered by the fusion of hydrogen into helium, and in a star similar to our Sun, as the source of usable hydrogen starts to run out, it starts to contract. This contraction increases temperatures within the star leading—in a process first suggested by Fred Hoyle—to the fusion of helium atoms. The energy released by this process causes the star to increase greatly in size—since it also produces redder light, such a star is referred to as a red giant.[2] Hoyle described what would happen in a radio broadcast and associated best-selling popular science book in the middle of the twentieth century, writing: 'As the sun steadily grills the earth it will swell, at first slowly and then with increasing rapidity, until it swallows the inner planets one by one'.[29] At the time it was thought that the Sun may expand to the orbit of Mars or beyond—but astronomers now think that the Earth's orbit is near the likely limit of expansion.[30] Even if Earth avoids being engulfed, it will become far too hot for any life to survive—'grilled' in Hoyle's culinary metaphor. This ultimate biosphere-ending disaster is not imminent; indeed it is likely to take place about 6 billion years from now.[2]

Long-term changes in the Sun present other problems to understanding the geological history (and fate) of life on Earth. The basic life history of stars similar to our Sun is well known and astronomers are clear that early in their life they produce less energy (making them 'fainter'). The astronomer Carl Sagan—who was also one of the most prominent popularizers of astronomy during the 1970s and 1980s when we were developing our scientific interests—realized that this raised an important question for the history of life, and had several goes at solving the problem. He first addressed the question in a paper in *Science* in 1972 co-authored with George Mullen.[31] They pointed out that the surface temperature of the Earth today is compatible with the presence of liquid water (obviously important for life) but solar luminosity was much lower in the past, so low that all water should have been frozen at times long after life is known to have evolved. Yet we have plentiful geological evidence for liquid water during this early period, some of which now suggests oceans may even have existed 4.4 billion years ago.[23]

How can this discrepancy be explained? Sagan and Mullen suggested that a greenhouse effect based on atmospheric ammonia could explain this apparent paradox, with the greenhouse atmosphere keeping the early Earth warm, despite the faint young Sun. Sagan returned to this idea towards the end of his life, refining the explanation in a posthumously published paper co-authored with Christopher Chyba.[32] However, while most astronomers and geologists have accepted Sagan's point that there is an apparent paradox created by the presence of liquid water on Earth associated with a faint young sun, and also that greenhouse gases provide a good solution to the problem, most have not accepted his proposed ammonia-based solution. A more plausible alternative to explain the continual presence of water, for *at least* the past 3.8 billion years,[33] is thought to be high levels of carbon dioxide in the atmosphere of the early Earth, along with potentially high levels of methane, which is another greenhouse gas.[12]

Clearly, astronomical events such as quasars, supernovae, asteroids, or the red giant expansion of the sun have the potential to end the biosphere—although the details of the mechanisms may be more suited to a book on 'Big questions in astronomy and cosmology'. Nevertheless, in the early 1980s, modelling studies suggested that interactions between the behaviour of the Sun, the long-term carbon cycle on Earth, and photosynthesis could end the biosphere as we know it, long before the demise of our star. These are more ecological ideas, directly relevant to the approach taken in this book, and we focus the rest of this chapter on discussing the research inspired by this approach. To do this, first we need to set some context by outlining the nature of the long-term carbon cycle.

The long-term carbon cycle

If you look in a university-level ecology textbook at a diagram of the carbon cycle, you will normally see a collection of arrows showing the flux (movement) of carbon between various reservoirs (such as the atmosphere, ocean sediments, or life). Various fluxes will be illustrated, such as photosynthesis, human burning of fossil fuels, and carbon dioxide dissolving in water.[34,35] However, for anyone used to thinking about the carbon cycle over geological time spans, one key process is strikingly absent from these ecology texts. This missing process is the weathering of silicate rocks, which on a geological timescale removes large quantities of carbon dioxide from the Earth's atmosphere and places it into carbonates and silica oxides. Indeed on a long timescale this is more important than the burial of organic carbon from photosynthesis in controlling the amounts of carbon dioxide in the atmosphere (Fig. 10.2).

In the 'short-term' carbon cycle of the ecology textbooks, the fluxes are large compared to the size of many of the reservoirs, such as the atmosphere or soil carbon. This means that even a relatively minor proportionate increase or decrease in some of these fluxes would potentially lead to major changes in the amount of carbon dioxide in the atmosphere. Since we know that atmospheric carbon dioxide does not change dramatically on the timescale of the short-term carbon cycle (measured in hundreds of years), this cycle must be in a surprisingly exact balance[13]—a balance we are now modifying by increasing the flux of carbon entering the atmosphere from burning of fossil fuels and vegetation change (see Chapter 9).

In contrast, in the longer-term 'geological' carbon cycle we describe, the fluxes are too small to affect the size of the reservoirs on a short timescale; therefore, the geological processes below affecting the carbon in rocks are only influential on long timescales (thousands of years or longer), as it takes time for these small fluxes to alter the size of a reservoir.[13]

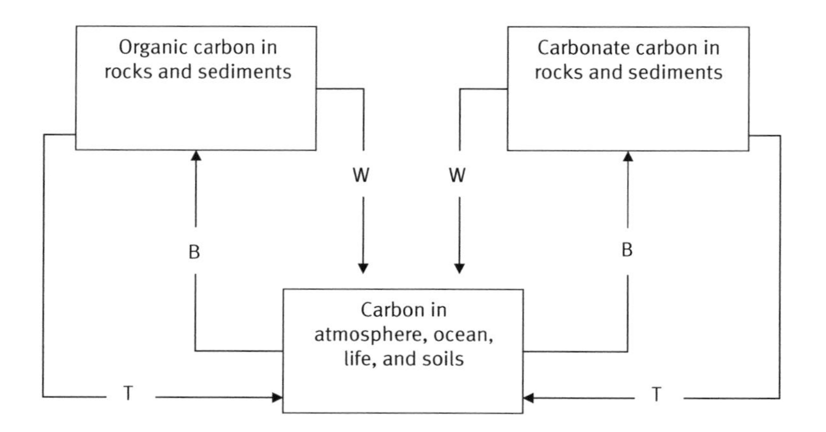

Figure 10.2 Schematic representation of the long-term carbon cycle—mainly based on the work of Robert Berner.[40] The fluxes (arrows) marked 'B' are the burial of carbon in sediments, those marked 'W' are weathering, while 'T' is the flux of carbon back out of geological sediments due to thermal processes such as volcanism. Most of the 'short-term carbon cycle' usually illustrated in ecology textbooks is concerned with the movement of carbon within the box labelled 'Carbon in atmosphere, ocean, life, and soils'.

The Earth's crust is composed of two main types of mineral, namely carbonates and silicates. The weathering of silicate rocks involves the minerals reacting with carbon dioxide from the atmosphere to produce carbonates and soluble silica oxides—these compounds can then make their way into ocean sediments and ultimately into more rock. The carbon dioxide locked up in the products of this weathering are eventually released back to the atmosphere when the rocks are subjected to large amounts of heat, as arises in volcanic and other thermal processes. Unlike silicate weathering, carbonate weathering does not provide a long-term sink for carbon: after the weathering products are transported to the oceans, new carbonates are precipitated releasing carbon dioxide and leading to no net change in atmospheric carbon dioxide.[13]

At the start of the 1980s Walker, Hays, and Kasting[36] realized that the long-term carbon cycle potentially provided a way of stabilizing the Earth's surface temperature over geological time. They pointed out that if temperature increases, then silicate weathering proceeds more quickly. This is because, like most chemical reactions, silicate weathering proceeds more rapidly at higher temperatures; in addition, in a warmer world more water evaporates from the oceans and when some of this falls on land as rain it also tends to increase weathering rates. If weathering increases then more carbon dioxide (a key greenhouse gas) is removed from the atmosphere, so cooling the Earth by reducing the greenhouse effect. Conversely, if global temperature falls, then the rate of silicate weathering declines. This reduces the rate that carbon dioxide is removed from the atmosphere—however, as it is still being added from volcanic processes this leads to a slow increase in atmospheric carbon dioxide over time, producing an increased greenhouse effect counteracting the effects of the initial cooling. Such a system can in theory operate similar to a thermostat, keeping the Earth's temperature *approximately* constant over long expanses of geological time.

The role of life in silicate weathering

In their original 1981 paper, Walker, Hays, and Kasting ignored the role of life in an attempt to keep their model simple—while acknowledging that land plants were potentially important.[36] Other scientists were quick to point out that under modern conditions plants can make a big impact on rates of rock weathering, and are also affected by temperature changes and carbon dioxide levels, and so must play a major role in the general feedback processes suggested by Walker and colleagues—the difficulty was representing these effects in a quantitative model. Jim Lovelock, working with Andy Watson[37] and Mike Whitfield,[38] made initial attempts at adding life to these models. It turns out that these attempts to introduce life into Earth's 'thermostat' has led to a very different answer to the question 'How will the biosphere end?'—but before we consider these ideas, we need to describe a bit more background on the interactions between life and weathering rates over geological timescales.

Biology potentially increases the rate of weathering by a number of mechanisms. For example, respiration by soil-living microorganisms raises the concentration of carbon dioxide in the soil (crucial for silicate-weathering reactions), while plants, along with their associated mycorrhizal fungi, produce a range of organic acids that also contribute to weathering.[39,40] In addition, plants stabilize soils, reducing erosion, and allowing deep soils to develop. For example, David Schwartzman[41] has calculated that a hypothetical soil 1 m deep composed of particles that were cubes of 1 mm^3 has 6,000 times the potential reactive surface compared to a soil-free impermeable bare rock surface (these unnaturally uniform and regular soil particles allow for easier calculations). So there is much more mineral material available to react with carbon dioxide in a soil compared to a rock surface.

In 1989, David Schwartzman and Tyler Volk[39] published a more detailed model of the role of life, in an influential paper published in the journal *Nature*. Schwartzman and Volk's[39] paper tried to answer two related questions—what has been the effect of life on weathering rates over geological time and how has this affected the Earth's climate through changes in the greenhouse gas carbon dioxide? One of the big difficulties in modelling such questions is attaching a realistic number to the effect of life on weathering rates. Schwartzman and Volk attempted to address this by running their calculations for several different plausible values for the effect of life on weathering (namely that life on land increases weathering rates by 10, 100, or 1,000 times over what would be seen without life). Their model suggested that without life, the temperature of the Earth would be between 15°C and 45°C warmer than it is today as a consequence of higher carbon dioxide levels—and they thought the higher temperature estimates were more likely to be correct.

Although the main mechanisms by which life enhances weathering are known, putting numbers to them is difficult and controversial. It appears that even a covering of lichens on a rock surface (Fig. 10.3) can speed up weathering—in one study the amount of magnesium being weathered from the rock was 16 times higher with lichens present compared to bare rock.[42] However, once there are large plants such as trees, then deep soils can form and potentially weathering rates could be even higher. One of the challenges in conducting such studies is that it is very difficult to find rocks with no life growing on them—even if it is only a thin microbial or lichen layer—so many studies compare rocks with minimal 'plant' cover (lichens and mosses) with areas with more extensive vegetation. Several of these studies suggest that this additional vegetation

Figure 10.3 Lichens colonizing a rock surface, the classic example of early stages in primary succession familiar from introductory ecology textbooks. In this case, the rocks are part of a prehistoric archaeological monument, the Rollright Stones, in Oxfordshire, England. The age of lichens can be very approximately estimated from their size and these are the oldest described from Britain—it is estimated that some of them started growing around AD 1200.[54] Their slow growth means that any effect on weathering in a given year is modest. However, over geological time their effect can be substantial. It is often assumed that some of the first life to colonize the land (and so start to increase weathering rates) may have been lichens[55]—however, their fossil record is poor and the date of the first lichens is very uncertain. Photo: DMW.

increases weathering by at least three times over what is seen with only 'lower' plants.[40] For example, an experimental study using boxes of sandy sediment in New Hampshire, USA, compared boxes 'kept mostly free of vegetation' with ones containing red pine saplings. The pines increased the rate of weathering of calcium by a factor of 10 and the weathering of magnesium by 18, over the course of several years.[43] These empirical studies suggest that the higher values for the biological enhancement of weathering used by Schwartzman and Volk in their much-cited *Nature* paper may be a bit high, but their lower factor of 10 seems justified, indeed it may be a bit conservative. While the quantitative details are still uncertain, the main conclusion that life has increased weathering rates and so reduced the amount of carbon dioxide in the Earth's atmosphere appears secure. As we will now see, this has implications for understanding both the past history of the Earth's climate and the future long-term survival of the biosphere.

The geological carbon cycle and the long-term future of the biosphere

The ecological disaster of our Sun becoming a red giant is in the far distant future, as it is predicted that our Sun will not run short on hydrogen for another 6 billion years[2]; this is a long time even by the standards of geology (the Earth is just over 4.5 billion years old). However in 1982, Jim Lovelock and Mike Whitfield published a paper suggesting that an ecological view of the long-term carbon cycle predicted the end of the biosphere in only 100 million years time.[38] The really original step in this paper was the realization that one could use the long-term carbon cycle to ask such a question in a quantitative way. These researchers used simple mathematical models to look at the effects of the steadily increasing luminosity of our Sun. This change in the Sun was predicted to gradually raise the temperature on Earth, so increasing the rate of silicate weathering and thereby reducing carbon dioxide concentrations in the atmosphere. In turn, the lower atmospheric carbon dioxide was predicted to gradually reduce the Earth's temperature by reducing the greenhouse effect. The predicted net outcome of all this is a relatively stable temperature on Earth, at least while the supplies of carbon dioxide last and while the global 'thermostat' is otherwise functioning properly. However, there are inevitable tolerance ranges to the 'thermostat'. Their models suggested that within around 100 million years atmospheric carbon dioxide levels would have been lowered to a point where photosynthesis became impossible—hence the title of their paper 'Life span of the biosphere'.

Even assuming these simple models were entirely realistic (a point we will consider later), the title of the paper really claims too much for these calculations. Lovelock[44] now admits that their paper should really have been called 'The life span of the biosphere as we know it': a less snappy title (except, perhaps, if you are a Star Trek fan) but a more realistic claim. First, there is the possibility that declining carbon dioxide levels would produce a strong selection pressure for the evolution of forms of photosynthesis that can operate in very low carbon dioxide levels. However, even without photosynthesis much microbial life could continue, albeit at much lower biomasses because of the greatly reduced amounts of energy entering food chains. Indeed, although details of the early history of life on Earth are hugely controversial, the general assumption of most scientists is that the first life was not photosynthetic. This pre-photosynthetic ecology was probably based on chemicals such as hydrogen, hydrogen sulphide, and methane, perhaps produced at hydrothermal vents and/or by the effects of sunlight on atmospheric chemicals.[45] So, photosynthesis is not required for a working biosphere, but it probably is necessary for a 'lively' highly productive biosphere of the type we see today (technically this probably requires the typical oxygen-producing photosynthesis beloved of biology textbooks, rather than the less-efficient non-oxygen-producing photosynthesis used by some microbes).[27]

A decade after the Lovelock and Whitfield paper, Ken Caldeira and James Kasting[46] returned to the problem caused by the change in the Sun's luminosity with the aid of a more complex model of the Earth's carbon cycle and climate along with more realistic biological assumptions. Key improvements on Lovelock and Whitfield's semi-quantitative first attempt at the problem were a better mathematical description of the 'greenhouse effect', a better attempt to quantify the biological enhancement of weathering, and a realization that some types of plants (plants utilizing so-called C_4 photosynthesis) can photosynthesize at lower levels of carbon dioxide than those used by Lovelock and Whitfield (who used minimum carbon dioxide levels for photosynthesis based on the commoner C_3 plants). On the basis of these improvements, they calculated that a biosphere with photosynthetic plants could survive for at least another 0.9–1.5 billion years, with its end brought about either by carbon dioxide starvation or high temperatures—which of these was most important depended on the exact assumptions made about ocean chemistry in the model.[46] Their model also suggested that in around 2.5 billion years, temperatures would have risen to a point where the Earth would lose all its water.[46]

The relationship between the Lovelock and Whitfield paper from the early 1980s and the Caldeira and Kasting study from 10 years later is typical of the history of many attempts to model phenomena in ecology and evolution. Lovelock and Whitfield identified the problem and had the first, semi-quantitative, attempt at modelling it. With the advantage of having the basic question already set out in the Lovelock and Whitfield paper—along with a decade's worth of additional knowledge of carbon and climate modelling—Caldeira and Kasting were able to produce a somewhat more realistic quantitative model that suggested the biosphere as we know it will survive for much longer than was suggested in the original paper. However, there are also aspects of Caldeira and Kasting's model that subsequent researchers have attempted to refine. For example, the way Caldeira and Kasting modelled soil carbon dioxide in relation to the atmospheric level of this gas effectively gives a figure for the biotic enhancement of weathering of around 1.5 times larger than the rate without life,[47] which is rather low compared to the data we discussed earlier.

In 2001, Tim Lenton and Werner von Bloh[47] published a model with a more realistic description of the biotic enhancement of weathering that varies with conditions on Earth—by including feedbacks between carbon dioxide levels and temperature with plant productivity (so plant production was higher at greater carbon dioxide levels and temperatures of around 25°C).[47] These improvements to the model increased the life span of a biosphere with terrestrial plants because as carbon dioxide levels fell, plant production declined so reducing silicate weathering—while all the time more carbon dioxide was being added to the atmosphere from volcanoes. So, while Caldeira and Kasting[46] suggested that carbon dioxide would become limiting after approximately 0.9 billion years, Lenton and von Bloh[47] predicted that something similar to the current biosphere could survive for 1.2 billion years (Caldeira and Kasting's model only predicted such long-term survival with particular assumptions about the behaviour of ocean chemistry). In addition, the much stronger feedbacks between vegetation and carbon dioxide in Lenton and von Bloh's model lead to a final dramatic collapse, rather than a slow long-term decline of plant life, which was seen in the Caldeira and Kasting model of 1992. This is because in the 1992 model the plants track the slowly decreasing carbon dioxide levels, while in Lenton and von Bloh's model the additional feedbacks allow plant-friendly amounts of carbon dioxide to persist for longer, before finally dropping dramatically when the feedbacks are pushed too far and the system can no longer self-regulate.

So how will the biosphere end?

Once atmospheric carbon dioxide levels have been reduced to near zero (somewhat ironic given the contemporary rise in the gas) then the feedbacks, which have helped maintain reasonably stable temperatures on Earth throughout the history of life, can no longer operate. This means that as the Sun continues to warm, the temperature on Earth will rise. Eventually, whole groups of organisms would be expected to go extinct as the temperatures become too hot for them. Today, the most heat-tolerant plants and animals can survive temperatures up to approximately 50°C, some single-celled eukaryotes can survive up to 60°C, the photosynthetic cyanobacteria do a bit better, surviving at 75°C, while the most heat-tolerant bacteria (archaea) can manage up to at least 113°C (based on various sources compiled by Wilkinson[27]), with temperatures of 121°C recently reported.[48] So as temperatures rise, multicellular eukaryotes (plants, fungi, and animals) would probably be the first to become extinct, followed by single-celled eukaryotes and finally by prokaryotes; this order of extinction is the reverse of the order in which these groups originally appear in the fossil record.[30] Recently, a long-term carbon cycle model has been used to estimate the potential times of these extinctions. It suggested that increases in temperature would cause multicellular eukaryotes to become extinct in 0.8–1.2 billion years and single-celled eukaryotes in 1.3–1.5 billion years. Reductions in carbon dioxide (rather than temperature rises) led to the extinction of photosynthetic bacteria at 1.6 billion years.[49] A biosphere of non-photosynthetic prokaryotes may survive significantly beyond this point but without photosynthesis this will be nothing like the biosphere as we know it. Ultimately at some point even these will become extinct, perhaps directly through increased temperature making their biochemistry inoperable or indirectly through a loss of water from the Earth. However, by the time the Sun expands to reach Earth orbit in some 6 billion years time, life will have gone from our planet.

Biological uncertainties

There is a problem with all these published studies of carbon cycling and the end of the biosphere as we know it, and it is the same problem that bedevils most of science fiction: a potential failure of imagination. In the approximately 1 billion years before the modelled end of the biosphere, there is scope for substantial evolution—potentially undermining some of the predictions about what levels of carbon dioxide a plant may need. For example, one billion years ago all life on Earth was microbial, and look what we now have living around us. As Tyler Volk[50] has pointed out when thinking about biogeochemical cycles and the end of the biosphere we need to think not just about 'life as we know it' but much more challengingly 'life as we do not know it'. Even so, it is hard to envisage life utilizing sunlight (photosynthesis) at vanishingly low carbon dioxide levels—evolution cannot provide chemically impossible solutions, and without extensive photosynthesis a productive biosphere seems impossible.

If the end of the biosphere is defined as the end of all life (the biosphere as defined by Suess[6]), rather than the end of a productive photosynthetically driven ecology, then there is another problem. We have already mentioned the difficulty in killing many microbes, especially prokaryotes, which can, for example, be found surviving in such extreme conditions as hydrothermal vents on the ocean floor with water temperatures around 120°C and in highly radioactive conditions in nuclear power plants.[48] However, prokaryotes can also remain in a dormant state for long periods of time, so while not ecologically active they are not dead—or at least are able to come back to life. How long dormant bacteria can survive is a highly controversial topic; however, there is reasonable evidence for survival for tens to possibly a few hundred million years in geological sediments followed by successful reproduction[27] (see also Chapter 1). This makes it very hard to define a planet as completely 'dead' (lifeless)—be that planet the Earth of the future, or the Mars of today.

Can intelligent life prolong the life span of the biosphere?

A final possibility is that humans, or whatever we evolve into, could intervene in the Earth's climate and biogeochemical cycles to postpone the end of a productive biosphere. The Sun will eventually grill us, but who knows, we may be able to keep the thermostat functioning for a few million years longer. The idea may not be so far-fetched: after all, we are already making serious changes to the short-term carbon cycle with potentially grave climatic implications, could we not do something similar in the far future, this time to save rather than challenge the biosphere? For example, it has recently been suggested[51] that we could counter our contemporary *increases* in atmospheric carbon dioxide by effectively increasing the rate of silicate weathering, so it would function to reduce carbon dioxide levels on short timescales of use to ourselves. This could conceivably be done by electrochemically (assuming the electricity is derived from sources other than fossil fuels) removing hydrochloric acid from sea water and reacting the acid with silicate rocks on a grand scale. This example illustrates the general idea of deliberately intervening in biogeochemical cycles, but as in the very long term the problem is too little carbon dioxide then this particular approach would not help in increasing the life of the biosphere.

Attempts in the far distant future to raise carbon dioxide levels in the atmosphere to facilitate photosynthesis are likely to create more problems than they solve. Increasing the amount of greenhouse gases in the context of an ever-warming Sun would potentially produce temperatures too high for a productive biosphere able to support human-like species. Perhaps more useful would be engineering solutions that reduce the amount of solar radiation reaching Earth—so reducing temperatures, which would lead to lower rates of silicate weathering and so the survival of photosynthetically useful levels of carbon dioxide for longer into the future. Such schemes have already been discussed in the context of reducing current human-caused 'global warming'. One especially ambitious possibility is a 2,000 km diameter mirror in space that, correctly positioned, could prevent around 2% of solar energy from reaching Earth.[52]

Humans could also potentially protect the Earth from biosphere-ending impacts of astronomical objects. This could be done by destroying them with nuclear weapons, the option preferred by science fiction film makers, or less dramatically by altering their trajectories so they miss Earth.[53] If you are an optimist you can argue that because of such possibilities, the evolution of intelligent life on a planet increases the potential life span of its biosphere—a pessimist would look at the example of humans on Earth, with nuclear weapons and climate-altering changes to the planet, and wonder if such an optimist has any grasp of reality. Another possibility is that our technological society will free us from having our fate chained to that of the Earth or even any particular planet. Currently, these are ideas better treated by science fiction than science.

Clearly, confident predictions of potential technological fixes in the far future are impossible. In the Hoyles' novel we quoted at the opening of this chapter, it becomes apparent, as the story unfolds, that some advanced extraterrestrial intelligence must have intervened to prevent the total destruction of all life on Earth by the quasar. Without intelligent intervention (from our descendants or outer space?) then the last microbial life on Earth will probably go with the aftermath of the loss of the oceans as the Sun becomes even hotter. The biosphere as we know it will probably have gone long before this due to a mixture of increasing temperatures and plummeting carbon dioxide levels—in perhaps 1–1.5 billion years from now. What fraction of this maximum time period humans (or the descendents of humans) will be around for is quite another matter.

11

General Conclusions

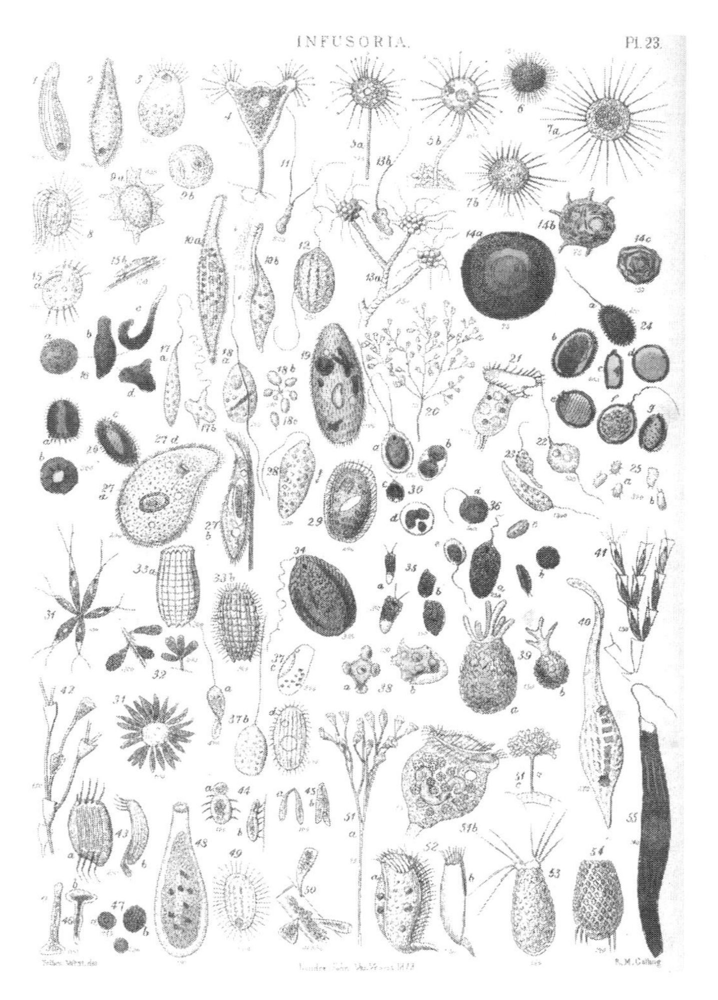

Figure 11.1 The diversity of eukaryotic microbes, as illustrated in J.W. Griffiths and A. Henfery's *Micrographic Dictionary* of 1875 (3rd edition, published by John van Voorst, London). The figure mainly illustrates various types of protists (under the archaic name 'Infusoria'), including ciliates, flagellates, and testate amoebae.

> *There ain't no reason things are this way, its how they've always been and they intend to stay, I can't explain why we live this way, we do it every day.*
>
> *—Brett Dennen, 'Ain't no reason' from his album So Much More (2006).*

Brett Dennen is a fine musician, but listening to his lyrics, one might be tempted to think that because we 'do [or see] it every day', then it does not deserve an explanation. Our book has dealt with a variety of everyday phenomena such as ageing, sex, species, a green world, and a blue sea, and we hope that by now our readers will agree that there *is* a reason why things are this way. Indeed, the fact that these phenomena are so commonplace makes the questions all the more important. The exciting thing is that while considerable progress has been made in each of the areas we address, we still do not have a complete answer to any of the questions we have posed.

We use this short concluding chapter to pull together some common threads and to discuss some of the interrelationships between our answers. First and foremost, even the most casual reader will note that there is a close interrelationship between the ecological and evolutionary explanations we have presented. Taking the perspective of evolutionary biology, almost all of the evolutionary explanations we have proposed include an important ecological component. For example, ageing is now widely seen to arise as a consequence of there being relatively weak natural selection late in an organism's life. Yet the primary reason for this 'selective shadow' is that predators and parasites are likely to have killed the organism long before it reaches an advanced stage of maturity. Likewise, one explanation for the evolution of sex is that the variation it generates allows at least some of the offspring to better compete with members of the same species, or to avoid parasitism.

In a similar vein, many of the ecological phenomena we have sought to explain have evolutionary origins. For example, tropical areas may have more species because rates of speciation are greater in the tropics, or because rates of extinction are greater at high latitudes, or both. Likewise, plants have evolved secondary compounds to deter herbivory, and the presence of these compounds may go some way towards understanding why the world remains green. Even our chapter on chaos appealed to evolutionary biology to help understand why demographic parameters were so rarely high enough to push the consequent dynamics into the chaotic regime. It may seem entirely natural to combine ecology and evolution in this way as many scientists do, but it is worth noting that in many biology degrees ecology is taught in one course, while evolution is taught somewhere else.

Although we have addressed each question in turn, it is also clear that they are not independent in that their solutions share many other underlying concepts. One common concept, invoked in almost all of our evolutionary chapters, was that of kin selection, where a particular form of a gene can be selected for not because of its beneficial effects on the carrier, but because of its beneficial effects on others who share the same form of the gene. This deceptively simple idea was relatively slow to be formally recognized (it took until the middle of the past century), but thanks to the genius of researchers like Bill Hamilton, it is now believed to underlie many examples of cooperation observed in the natural world, and it helps explain a wide range of puzzling phenomena, including why individuals may occasionally commit suicide rather than inflict their relatives with a harmful parasite. Conversely, we now recognize that arguments based on higher levels of selection *usually* depend heavily on an array of unlikely assumptions. This helps to rule out several seemingly attractive explanations for observed phenomena; for example, we now appreciate that the planet does not remain green because herbivores refrain from consuming their food supply, and individuals do not age because it relieves other members of the community from the burden of looking after the injured.

Another concept we have repeatedly invoked is that of the trade-off. As we have seen, one popular evolutionary argument for ageing is that it arises as a consequence of genes with antagonistic effects, rather like performance-enhancing drugs that can improve athletic performance but then cause all sorts of deleterious effects years later. Yet we also invoke trade-offs to understand how several related species can coexist, to understand why individuals do not tend to evolve such high rates of reproduction that the population ends up chaotic, and to understand how sex can produce offspring that have a competitive advantage that varies with the nature of the environment. So, if you want to study evolution and ecology, there is no avoiding trade-offs. Quite simply, the idea is universal, and time and time again one will see the same type of 'trade-off' argument applied at different levels of biological organization.

An understanding of population dynamics also plays a fundamental role in many of our chapters. Although it is tempting to think the world is green because individual herbivores decide not to consume all the plants they are feeding on, in reality the number of mouths may play just as important a role as what individual mouths do. In this way, the 'green world' problem is a problem of population dynamics—essentially, why is there a non-zero equilibrium? Likewise, we note that humans had an increasing impact on their environment as their population size increased; something that happened at an increasing pace after the invention of agriculture. Although most politicians understandably shy away from discussing population size, it is plain that most of our current environmental problems would be much less pressing if there were far fewer of us on the planet.

One of the problems of a chapter-based book, and formal education in general, is that it necessarily compartmentalizes knowledge, but this is not an approach we want to encourage. Our chapters share many common concepts, but they also share explanations, albeit at different levels. For example, both ageing and sex may have closer connections than may first appear, with both phenomena linked directly and indirectly to a form of rejuvenation—in both cases, parents might be considered 'damaged goods'. Our chapters on when we started to change things, and why is the sea blue, introduced the important idea of biogeochemical cycles, yet the insights of Lovelock and Whitfield indicate that the same processes may be extrapolated to help understand how the biosphere will end. Likewise, to understand why the tropics are so diverse, it is crucial to know how species are actually formed, as many theories depend explicitly on assumptions relating to the mode and frequency of speciation.

Some of the big questions facing researchers have extra resonance because of the challenges facing humanity. Our book has not dealt directly with conservation issues, but we hope that some of our chapters will have given some background into the timing and type of effects our species has had on our planet, as well as the more 'natural' influence of biogeochemical cycles. In doing so, we hope that we have provided at least a few insights into our current predicament, such as to how concentrations of greenhouse gases and temperature have fluctuated over the geological past (it has been a whole lot warmer, and carbon dioxide levels have been a whole lot higher in the past), without understating the current level of crisis. Ironically, we are facing almost unparalleled loss of biodiversity, before we fully understand why areas such as the tropics have so many species. Likewise, we are still discovering new groups of marine plankton, and highly rich areas of biodiversity of Antarctica, while at the same time pushing many of the world's fisheries to extinction and treating the seas as garbage dumps.

Finally, we can ask why it is that so many of the questions we have posed do not yet have complete answers. One reason may be that ecology and evolution are relatively young and poorly funded sciences, and that it takes time and effort to address problems, coupled with occasional good luck in identifying a key test, or natural experiment. This slow development may have been compounded by the fact that researchers were slow to recognize certain problems (the 'green world' problem was only explicitly articulated in 1960 and very few ecologists considered chaos before the 1980s). However, above and beyond this is the fact that these questions are plainly difficult to answer. Much of this difficulty may come down to the complexity of the phenomena we are attempting to explain. When multiple variables (such as temperature, land area, and primary productivity) are associated with species diversity (one could equally think of multiple correlates of sex, or ageing, say), then it is really hard to know which variables are the underlying primary cause, and which variables simply co-vary with the causal agent. Moreover, as our chapters on the evolution of ageing, sex, and the green world serve to demonstrate, very often there may be more than one plausible explanation, with each process playing some role. The difficulty then is more quantitative than qualitative—evaluating which agent is most important. Without full knowledge of the parameters involved, these quantitative questions are intrinsically hard questions to answer.

With any luck, some of the ideas discussed in this book will continue to be both valid and relevant for many years to come. Certain ideas we have favoured will undoubtedly require considerable revision, or may be rejected outright. For example, Bill Ruddiman's ideas on an early start to our modifications of the global environment open up fascinating new areas for research, but may well turn out to be wrong in detail. Nevertheless, our book will have served its purpose if we have stimulated some readers to think about the questions raised, look at the solutions that have been offered and, if they do not buy into them, propose their own explanations.

Species list

For ease of communication, we have tended to use common names, when available and reasonably consistent, in our text. Here we provide the formal scientific name for specific species so far described only by their common name in our book.

Abyssinian Mustard *Brassica carinata*
African ('Savannah') Elephant *Loxodonta africana*
African Lion *Panthera leo* (a variety of subspecies)
American Crow *Corvus brachyrhynchos*
American Lion *Panthera leo* (subspecies *atrox*; some consider this a distinct species)
American Lobster *Homarus americanus*
Antler Fly *Protopiophila litigate*
Apple-blossom Thrip (Plague Thrip) *Thrips imaginis*
Azure-winged Magpie *Cyanopica cyanus*
Baboon *Papio* spp.
Badger (American) *Taxidea taxus*
Balsam Fir *Abies balsamea*
Banana *Musa* spp.
Barn Swallow *Hirundo rustica*
Belding's Ground Squirrel *Spermophilus beldingi*
Bighorn Sheep *Ovis canadensis*
Blackberry (Common) *Rubus fruticosus*
Black Mustard *Brassica nigra*
Blowfly (Australian Sheep) *Lucilia cuprina*
Blowfly (Green Bottle) *Lucilia sericata*
Bobwhite Quail *Colinus virginianus*
Bristlecone Pine (Great Basin) *Pinus longaeva*
Brown Bear *Ursus actos*
Brown Hyena *Parahyaena brunnea*
Cabbage *Brassica oleracea*
Canadian Lynx *Lynx canadensis*
Capuchin Monkey (Brown) *Cebus paella*
Chimpanzee (Common) *Pan troglodytes*
Cinnabar Moth *Tyria jacobaeae*
Columbian Mammoth *Mammuthus columbi*
Common Duckweed *Lemma minor*
Cory's Shearwater *Calonectris diomedea*
Cotton-top Tamarin *Saguinus oedipus*
Coyote *Canis latrans*
Creosote Bush *Larrea tridentata*
Cuckoo *Cuculus canorus*
Damselfish (Australian) *Pomacentrus amboinensis*

Dandelion *Taraxacum* spp.
Dodo *Raphus cucullatus*
Dog (Domestic) *Canis lupus familiaris*
Emmer Wheat *Triticum dicoccum*
Fiddler Crab (Australian) *Uca mjoebergi*
Flax *Linum ustatissimum*
Florida Scrub Jay *Aphelocoma coerulescens*
Fruit Fly Diptera of family Drosophilidae
Furbelows *Saccorhiza polyschides*
Galapágos Giant Tortoise *Geochelone nigra*
Galapágos Penguin *Spheniscus mendiculus*
Giant Deer ('Irish Elk') *Megaloceros giganteus*
Great White Shark *Carcharodon carcharias*
Guillemot (Common) *Uria aalge*
Guppy (Trinidadian) *Poecilia reticulata*
Heather (Common) *Calluna vulgaris*
Heather Beetle *Lochmaea suturalis*
Honeybee (Western) *Apis mellifera*
Horse *Equus caballus*
Human *Homo sapiens*
Ibex *Capra* spp.
Impala *Aepyceros melampus*
Japanese Quail *Coturnix japonica*
Koala *Phascolarctos cinereus*
Komodo Dragon *Varanus komodoensis*
Leach's Storm Petrel *Oceanodroma leucorhoa*
Leaf Mustard *Brassica juncea*
Long-tailed Tit *Aegithalos caudatus*
Magellanic Penguin *Spheniscus magellanicus*
Mallard Duck *Anas platyrhynchos*
Meerkat *Suricata suricatta*
Migratory Locust *Locusta migratoria*
Monarch Butterfly *Danaus plexippus*
Moose (Elk in Europe) *Alces alces*
Mountain Gorilla *Gorilla beringei beringei*
Naked Mole Rat *Heterocephalus glaber*
New Zealand Mud Snail *Potamopyrgus antipodarum*
Oak *Quercus* spp.
Oil-seed Rape (Rapeseed) *Brassica napus*
Olive *Olea europaea*
Paradise Kingfishers *Tanysiptera* spp.
Pea Aphid *Acyrthosiphon pisum*
Quahog Clam (Oceanic) *Arctica islandica*
Ragwort *Senecio jacobaea*
Rainbow Trout *Oncorhynchus mykiss*

Red Deer *Cervus elaphus*
Red Pine *Pinus resinosa*
Red-winged Blackbird *Agelaius phoeniceus*
Sargasso Weed *Sargassum* spp
Sheep (Domestic) *Ovis aries*
Sooty Shearwater *Puffinus griseus*
Soybean *Glycine max*
Swan (Mute) *Cygnus olor*
Sycamore *Acer pseudoplatanus*
Three-spined Stickleback *Gasterosteus aculeatus*
Tiger *Panthera tigris*
Transvaal Stone plant *hithops ruschiorum*
Turnip *Brassica rapa*
Vampire Bat (Common) *Desmodus rotundus*
Water Hyacinth *Eichhorina crassipes*
Weasel *Mustela* spp
White-Fronted Bee-eater *Merops bullockoides*
Wildebeest (Blue) *Connochaetes taurinus*
Wolf (Grey) *Canis lupus*
Woolly Mammoth *Mammuthus primigenius*

A male African Lion in Limpopo Province, South Africa. Photo: DMW.

Glossary

This book clearly considers an extremely wide range of issues in ecology and evolution and as such it cuts across a large number of scientific disciplines. We therefore hope the following glossary will be of assistance, in giving a quick summary of the basic concepts involved. In so doing we are attempting to give an explanation of the terms as we have used them rather than a formal definition. If you find yourself having to look up terms repeatedly and the glossary does not provide satisfaction, then Calow (1999), Lincoln *et al.* (1998), or Hale *et al.* (2005) may address your needs. Some of the material has been extracted from Wilkinson (2006), but we have also been aided by some excellent source texts, notably Bell (1982) and Futuyma (1998). Note that '*q.v.*' means 'see also'—so 'biosphere *q.v.*' means that the word 'biosphere' is also defined elsewhere in the glossary, while '*c.f.*' means 'compare with' and draws attention to superficially similar terms that are also listed. A brief guide to the geological time scale is given at the end of the glossary.

Abiotic environment Traditionally ecology textbooks split the environment into 'biotic' (the biological aspects of the environment) and the 'abiotic' (the physical aspects). Abiotic aspects would include factors such as aspects of climate, soil pH, or oxygen concentration of the atmosphere; however, the approach taken in this book (e.g. Chapters 9 and 10) makes it clear that this classification is often unhelpful as most of the traditional abiotic aspects of the environment are affected by biology.

Adaptive radiation The evolution of ecological and phenotypic diversity from members of a single phylogenetic line within a diverging lineage.

Ageing (senescence) Decrease in an individual's performance, survivorship, and/or fecundity as age increases.

Albedo The reflectivity of an object. For example, the albedo of fresh snow is around 0.95 whereas a flat calm ocean can be as low as 0.20.

Alkaloid Poisonous, nitrogen-containing compounds found in some plants.

Allele One of several alternative forms (nucleotide sequences) of a gene, usually recognized by their phenotypic effects.

Allopatric Pertaining to populations or species, the ranges of which do not overlap.

Allopatric speciation The origin of two or more species resulting from divergent evolution of populations that are geographically isolated from one another.

Alloploidy Having one or more sets of chromosomes from a parent of one species and another set or sets from a parent of another species.

Alternation of generations The alternation of ploidy (*q.v.*) enforced in all sexual organisms by the alternation of meiosis (*q.v.*) and syngamy (*q.v.*).

Altruism A behaviour that is costly to an individual but results in a benefit to others (*c.f.* cooperation).

Amphimixis Fusion (syngamy) between gametes produced by different individuals (sexual reproduction).

Anisogamy The occurrence of gametes of different size, shape, structure, or behaviour (c.*f.* isogamy), for example, sperm and eggs.

Anthropocene Informal name for the period of time during which humans have been having major global environmental effects—often assumed to start with the Industrial Revolution.

Apparency Term which tries to convey the idea that some plant species are more obvious (apparent) to their predators than others.

Apomixis A form of parthenogenetic reproduction in which offspring develop from *mitotically* produced cells that have not experienced meiosis or syngamy (*c.f.* automixis).

Asexual Lacking sexuality ('amictic').

Assortative mating A non-random mating pattern, mediated either directly through mate choice, or indirectly through habitat choice (for taxa that mate within their habitat).

Automixis Syngamy (fusion) between *meiotically* reduced nuclei descending immediately from the same zygote (*c.f.* apomixis).

Autoploidy The occurrence of two or more sets of chromosomes, all of which derive from members of the same species.

Autotroph An organism which can use non-organic sources of energy. The most well-known examples being green plants using solar energy.

Bdelloid rotifers A group of multicellular microscopic animals that seem to have experienced a period of evolution without sex, and probably without other forms of recombination, for more than 80 million years.

Biological Species Concept See 'Species'.

Biomass The total mass of living material. For example, the biomass of a forest would be the combined mass of every organism living in that habitat.

Biome An area of the world with similar ecologies. Examples include 'tropical rainforest' or 'temperate deciduous forest'.

Biosphere This term has been used in several different ways. Some authors use it to mean the 'totality of living things residing on the Earth', others use it to mean 'the space occupied by living things', while others use it to refer to 'life and life support systems'. In the context of this book all of these definitions are applicable—for example, our question 'How will the biosphere end' makes sense under all three definitions of the term.

Bottleneck A short-term severe reduction in population size, which can occur either through much reduced survival within a population or through colonization of a new habitat. This can lead to a population which is less genetically diverse than before the bottleneck.

Butterfly effect A metaphor for the extreme sensitivity to initial conditions that characterizes (and defines) chaotic systems, coined by Edward Lorenz. Thus, in a chaotic system the difference between a butterfly flapping or not flapping its wings (or frog croaking, etc.) could result in the difference between a hurricane arising or not arising.

By-product mutualism A cooperative relationship between species in which the species benefit from one another's products which would be made even in the absence of the relationship.

Carrying capacity The maximum number of individuals that an area can support at equilibrium. Often represented as 'K' in the logistic equation (*q.v.*).

Chaos Complex dynamics which are bounded in magnitude, yet neither at equilibrium nor in a predictable cycle, and shows extreme sensitivity to initial conditions (*q.v.* butterfly effect).

Character displacement A pattern of geographic variation, in which a character (such as beak size) differs more between sympatric than between allopatric populations of two species (*q.v.* reinforcement).

Cheats Individuals who do not cooperate (or cooperate less than their 'fair' share), but are potentially able to gain the benefit of others cooperating.

Chromosome In eukaryotes (*q.v.*), a visible microscopic structure within the nucleus bearing genetic information arranged in a linear sequence. In prokaryotes (*q.v.*), it is a circular structure not localized within a nuclear membrane.

Ciliates Informal name for a phylum of 'protozoa' (Protists *q.v.*)—the Ciliophora.

Clade A monophyletic group (*q.v.*) of organisms.

Clonal interference The reduced competitive advantage of a clone that carries a beneficial mutation owing to the simultaneous presence of one or more other clones that carry different beneficial mutations.

Clone An assemblage of genetically identical organisms (in strict use, they must be identical by decent for every allele at every locus).

Conspecific Belonging to the same species.

Cooperation A behaviour which provides a benefit to another individual (recipient), and which is selected for because of its beneficial effect on the recipient (*c.f.* altruism).

Cosmic rays Elementary particles (e.g. protons, electrons, etc.) travelling through space.

Cosmopolitan An organism that is widely distributed, usually applied to species found on most continents, or most oceans if marine.

Crossing over The exchange of genetic material between non-sister chromatids of homologous chromosomes by symmetrical breakage and reunion during meiosis (*q.v.* recombination).

Cyanobacteria Oxygenic photosynthetic bacteria, have often been referred to as blue green bacteria (or 'algae') in the past.

Cyanogenic glycosides Hydrogen cyanide conjugated with glucose. Many plant species store the poison cyanide in this form; upon insect attack chemical reactions can release the cyanide as a defence measure.

Deme A local population of organisms of a sexual species that actively interbreed with one another and share a distinct gene pool.

Diatoms Aquatic single-celled eukaryotes with silica-rich shells.

Diploid Having two (hence 'di') sets of homologous (*q.v.*) chromosomes, one of maternal and the other of paternal descent.

Disruptive selection Natural selection within a single population against intermediates and favouring increased variance, such as two different (usually extreme) phenotypes.

Ecological engineering It has long been known that organisms modify their environment; however, during the 1980s Clive Jones and colleagues formalized this into the concept of 'ecological engineers'. These organisms have a particular effect on their environment, either by their physical structure (e.g. trees) or by their behaviour (e.g. beavers). Extinction of such species may have major effects on the environment.

Ecosystem services Services of use to humans and other organisms provided by ecosystems, for example, oxygen, food, 'clean' water, and so on.

Emergence The formation of global patterns from local interactions; density-dependent regulation is the most well-known ecological example. This concept was first named by the biologist G.H. Lewes (now better known as the partner of the novelist George Eliot) in the nineteenth century, but attracted relatively little interest until the late twentieth century.

Endemic Organism native to a particular, restricted area and found only in that place.

Epistasis An interactive effect in which the fitness advantage (or disadvantage) provided by a given allelic form of a gene is influenced by the allelic forms of other genes (*q.v.* Kondrashov's Hatchet).

Eukaryote An organism with linear chromosomes localized within a nuclear membrane, typically showing both mitosis and meiosis. One of the two great divisions ('Empires' or 'Super Kingdoms') of life on Earth, the other is the Prokaryotes (*q.v*).

Extinction The death of all individuals in a local population, a species, or a higher taxon.

Fecundity The quantity of gametes (usually fertilized eggs) produced.

Female The gender producing the larger gametic type (an ovule or ovum) in an anisogamous organism *c.f.* male, which produces the smaller gametic type (sperm or pollen cell).

Fertilization The union of gametes—'syngamy'.

Fitness A rate of increase; usually the rate of increase of some type relative to that of an alternative type in the same population over a specified period of time (*c.f.* inclusive fitness).

Fixation An end point in which all members of a population carry the same allelic form for a given gene, so that its relative frequency is 100%.

Flagellates Protists (protozoa) in the phylum Discomitochondria.

Food chain A linear sequence describing feeding relationships. For example, Plant → Caterpillar → Tit (Chickadee) → Hawk, where the arrows show feeding relationships. So in this example caterpillars eat plants but tits eat caterpillars.

Foraminifera Informally called 'forams'. Marine shell-building eukaryotic microorganisms. Most are benthic, living attached to surfaces or in sediments, but a few groups are free-swimming planktonic organisms.

Fossil Any recognizable trace of an ancient organism preserved in a geologic deposit.

Founder effect The genetic consequences of starting a new population from a small number of individuals. The newly founded population is likely to have quite different gene frequencies than the source population because of sampling variation and subsequent genetic drift, and have less genetic variation than the source population.

Frequency-dependent selection The term given to the outcome of natural selection in which the fitness (*q.v.*) of a phenotype is dependent on its frequency relative to other phenotypes in a given population.

Gaia The idea, suggested by James Lovelock, that organisms and their material environment evolve as a single-coupled system, from which emerges the sustained self-regulation of climate and chemistry at a habitable state for whatever is the current biota.

Gamete A cell, such as an egg or sperm, capable of undergoing syngamy with another cell to form a zygote.

Gender The set of individuals or gametes of the same species incapable of fertilizing one another.

Gene A unit of heredity, usually a sequence of DNA that encodes a protein or other product that influences the development of one or more characters.

Gene flow The movement of genes from one population into another (usually of the same species) resulting from movement of individuals or their gametes.

Genetic drift The occurrence of changes in gene frequency brought about not by natural selection, but by chance. Its effects are particularly strong in small populations. Also known simply as 'drift'.

Genome The totality of genetic material transmitted to progeny.

Genotype The genetic make-up of an individual. More formally the allelic state of any specified number of loci in a given individual (*q.v.* phenotype).

Germ line The lineage of generative cells and their descendants that give rise to gametes.

Group selection The differential rate of origination or extinction of whole groups (colonies, populations, or even species) on the basis of differences among them in one or more characteristics (*c.f.* kin selection).

Guild A group of species who all make their living in a similar way. Named for a perceived analogy to medieval guilds of tradesmen; however, the guilds of medieval England were rather more complex institutions than this analogy suggests.

Haldane's rule Named after J.B.S. Haldane, it is the observation that, if hybrids of only one sex are sterile or inviable in a species cross, that sex is nearly always the one having heterogametic (e.g. XY) sex chromosomes.

Hamilton's rule Named after W.D. Hamilton, it is an inequality ($rb - c > 0$) that predicts when a trait is favoured by kin selection, where *c* is the fitness cost to the actor of performing the behaviour, *b* is the benefit to the individual who the behaviour is directed towards, and *r* is a measure of the genetic relatedness between those individuals.

Haploid Possessing a single set of chromosomes, none of which are homologous—hence a single gene copy at each locus (*q.v.* diploid, triploid, etc.).

Hayflick limit Named after its discoverer, Leonard Hayflick, it is the finite number of divisions of which a cell is capable, widely associated with the shortening of telomeres (*q.v.*).

Heterozygous A given locus in a diploid (or polyploid) individual is heterozygous if it bears two different alleles (or more).

Holocene The geological 'series' in which we are currently living. The most recent part of the Quaternary (*q.v.*) comprising the last 11,600 calendar years (approximately 10,000 radiocarbon years) since the end of the last glaciation.

Homologous chromosomes The paternally and maternally derived chromosomes which bear the same sequence of loci pair during meiosis and may cross over.

Homozygous A given locus in a diploid (or polyploid) individual is homozygous if two alleles are identical.

Host The organism upon which a parasite (*q.v.*) lives, and usually feeds.

Hybrid zone A region in which genetically distinct populations come into contact and produce at least some offspring of mixed ancestry.

Incipient species Two or more diverged populations that are substantially, but not completely, reproductively isolated.

Inclusive fitness The sum of an individual's fitness quantified as the reproductive success of the individual and its relatives, with the relatives devalued in proportion to their genetic distance.

Isogamy The state in which gametes of different gender have the same size (*q.v.* anisogamy).

Kin discrimination When behaviours are directed towards or against individuals depending on their relatedness of the recipient the actor.

Kin selection A process by which traits are favoured because of their beneficial effects on the fitness of relatives (*c.f.* group selection).

Kondrashov's hatchet A theory for the evolution of sex based on 'synergistic' epistasis (*q.v.*). If rates of deleterious mutations per diploid genome per generation are high, and deleterious mutations interact synergistically (such that their combined disadvantage is more than their sum), then sexual forms can overcome their inherent two-fold disadvantage (*q.v.*).

Leucocytes White blood cells, involved in defence against parasites.

Lignin Complex polymer found in plant cell walls, which provides mechanical support for the cell. Lignin is the second most abundant polymer in plants after cellulose.

Lineage A chronological sequence of populations with ancestor–descendant relationships.

Little Ice Age Unusually cold period from approximately the middle of the sixteenth century until the mid-nineteenth century.

Locus A site on a chromosome occupied by a specific gene; more loosely, the gene itself, in all its allelic states (plural loci).

Logistic equation One of the simplest equations to relate the rate of growth of a population to population density. As density increases, the rate of growth gradually declines, until population growth stops and finally goes negative. In its simplest form, the discrete time logistic can be represented as $x_{t+1} = rx_t(1-x_t)$ where x_t is an index of population density and r is the intrinsic population growth rate.

Macroparasites Multicellular parasites, such as nematodes and lice. They reproduce much more slowly than most microparasites (*q.v.*).

Mass extinction A very high number of extinctions (with a concomitant decline in diversity) over a geologically short interval of time (years to many thousands of years).

Meiosis A sequence of nuclear divisions during which ploidy is halved, typically involving genetic recombination through crossing over between homologous chromosomes (*c.f.* mitosis).

Microparasites Small parasitic organisms such as viruses, bacteria, and protozoa. Often capable of extreme multiplication within the host (*q.v.*) as they have very short generation times. *c.f.* macroparasites.

Milankovitch cycles The cyclical changes in the Earth's climate which, in the recent geological past, have been involved in driving the series of successive glacials ('ice ages') and interglacial (warm) conditions. Named after Mulutin Milankovitch (1879–1958) one of the key theorists involved in the mathematical understanding of how changes in the Earth's orbit cause these cycles.

Mitosis A single nuclear division, resulting in the exact replication of the genome (*q.v.* meiosis).

Mixis The rearrangement of genetic material through meiosis and/or syngamy (usually both), almost always resulting in the production of one or more new organisms differing genetically from one another and from their parents.

Monophyletic Derived from the same ancestral taxon. Sometimes used in a restricted sense to include the ancestral species and all descendant species.

Morphological species concept A species defined by being sufficiently morphologically distinct from all others. The definition can include multivariate tests of the statistical distance between species centroids in relation to intraspecific variation about the centroids.

Muller's ratchet The process by which the genome of an asexual population accumulates deleterious mutations in an irreversible manner, owing to the chance loss of individuals with the lowest number of mutations.

Mutation An error in replication of a nucleotide sequence (and therefore heritable), or any other alteration of the genome that is not simply the product of recombination.

Mutualism A mutually beneficial relationship, usually used to describe the relationship between different species. See also symbiosis (*q.v.*).

Mycorrhizae An association between plant roots and fungi which is normally beneficial to both species.

Neutral alleles Allelic variants at a locus that have no effect on the fitness of the bearer. Such alleles are said to be 'selectively neutral'.

Niche The range of combinations of all relevant environmental variables under which a species or population can persist. Perhaps best left vague, textbooks often write of a range of different definitions such as Eltonian niches or Hutchsonian niches.

Niche construction The process by which organisms modify their own, or other organisms' environments. This approach was developed during the 1990s by John Odling-Smee, Kevin Laland, and Marcus Feldman. It emphasizes the evolutionary effects of this process and has its theoretical origins in ideas from population genetics.

Parasite An organism which lives in, or on, another organism and in so doing obtains resources at its host's (*q.v.*) expense. It is remarkably difficult to produce a rigorous definition of parasite which successfully covers the wide range of examples seen in nature. See also microparasite and macroparasite.

Parasitoid An organism which lays its eggs inside or on another organism; the growth of the parasitoid ultimately kills the host. The classic examples include species of wasp (Hymenoptera) and fly (Diptera).

Parthenogenesis ('Virgin birth') The production of eggs which develop without fertilization by another individual.

Phenotype Any set of measurable characteristics of an organism manifested throughout its life (and therefore a product of both genotype and environment). Used conventionally in opposition to genotype (*c.f.*).

Phenotypic divergence Divergence of means of particular characters (such as beak length) between two or more populations.

Phylogeny The evolutionary relationships among a group of organisms.

Planktonic Living in open water.

Pleiotropy Multiple effects of alleles on more than one character. Pleiotropy leads to correlated responses to selection because allele frequency change caused by selection alters the value of all of the traits affected by a pleiotropic allele. Likewise, an allelic form of a gene with pleiotropic effects can be selected for due to its beneficial effects early in life even if this comes at the expense of deleterious effects late in life ('antagonistic pleiotropy').

Pleistocene Geological 'series' comprising most of the Quaternary (*q.v.*), its start date is controversial but in this book it is defined as starting 2.6 million years and ending 11,600 years ago.

Ploidy Number of sets of chromosomes in the genome of an organism.

Polymorphism Occurrence of different phenotypes among members of the same population.

Polyploid Possessing several haploid complements (hence triploid (3), tetraploid (4), etc.).

Post-zygotic isolation Reproductive isolation that occurs after fertilization, such as hybrid sterility and hybrid inviability.

Predator An organism which consumes another killing it in the process. It is remarkably difficult to produce a rigorous definition of predator which successfully distinguishes it from parasite (*c.f.*).

Pre-zygotic isolation Reproductive isolation that occurs before fertilization. Includes species differences in traits such as sexual behaviour, habitat preference, seasonal breeding (all pre-mating), and gametic compatibility (post-mating but pre-zygotic).

Primary Production The rate of production of biomass (*q.v.*) by autotrophs (*q.v.*).

Prisoner's Dilemma In its simplest form it is a two-player game in which players simultaneously decide whether to cooperate (C) or defect (D). The relative sizes of the pay-offs define the game, in that mutual cooperation pays more than mutual defection but defecting while your partner cooperates provides the highest pay-off and cooperating while your partner defects provides the lowest pay-off. The game has long been thought to epitomize the relative pay-offs involved in cooperation between non-relatives since it captures both the temptation to defect and the low pay-off for being a sucker (*c.f.* the 'tragedy of the commons', which arises in a multiplayer version of this game).

Prokaryote An organism (including bacteria) with a circular chromosome not localized in a nuclear envelope. One of the two great divisions ('Empires' or 'Super Kingdoms') of life on Earth, the other is the eukaryotes (*q.v.*).

Protists Organisms in the eukaryotic kingdom Protista, includes the traditional protozoa and a collection of other organisms which do not neatly fit into the plants, animals, or fungi. This kingdom is not monophyletic (*q.v.*).

Public good A resource that is costly to produce, and provides a benefit to all the individuals in the local group or population. Public goods systems are often open to exploitation by cheats who benefit, but do not pay the cost.

Quaternary The most recent geological period, in which we are still living. Its formal status and start date are both controversial; however, in this book it is defined as starting 2.6 million years ago. It is commonly subdivided into two 'series' the Pleistocene (*q.v.*) and the Holocene (*q.v.*).

Rapoport's rule The suggestion that species at higher latitudes tend to have larger range sizes and wider ecological tolerances. Named after the Mexican ecologist Eduardo Rapoport who suggested the idea in 1975.

Reactive oxygen species Oxygen-containing molecules with an unpaired electron (free radicals).

Recessive gene A gene that does not express its effect when it is present in the heterozygous state.

Reciprocal altruism A relationship in which altruism is maintained through application of the simple principle of 'you scratch my back, I'll scratch yours'. Tit-for-tat (cooperate at first and thereafter do what your partner did to you the previous round) is one of a range of competitively successful strategies which employs this principle.

Recombination The change in the relationship between loci on the same chromosome caused by crossing over (*q.v.*).

Red Queen A metaphor, derived from Lewis Carroll's 'Through the Looking-Glass' in which the Red Queen said, 'It takes all the running you can do, to keep in the same place'. Following Leigh Van Valen, it refers to coevolving systems, such as hosts and parasites, in which both parties are continually selected to deal with the other's counter-measures without any given species necessarily gaining the upper hand.

Reinforcement The adaptive strengthening of pre-zygotic isolating mechanisms in a zone of secondary contact between two distinct taxa due to the selective disadvantage of hybridization (*q.v.* character displacement).

Relatedness A measure of genetic similarity.

Reproductive isolation Absence (or severe restriction) of gene flow between populations whose members are in contact with one another.

Ribosomal RNA RNA transcribed from nuclear DNA which, along with proteins, forms cellular particles called ribosomes which are involved in protein synthesis.

Ring species Two or more reproductively isolated forms connected by a chain of interbreeding populations, typically wrapped around a geographical barrier.

Secondary plant compounds Chemicals produced by plants which are clearly not used by the plant in primary metabolism—that is, in the main energy supplying chemical processes in a cell.

Segregation The separation of pairs of alleles at meiosis and their passage into different haploid cells.

Selection Shorthand for 'natural selection', that is, consistent differences in the rate of survival or reproduction between different genotypes or alleles due to differences in the phenotypes they produce.

Selfishness A behaviour which is beneficial to the actor and costly to the recipient.

Senescence The biological process of ageing (*q.v.*).

Sex A process of genetic reorganization through meiosis and syngamy (gamete fusion), usually associated with the formation of several or many reproductive propagules (sexual reproduction).

Sexual selection Differential reproductive success resulting from competition for fertilization. Competition for fertilization occurs through direct competition between members of the same sex or through the attraction of one sex to the other.

Soma Pertaining to the body or any non-germinal cell, tissue, structure, or process (*c.f.* germ line).

Speciation The phenomenon of species formation.

Species Many different definitions exist, dependent on what can be measured and what use the concept is put to. The most conventional definition is based on the 'Biological Species Concept' (*c.f.* Morphological Species Concept) described by Ernst Mayr. Thus, species are groups of actually or potentially interbreeding natural populations, which are reproductively isolated from other such groups. Most commentators argue that speciation is characterized by substantial but not necessarily complete reproductive isolation.

Spite A behaviour which is costly to both the actor and the recipient (*c.f.* selfishness).

Stomata Opening in the surface of a plant (often in the leaves) which allows gaseous exchange between the plants interior and the surrounding atmosphere.

Strange attractor An unusual geometrical object, typically with self-similarity at a variety of spatial scales, generated when chaotic dynamics (*q.v.*) is displayed in phase space (i.e. a representation of the dynamical relationship between variables where time is implicit).

Supernova An 'exploding' star caused by the contraction of its core.

Symbiosis In its original usage this described two organisms living very closely together. Such a relationship could be mutually beneficial, harmful to one of the partners or neutral with respect to benefits. This is our preferred usage; however, some authors use it as a synonym for mutualism (*q.v.*).

Sympatric speciation Speciation occurring within a single geographical area.

Sympatry Having overlapping or coincident geographical distributions.

Syngamy Fusion of gametes (sex cells).

Taxon (pl. taxa) The named taxonomic unit to which individuals, or sets of species, are assigned. Higher taxa are those above the species level.

Taxonomy The study of the classification of organisms. Scientists who work in this area are called taxonomists.

Telomere (telos—end, meres—part) A region of repetitive DNA at the end of chromosomes involved in facilitating chromosomal replication and stability (*q.v.* Hayflick limit).

Testate amoebae Shell-building protozoa which are often common in soils and freshwater, especially in sediments of high organic matter content. This group is not monophyletic.

Tragedy of the commons A situation in which individuals would do better if they all cooperate, compared to them all defecting, but cooperation is unstable because each individual gains by selfishly pursuing their own short-term interests (*c.f.* Prisoner's Dilemma).

Two-fold cost One of the central problems that must be overcome to help explain the maintenance of sex, highlighting the apparent inefficiency of producing males. All else being equal, in a sexual population any mutation that suppressed sex would rapidly spread, because asexual females would each produce asexual offspring that are directly capable of reproducing whereas sexual females would produce a mixture of males and females.

Vegetative reproduction Any mode of reproduction not involving the production of eggs.

Zygote A single-celled individual (the fertilized egg) formed by union of gametes and their nuclei.

The Geological Time Scale

The geological time scale is split into four 'eons' which are subdivided into 'eras'; these eras are further subdivided into the more familiar geological 'periods' (systems), which can in turn be further subdivided into 'series' (epochs). In assigning dates to this time scale, we have followed the 2004 recommendations of the International Commission on Stratigraphy. The one exception to this is that we have retained the Quaternary as a full geological period starting 2.6 million years ago.

Eons

Hadean before approximately 3,800 million years ago
Archaean from approximately 3,800 to 2,500 million years ago
Proterozoic from 2,500 to 542 million years ago
Phanerozoic from 542 million years ago until the present.

The Phanerozoic in more detail

Within this book we only use formal names of eras and periods (systems) within the most recent geological era the Phanerozoic. Note that the well-known period of the 'Tertiary' no longer has formal status and is split into the Palaeogene and the Neogene—this does not stop most geologists talking about the 'Cretaceous-Tertiary (K-T) boundary' when discussing mass extinctions as old habits die hard. The subdivisions of the Phanerozoic are shown in the following table:

Era	Period (system)	Age in years (millions of years)
Cenozoic	Quaternary	0–2.6*
Cenozoic	Neogene	2.6–23.03
Cenozoic	Palaeogene	23.03–65.5
Mesozoic	Cretaceous	65.5–145.5
Mesozoic	Jurassic	145.5–199.6
Mesozoic	Triassic	199.6–251.0
Palaeozoic	Permian	251.0–299.0
Palaeozoic	Carboniferous	299.0–359.2
Palaeozoic	Devonian	359.2–416.0
Palaeozoic	Silurian	416.0–443.7
Palaeozoic	Ordovician	443.7–488.3
Palaeozoic	Cambrian	488.3–542.0

*Note that many authors use 1.8 million as the base of the Quaternary; however, 2.6 seems to be gaining increasing support. This period is subdivided into two 'series', the Pleistocene and the Holocene (see main glossary).

References

Bell, G. (1982) *The Masterpiece of Nature: The Evolution and Genetics of Sexuality*. Croom Helm.
Calow, P. (1999) *The Encyclopaedia of Ecology and Environmental Management*. Blackwell Science.

Futuyma, D.J. (1998) *Evolutionary Biology*. Third Edition, Sinauer.

Hale, W.G., Saunders, V.A., and Margham, J.P. (2005) *Collins Dictionary of Biology*. Third Edition, Harper Collins.

Lincoln, R.J., Boxshall, G.A., and Clark, P.F. (1998) *A Dictionary of Ecology, Evolution and Systematics*. Second Edition, Cambridge University Press.

Wilkinson, D.M. (2006) *Fundamental Processes in Ecology; An Earth Systems Approach*. Oxford University Press.

Geological time as exposed in the cliffs of the Grand Canyon in Arizona, USA. The rocks in the Canyon cover nearly two billion years of Earth history. Photo: DMW.

Bibliography

Chapter 1: Why Do We Age?

Ageing research is an enormous field—for example, a 'Web of Science' search of published journal articles that include the term 'senescence' reports over 17,000 papers published since 1946. We list a selection of books and journal articles that may help in providing a more comprehensive overview of the field; recent reviews are given priority. We then provide technical references supporting specific arguments.

Suggested general reading

Bell, G. (1987) *Sex and Death in Protozoa: The History of an Obsession*. Cambridge University Press.

Bonsall, M.B. (2006) Longevity and ageing: appraising the evolutionary consequences of growing old. *Philosophical Transactions of the Royal Society B-Biological Sciences* 361, 119–135.

Charlesworth, B. (1980) *Evolution in Age-Structured Populations*. Cambridge University Press.

Comfort, A. (1979) *The Biology of Senescence*. Churchill Livingstone.

Hughes, K.A. and Reynolds, R.M. (2005) Evolutionary and mechanistic theories of aging. *Annual Review of Entomology* 50, 421–445.

Kirkwood, T. (2001) *The End of Age: Why Everything About Ageing is Changing*. Profile Books.

Kirkwood, T.B.L. and Austad, S.N. (2000) Why do we age? *Nature* 408, 233–238.

Monaghan, P. *et al.* (2008) The evolutionary ecology of senescence. *Functional Ecology* 22, 371–378.

Partridge, L. and Barton, N.H. (1993) Optimality, mutation and the evolution of aging. *Nature* 362, 305–311.

Partridge, L. and Gems, D. (2002) The evolution of longevity. *Current Biology* 12, R544–R546.

Ricklefs, R.E. and Finch, C.B. (1995) *Aging: A Natural History*. Scientific American Library.

Rose, M.R. (1991) *Evolutionary Biology of Aging*. Oxford University Press.

Rose, M.R. (2005) *The Long Tomorrow: How Advances in Evolutionary Biology Can Help Us Postpone Aging*. Oxford University Press.

Volk, T. (2002) *What is Death?* John Wiley & Sons.

References

1. Stoker, B. (1897) *Dracula*. p. 148, Penguin Classics.
2. DePinho, R.A. (2000) The age of cancer. *Nature* 408, 248–254.
3. Nesse, R.M. and Williams, G.C. (1996) *Why We Get Sick: the New Science of Darwinian Medicine*. Vintage Books.
4. Wortley, J. (2001) Geriatric pathology À L'ancienne. *International Journal of Aging and Human Development* 53, 167–179.
5. Williams, G.C. (1957) Pleiotropy, natural-selection, and the evolution of senescence. *Evolution* 11, 398–411.
6. Banks, M.J. and Thompson, D.J. (1985) Lifetime mating success in the damselfly *Coenagrion puella*. *Animal Behaviour* 33, 1175–1183.
7. Finkel, T. and Holbrook, N.J. (2000) Oxidants, oxidative stress and the biology of ageing. *Nature* 408, 239–247.

8. Hughes, K.A. and Reynolds, R.M. (2005) Evolutionary and mechanistic theories of aging. *Annual Review of Entomology* 50, 421–445.
9. Levine, R.L. (2002) Carbonyl modified proteins in cellular regulation, ageing and disease. *Free Radical Biology and Medicine* 32, 790–796.
10. Arking, A. (1998) *Biology of Aging: Observations and Principles*. Sinauer Associates.
11. Lanner, R.M. and Connor, K.F. (2001) Does bristlecone pine senesce? *Experimental Gerontology* 36, 675–685.
12. Cano, R. and Monica, K.B. (1995) Revival and identification of bacterial spores in 25- to 40-million year-old Dominican amber. *Science* 268, 1060–1064.
13. http://news.bbc.co.uk/1/hi/sci/tech/7066389.stm (last accessed 27 May 2008).
14. http://news.bbc.co.uk/1/hi/world/asia-pacific/5109342.stm (last accessed 27 May 2008).
15. http://www.nature.ca/ukaliq/elem/pop/P0101-e.html (last accessed 27 May 2008).
16. Corbet, G.B. and Southern, H.N. (1977) *The Handbook of British Mammals*. Second Edition, Blackwell.
17. Comfort, A. (1979) *The Biology of Senescence*. Churchill Livingstone.
18. Herman, W.S. and Tatar, M. (2001) Juvenile hormone regulation of longevity in the migratory monarch butterfly. *Proceedings of the Royal Society B* 268, 2509–2514.
19. Law, R. *et al.* (1977) Life-history variation in *Poa annua*. *Evolution* 31, 233–246.
20. Tsubaki, Y. *et al.* (1997) Differences in adult and reproductive lifespan in the two male forms of *Mnais pruinosa costalis* Selys (Odonata: Calopterygidae). *Researches in Population Ecology* 39, 149–155.
21. Loeb, J. and Northrop, J.H. (1916) Is there a temperature coefficient for the duration of life. *Proceedings of the National Academy of Sciences USA* 2, 456–457.
22. Pearl, R. (1928) *The Rate of Living*. Alfred A. Knopf.
23. Schaffer, D. (2005) No old man ever forgot where he buried his treasure: concepts of cognitive impairment in old age circa 1700. *Journal of the American Geriatrics Society* 53, 2023–2027.
24. Bonner, J.T. (2006) *Why Size Matters: From Bacteria to Blue Whales*. Princeton University Press.
25. Partridge, L. and Gems, D. (2002) The evolution of longevity. *Current Biology* 12, R544–R546.
26. Brunet-Rossinni, A.K. and Austad, S.N. (2004) Ageing studies on bats: a review. *Biogerontology* 5, 211–222.
27. Speakman, J.R. *et al.* (2002) Living fast, dying when? The link between aging and energetics. *The Journal of Nutrition* 132 (supplement), 1583S–1597S.
28. Hayflick, L. (2000) The future of ageing. *Nature* 408, 267–269.
29. Kirkwood, T.B.L. and Austad, S.N. (2000) Why do we age? *Nature* 408, 233–238.
30. Cosgrove, J. (2006) On the record. *Nature* 439, 640.
31. Anderson, R.C. *et al.* (2002) Octopus senescence: the beginning of the end. *Journal of Applied Animal Welfare Science* 5, 275–283.
32. Silvertown, J. *et al.* (2001) Evolution of senescence in iteroparous perennial plants. *Evolutionary Ecology Research* 3, 393–412.
33. Ricklefs, R.E. (1998) Evolutionary theories of aging: confirmation of a fundamental prediction, with implications for the genetic basis and evolution of life span. *American Naturalist* 152, 24–44.
34. Gaillard, J.M. *et al.* (1994) Senescence in natural populations of mammals—a reanalysis. *Evolution* 48, 509–516.
35. Ackermann, M. *et al.* (2003) Senescence in a bacterium with asymmetric division. *Science* 300, 1920.

36. Stephens, C. (2005) Senescence: even bacteria get old. *Current Biology* 15, R308–R310.
37. Ricklefs, R.E. (2007) Tyrannosaur ageing. *Biology Letters*, 3, 214–217.
38. Bonduriansky, R. and Brassil, C.E. (2002) Rapid and costly ageing in wild male flies. *Nature* 420, 377.
39. Banks, M.J. and Thompson, D.J. (1987) Lifetime reproductive success of females of the damselfly *Coenagrion puella*. *Journal of Animal Ecology* 56, 815–832.
40. Gaillard, J.-M. *et al.* (2003) Ecological correlates of lifespan in populations of large herbivorous mammals. In *Life Span: Evolutionary, Ecological, and Demographic Perspectives* (Carey, J.R. and Tuljapurkar, S. eds), Population Council.
41. Hendry, A.P. *et al.* (2004) Adaptive variation in senescence: reproductive lifespan in a wild salmon population. *Proceedings of the Royal Society B* 271, 259–266.
42. Morbey, Y.E. *et al.* (2005) Rapid senescence in Pacific salmon. *American Naturalist* 166, 556–568.
43. Oakwood, M. *et al.* (2001) Semelparity in a large marsupial. *Proceedings of the Royal Society B* 268, 407–411.
44. Partridge, L. and Barton, N.H. (1993) Optimality, mutation and the evolution of aging. *Nature* 362, 305–311.
45. Stewart, E.J. *et al.* (2005) Aging and death in an organism that reproduces by morphologically symmetric division. *PLoS Biology* 3, e45.
46. Jazwinski, S.M. (2002) Growing old: metabolic control and yeast aging. *Annual Review of Microbiology* 56, 769–792.
47. Ackermann, M. *et al.* (2007) On the evolutionary origin of aging. *Aging Cell* 6, 235–244.
48. Rose, M.R. (1991) *Evolutionary Biology of Aging*. Oxford University Press.
49. Bell, G. (1987) *Sex and Death in Protozoa: The History of an Obsession*. Cambridge University Press.
50. Norwood, T.H. and Gray, M. (1996) The role of DNA damage in cellular aging: is it time for a reassessment? *Experimental Gerontology* 31, 61–68.
51. Olovnikov, A.M. (1996) Telomeres, telomerase, and aging: origin of the theory. *Experimental Gerontology* 31, 443–448.
52. Greider, C.W. (1998) Telomeres and senescence: the history, the experiment, the future. *Current Biology* 8, R178–R181.
53. Guarente, L. and Kenyon, C. (2000) Genetic pathways that regulate ageing in model organisms. *Nature* 408, 255–262.
54. Haussmann, M.F. *et al.* (2003) Telomeres shorten more slowly in long-lived birds and mammals than in short-lived ones. *Prroceedings of the Royal Society B* 270, 1387–1392.
55. Campisi, J. (2003) Cellular senescence and apoptosis: how cellular responses might influence aging phenotypes. *Experimental Gerontology* 38, 5–11.
56. Rose, M.R. *et al.* (2005) The effects of evolution are local: evidence from experimental evolution in *Drosophila*. *Integrative and Comparative Biology* 45, 486–491.
57. Weinert, B.T. and Timiras, P.S. (2003) Theories of aging. *Journal of Applied Physiology* 95, 1706–1716.
58. Maynard Smith, J. (1962) Review lectures on senescence. I. The causes of aging. *Proceedings of the Royal Society B* 157, 115–127.
59. Dawkins, R. (1995) *River Out of Eden*. Weidenfeld & Nicholson.
60. Medvedev, Z.A. (1990) An attempt at a rational classification of theories of ageing. *Biological Reviews of the Cambridge Philosophical Society* 65, 375–398.

61. Wallace, A.R. (1865–1870) The action of natural selection in producing old age, decay, and death. Note reproduced in *Essays Upon Heredity and Kindred Biological Problems by August Weismann* (1889).
62. Weismann, A. (1889) *Essays Upon Heredity and Kindred Biological Problems*. Clarendon Press.
63. Nuland, S.B. (1994) *How We Die: Reflections on Life's Final Chapter*. Vintage Books.
64. Mitteldorf, J. (2004) Ageing selected for its own sake. *Evolutionary Ecology Research* 6, 937–953.
65. Skulachev, V.P. (2001) The programmed death phenomena, aging and the Samuri law of biology. *Experimental Gerontology* 36, 995–1024.
66. Leslie, M. (2001) Aging research grows up. *Science's SAGE KE, 2001, 3 October*, p. oa1.
67. Mitteldorf, J. (2006) Chaotic population dynamics and the evolution of ageing. *Evolutionary Ecology Research* 8, 561–574.
68. Kirchner, J.W. and Roy, B.A. (1999) The evolutionary advantages of dying young: epidemiological implications of longevity in metapopulations. *American Naturalist* 154, 140–159.
69. Travis, J.M.J. (2004) The evolution of programmed death in a spatially structured population. *Journals of Gerontology Series A-Biological Sciences and Medical Sciences* 59, 301–305.
70. McAllistair, M.K. and Roitberg, B.D. (1987) Adaptive suicidal behaviour in pea aphids. *Nature* 328, 797–799.
71. Engelberg-Kulka, H. and Glaser, G. (1999) Addiction modules and programmed cell death and antideath in bacterial cultures. *Annual Review of Microbiology* 53, 43–70.
72. Crespi, B.J. (2001) The evolution of social behavior in microorganisms. *Trends in Ecology and Evolution* 16, 178–183.
73. Blest, A.D. (1963) Longevity, palatability and natural selection in five species of New World saturniid moth. *Nature* 197, 1183–1186.
74. Hamilton, W.D. (1964) Genetical evolution of social behaviour 2. *Journal of Theoretical Biology* 7, 17–89.
75. Andres Blanco, M. and Sherman, P.W. (2005) Maximum longevities of chemically protected and non-protected fishes, reptiles, and amphibians support evolutionary hypotheses of aging. *Mechanisms of Ageing and Development* 126, 794–803.
76. Lahdenpera, M. *et al.* (2004) Fitness benefits of prolonged post-reproductive lifespan in women. *Nature* 428, 178–181.
77. Amdam, G.V. and Page, R.P. (2005) Intergenerational transfers may have decoupled physiological and chronological age in a eusocial insect. *Ageing Research Reviews* 4, 398–408.
78. Lee, R.D. (2003) Rethinking the evolutionary theory of aging: transfers, not births, shape social species. *Proceedings of the National Academy of Sciences USA* 100, 9637–9642.
79. Barrett, P.H. *et al.* (1987) *Charles Darwin's Notebooks 1836–1844*. Cornell University Press.
80. Darwin, C. (1872) *Origin of Species by Means of Natural Selection or the Preservation of Favoured Races in the Struggle for Life*. Sixth Edition, John Murray.
81. Medawar, P.B. (1952) *An Unsolved Problem of Biology*. H.K. Lewis.
82. Medawar, P.B. (1946) Old age and natural death. *Modern Quarterly* 1, 30–56.
83. Haldane, J.B.S. (1941) *New Paths in Genetics*. George Allen & Uwin.
84. Hamilton, W.D. (1966) Moulding of senescence by natural selection. *Journal of Theoretical Biology* 12, 12–45.
85. Charlesworth, B. (1980) *Evolution in Age-Structured Populations*. Cambridge University Press.
86. Kirkwood, T.B.L. (1977) Evolution of aging. *Nature* 270, 301–304.
87. Gavrilov, L.A. and Gavrilova, N.S. (2001) The reliability theory of aging and longevity. *Journal of Theoretical Biology* 213, 527–545.

88. Gavrilov, L.A. and Gavrilova, N.S. (2006) Reliability theory of aging and longevity. In *Handbook of the Biology of Aging*, Sixth Edition (Masoro, E.J. and Austad, S.N. eds), pp. 4–42, Academic Press.
89. Cournil, A. (2006) Longevity: genetics. In *Encyclopedia of Life Sciences*, pp. 1–4, John Wiley & Sons.
90. Cournil, A. and Kirkwood, T.B.L. (2001) If you would live long, choose your parents well. *Trends in Genetics* 17, 233–235.
91. Rose, M.R. (1984) Laboratory evolution of postponed senescence in *Drosophila melanogaster*. *Evolution* 38, 1004–1010.
92. Luckinbill, L.S. *et al.* (1984) Selection for delayed senescence in *Drosophila melanogaster*. *Evolution* 38, 996–1003.
93. Zwaan, B. *et al.* (1995) Direct selection on life-span in *Drosophila melanogaster*. *Evolution* 49, 649–659.
94. Stearns, S.C. *et al.* (2000) Experimental evolution of aging, growth, and reproduction in fruitflies. *Proceedings of the National Academy of Sciences USA* 97, 3309–3313.
95. Abrams, P.A. (1993) Does increased mortality favor the evolution of more rapid senescence? *Evolution* 47, 877–887.
96. Williams, P.D. and Day, T. (2003) Antagonistic pleiotropy, mortality source interactions, and the evolutionary theory of senescence. *Evolution* 57, 1478–1488.
97. Reznick, D.N. *et al.* (2004) Effect of extrinsic mortality on the evolution of senescence in guppies. *Nature* 431, 1095–1099.
98. Kirkwood, T.B.L. (2001) Sex and ageing. *Experimental Gerontology* 36, 413–418.
99. Charmantier, A. *et al.* (2006) Quantitative genetics of age at reproduction in wild swans: support for antagonistic pleiotropy models of senescence. *Proceedings of the National Academy of Sciences USA* 103, 6587–6592.
100. Leroi, A.M. *et al.* (1994) The evolution of phenotypic life-history trade-offs—an experimental-study using *Drosophila melanogaster*. *American Naturalist* 144, 661–676.
101. Kirkwood, T.B.L. (2005) Understanding the odd science of aging. *Cell* 120, 437–447.
102. Promislow, D.E.L. (2004) Protein networks, pleiotropy and the evolution of senescence. *Proceedings of the Royal Society B* 271, 1225–1234.
103. Pletcher, S.B. *et al.* (2002) Genome-wide transcript profiles in ageing and calorically restricted *Drosophila melanogaster*. *Current Biology* 12, 712–723.
104. Rose, M.R. and Long, A.D. (2002) Ageing: the many-headed monster. *Current Biology* 12, R311–R312.
105. Rose, M.R. (2005) *The Long Tomorrow: How Advances in Evolutionary Biology Can Help Us Postpone Aging*. Oxford University Press.
106. Partridge, L. *et al.* (2005) Dietary restriction, mortality trajectories, risk and damage. *Mechanisms of Ageing and Development* 126, 35–41.
107. Partridge, L. and Gems, D. (2006) Beyond the evolutionary theory of ageing, from functional genomics to evo-gero. *Trends in Ecology and Evolution* 21, 334–340.
108. Linnen, C. *et al.* (2001) Cultural artifacts: a comparison of senescence in natural, laboratory-adapted and artificially selected lines of *Drosophila melanogaster*. *Evolutionary Ecology Research* 3, 877–888.
109. Jemielity, S. *et al.* (2005) Long live the queen: studying aging in social insects. *Age* 27, 241–248.
110. Vaupel, J.W. *et al.* (2004) The case for negative senescence. *Theoretical Population Biology* 65, 339–351.

111. Promislow, D.E.L. and Pletcher, S.D. (2002) Advice to an aging scientist. *Mechanisms of Ageing and Development* 123, 841–850.

Chapter 2: Why Sex?

Suggested general reading

Bell, G. (1982) *The Masterpiece of Nature: The Evolution and Genetics of Sexuality*. Croom Helm.

Maynard Smith, J. (1978) *The Evolution of Sex*. Cambridge University Press.

Ridley, M. (1993) *The Red Queen: Sex and the Evolution of Human Nature*. Viking Books.

West, S.A. (2002) Sex. In *Encylopedia of Evolution* (Pagel, M. ed), pp. 1022–1030, Oxford University Press.

Williams, G.C. (1975) *Sex and Evolution*. Princeton University Press.

References

1. Hamilton, W.D. (1975) Gamblers since life began: barnacles, aphids, elms. *Quarterly Review of Biology* 50, 175–180.
2. Killick, S.C. *et al.* (2006) Testing the pluralist approach to sex: the influence of environment on synergistic interactions between mutation load and parasitism in *Daphnia magna*. *Journal of Evolutionary Biology* 19, 1603–1611.
3. Olsen, M.W. (1966) Segregation and replication of chromosomes in turkey parthenogenesis. *Nature* 212, 435–436.
4. Maynard Smith, J. (1971) The origin and maintenance of sex. In *Group Selection* (Williams, G.C. ed), pp. 163–171, Aldine-Atherton.
5. Maynard Smith, J. (1998) *Evolutionary Genetics*. Second Edition, Oxford University Press.
6. Otto, S.P. and Lenormand, T. (2002) Resolving the paradox of sex and recombination. *Nature Reviews Genetics* 3, 252–261.
7. Hughes, R.N. (1989) *A Functional Biology of Clonal Animals*. Chapman & Hall.
8. Cordero-Rivera, A. *et al.* (2005) Parthenogenetic *Ischnura hastata* (Say) widespread in the Azores (Zygoptera: Coenagrionidae). *Odonatologica* 34, 1–9.
9. Sherratt, T.N. and Beatty, C.D. (2005) Island of the clones. *Nature* 435, 1039–1040.
10. Cavalier-Smith, T. (2002) Origins of the machinery of recombination and sex. *Heredity* 88, 125–141.
11. Ramesh, M.A. *et al.* (2005) A phylogenomic inventory of meiotic genes: evidence for sex in *Giardia* and an early eukaryotic origin of meiosis. *Current Biology* 15, 185–191.
12. Judson, O.P. and Normark, B.B. (1996) Ancient asexual scandals. *Trends in Ecology and Evolution* 11, 41–46.
13. Hayden, E.C. (2008) Scandal! Sex-starved and still surviving! *Nature* 492, 678–680.
14. Dawkins, R. (2005) *The Ancestor's Tale: A Pilgrimage to the Dawn of Life*. Weidenfeld & Nicolson.
15. Bell, G. (1982) *The Masterpiece of Nature: The Evolution and Genetics of Sexuality*. Croom Helm.
16. Williams, G.C. (1975) *Sex and Evolution*. Princeton University Press.
17. Dagg, J.L. (2006) Could sex be maintained through harmful males? *Oikos* 112, 232–235.
18. Footman, T. ed (2001) *Guinness World Records*. Guiness Records.
19. Bernstein, H. *et al.* (1981) Evolution of sexual reproduction: importance of DNA repair, complementation, and variation. *American Naturalist* 117, 537–549.

20. Bernstein, H. *et al.* (1988) Is meiotic recombination an adaptation for repairing DNA, producing genetic variation, or both? In *Evolution of Sex: An Examination of Current Ideas* (Michod, R.E. and Levin, B.R. eds), pp. 106–125, Sinauer.
21. Maynard Smith, J. and Szathmary, E. (1997) *Major Transitions in Evolution*. Oxford University Press.
22. Jokela, J. *et al.* (1997) Evidence for a cost of sex in the freshwater snail *Potamopyrgus antipodarum*. *Ecology* 78, 452–460.
23. Kumpulainen, T. *et al.* (2004) Parasites and sexual reproduction in psychid moths. *Evolution* 58, 1511–1520.
24. Weismann, A. (1889) *Essays Upon Heredity and Kindred Biological Problems*. Clarendon Press.
25. Weismann, A. (1904) *The Evolution Theory*. Edward Arnold.
26. Muller, H.J. (1932) Some genetic aspects of sex. *American Naturalist* 66, 118–138.
27. Fisher, R.A. (1930) *The Genetical Theory of Natural Selection*. Oxford University Press.
28. Peck, J.R. (1994) A ruby in the rubbish: beneficial mutations and the evolution of sex. *Genetics* 137, 597–606.
29. Nunney, L. (1989) The maintenance of sex by group selection. *Evolution* 43, 245–257.
30. Bulmer, M. (1994) *Theoretical Evolutionary Ecology*. Sinauer Associates.
31. Stearns, S.C. and Hoekstra, R.F. (2005) *Evolution: An Introduction*. Oxford University Press.
32. Kondrashov, A.S. (1993) Classification of hypotheses on the advantage of amphimixis. *Journal of Heredity* 84, 372–387.
33. Silvertown, J.W. and Lovett Dorst, J. (1993) *Introduction to Plant Population Biology*. Blackwell.
34. Colgrave, N. (2002) Sex releases the speed limit on evolution. *Nature* 420, 664–666.
35. Muller, H.J. (1964) The relation of recombination to mutational advance. *Mutation Research* 1, 2–9.
36. Kondrashov, A.S. (1988) Deleterious mutations and the evolution of sexual reproduction. *Nature* 336, 435–440.
37. Charlesworth, B. (1990) Mutation–selection balance and the evolutionary advantage of sex and recombination. *Genetical Research* 55, 199–221.
38. Keightley, P.D. and Eyre-Walker, A. (2000) Deleterious mutations and the evolution of sex. *Science* 290, 331–333.
39. Haag-Liautard, C. *et al.* (2007) Direct estimation of per nucleotide and genomic deleterious mutation rates in *Drosophila*. *Nature* 445, 82–85.
40. Maynard Smith, J. (1971) What use is sex? *Journal of Theoretical Biology* 30, 319–335.
41. Ghiselin, M.T. (1974) *The Economy of Nature and the Evolution of Sex*. University of California Press.
42. Kelley, S.E. *et al.* (1988) A test of the short-term advantage of sexual reproduction. *Nature* 331, 714–716.
43. Koella, J.C. (1988) The tangled bank: the maintenance of sexual reproduction through competitive interactions. *Journal of Evolutionary Biology* 2, 95–116.
44. Doncaster, C.P. *et al.* (2000) The ecological cost of sex. *Nature* 404, 281–285.
45. Van Valen, L. (1973) A new evolutionary law. *Evolutionary Theory* 1, 1–30.
46. Wilkinson, D.M. (2000) Running with the Red Queen: reflections on 'sex versus non-sex versus parasite'. *Oikos* 91, 589–596.
47. Carius, H.J. *et al.* (2001) Genetic variation in a host–parasite association: potential for coevolution and frequency-dependent selection. *Evolution* 55, 1136–1145.
48. Lively, C.M. and Dybdahl, M.F. (2000) Parasite adaptation to locally common host genotypes. *Nature* 405, 679–681.

49. Baer, B. and Schmid-Hempel, P. (1999) Experimental variation in polyandry affects parasite loads and fitness in a bumble-bee. *Nature* 397, 151–154.
50. Jokela, J. (2001) Sex: advantage. In *Encyclopedia of Life Sciences*, pp. 1–6, John Wiley & Sons.
51. Agrawal, A.F. (2006) Similarity selection and the evolution of sex: revisiting the red queen. *PloS Biology* 4, 1364–1371.
52. Otto, S.P. and Nuismer, S.L. (2004) Species interactions and the evolution of sex. *Science* 304, 1018–1020.
53. Lively, C.M. (1987) Evidence from a New Zealand snail for the maintenance of sex by parasitism. *Nature* 328, 519–521.
54. Mee, J.A. and Rowe, L. (2006) A comparison of parasite loads on asexual and sexual *Phoxinus* (Pisces: Cyprinidae). *Canadian Journal of Zoology-Revue Canadienne De Zoologie* 84, 808–816.
55. Moritz, C. *et al.* (1991) Parasite loads in parthenogenetic and sexual lizards (*Heteronotia binoei*): support for the Red Queen hypothesis. *Proceedings of the Royal Society B* 224, 145–149.
56. Goddard, M.R. *et al.* (2005) Sex increases the efficacy of natural selection in experimental yeast populations. *Nature* 434, 636–640.
57. Burt, A. and Bell, G. (1987) Mammalian chiasma frequencies as a test of two theories of recombination. *Nature* 326, 803–805.
58. Rice, W.R. (2002) Experimental tests of the adaptive significance of sexual recombination. *Nature Reviews Genetics* 3, 241–251.
59. West, S.A. *et al.* (1999) A pluralist approach to sex and recombination. *Journal of Evolutionary Biology* 12, 1003–1012.
60. Kondrashov, A.S. (1999) Being too nice may be not too wise. *Journal of Evolutionary Biology* 12, 1031.
61. Kirkpatrick, M. and Jenkins, C.D. (1989) Genetic segregation and the maintenance of sexual reproduction. *Nature* 339, 300–301.

Chapter 3: Why Cooperate?

Suggested general reading

Axelrod, R. (1984) *The Evolution of Cooperation*. Basic Books.

Axelrod, R. and Hamilton, W.D. (1981) The evolution of cooperation. *Science* 211, 1390–1391.

Dugatkin, L.A. (1997) *Cooperation among Animals: An Evolutionary Perspective*. Oxford University Press.

Dugatkin, L.A. (2006) *The Altruism Equation: Seven Scientists Search for the Origins of Goodness*. Princeton University Press.

Fehr, E. and Gächter, S. (2002) Altruistic punishment in humans. *Nature* 415, 137–140.

Nowak, M.A. (2006) Five rules for the evolution of cooperation. *Science* 314, 1560–1563.

Nowak, M.A. and Sigmund, K. (2005) Evolution of indirect reciprocity. *Nature* 437, 1291–1298.

Ridley, M. (1997) *The Origins of Virtue*. Viking Books.

Trivers, R.L. (1971) Evolution of reciprocal altruism. *Quarterly Review of Biology* 46, 35–57.

References

1. Mandeville, B. (1732) *The Fable of the Bees or Private Vices, Publick Benefits*. Clarendon Press, Oxford (http://oll.libertyfund.org/EBooks/Mandeville_0014.01.pdf).
2. Nowak, M.A. (2006) Five rules for the evolution of cooperation. *Science* 314, 1560–1563.

3. West, S.A. *et al.* (2007) Social semantics: altruism, cooperation, mutualism, strong reciprocity and group selection. *Journal of Evolutionary Biology* 20, 415–432.
4. Clutton-Brock, T.H. *et al.* (2001) Cooperation, control, and concession in meerkat groups. *Science* 291, 478–481.
5. Wilkinson, G.S. (1984) Reciprocal food sharing in the vampire bat. *Nature* 308, 181–184.
6. Henzi, S.P. *et al.* (1997) Cohort size and the allocation of social effort by female mountain baboons. *Animal Behaviour* 54, 1235–1243.
7. Engelberg-Kulka, H. and Glaser, G. (1999) Addiction modules and programmed cell death and antideath in bacterial cultures. *Annual Review of Microbiology* 53, 43–70.
8. Crespi, B.J. (2001) The evolution of social behavior in microorganisms. *Trends in Ecology and Evolution* 16, 178–183.
9. May, R.M. (2006) Threats to tomorrow's world. Address of the President, Lord May of Oxford OM AC FRS, given at the anniversary meeting on 30 November 2005. *Notes and Records of the Royal Society* 60, 109–130.
10. Colman, A.M. (2006) The puzzle of cooperation (book review). *Nature* 440, 744–745.
11. Mesterton-Gibbons, M. and Dugatkin, L.A. (1992) Cooperation among unrelated individuals—evolutionary factors. *Quarterly Review of Biology* 67, 267–281.
12. Brembs, B. (1996) Chaos, cheating and cooperation: potential solutions to the Prisoner's Dilemma. *Oikos* 76, 14–24.
13. Dugatkin, L.A. (2002) Cooperation in animals: an evolutionary overview. *Biology and Philosophy* 17, 459–476.
14. Darwin, C. (1859) *On the Origin of Species by Means of Natural Selection, or the Preservation of Favoured Races in the Struggle for Life.* First Edition, John Murray, London.
15. Maynard Smith, J. (1975) Survival through suicide. *New Scientist* 28, 496.
16. Haldane, J.B.S. (1955) Population genetics. *New Biology* 18, 34–51.
17. Fisher, R.A. (1930) *The Genetical Theory of Natural Selection.* Clarendon Press.
18. Hamilton, W.D. (1964) Genetical evolution of social behaviour I. *Journal of Theoretical Biology* 7, 1–16.
19. Hamilton, W.D. (1964) Genetical evolution of social behaviour 2. *Journal of Theoretical Biology* 7, 17–89.
20. Hamilton, W.D. (1963) The evolution of altruistic behaviour. *American Naturalist* 97, 354–356.
21. Dugatkin, L.A. (2006) *The Altruism Equation: Seven Scientists Search for the Origins of Goodness.* Princeton University Press.
22. Rice, S.H. (2004) *Evolutionary Theory: Mathematical and Conceptual Foundation.* Sinauer.
23. West, S.A. *et al.* (2001) Testing Hamilton's rule with competition between relatives. *Nature* 409, 510–513.
24. Dawkins, R. (1979) 12 Misunderstandings of kin selection. *Zeitschrift Fur Tierpsychologie-Journal of Comparative Ethology* 51, 184–200.
25. Gribbin, J. and Cherfas, J. (1982) *Monkey Puzzle.* Bodley Head, UK.
26. Quotation by May, R.M. (2000) End of the beginning. *New Scientist* No. 2245, 1 July 2000.
27. Plomin, R. *et al.* (1997) *Behavioral Genetics: A Primer.* Third Edition, W.H. Freeman & Co.
28. Maynard Smith, J. and Szathmary, E. (1995) *The Major Transitions in Evolution.* Oxford University Press.
29. Dawkins, R. (1976) *The Selfish Gene.* Oxford University Press.
30. Keller, L. and Ross, K.G. (1998) Selfish genes: a green beard in the red fire ant. *Nature* 394, 573–575.

31. Roberts, G. and Sherratt, T.N. (2002) Behavioural evolution—does similarity breed cooperation? *Nature* 418, 499–500.
32. Hudson, R.E. *et al.* (2002) Altruism, cheating, and anticheater adaptations in cellular slime molds. *American Naturalist* 160, 31–43.
33. Fortunato, A. *et al.* (2003) Co-occurrence in nature of different clones of the social amoeba, *Dictyostelium discoideum*. *Molecular Ecology* 12, 1031–1038.
34. Sherman, P.W. (1977) Nepotism and evolution of alarm calls. *Science* 197, 1246–1253.
35. Sherman, P.W. (1985) Alarm calls of belding ground-squirrels to aerial predators—nepotism or self-preservation. *Behavioral Ecology and Sociobiology* 17, 313–323.
36. Pfennig, D.W. *et al.* (1993) Kin recognition and cannibalism in spadefoot toad tadpoles. *Animal Behaviour* 46, 87–94.
37. Wilson, E.O. (1995) *Naturalist*. Warner Books.
38. Queller, D.C. and Strassmann, J.E. (2002) Kin selection. *Current Biology* 12, R832–R832.
39. Queller, D.C. and Strassmann, J.E. (1998) Kin selection and social insects. *Bioscience* 48, 165–175.
40. Wenseleers, T. and Ratnieks, F.L.W. (2006) Enforced altruism in insect societies. *Nature* 444, 50.
41. Lewis, S. *et al.* (2007) Fitness increases with partner and neighbour allopreening. *Biology Letters* 3, 386–389.
42. Clutton-Brock, T. (2002) Behavioral ecology—breeding together: kin selection and mutualism in cooperative vertebrates. *Science* 296, 69–72.
43. Griffin, A.S. and West, S.A. (2003) Kin discrimination and the benefit of helping in cooperatively breeding vertebrates. *Science* 302, 634–636.
44. Wilkinson, G.S. (1988) Reciprocal altruism in bats and other mammals. *Ethology and Sociobiology* 9, 85–100.
45. Hart, B.L. and Hart, L.A. (1992) Reciprocal allogrooming in impala, *Aepyceros melampus*. *Animal Behaviour* 44, 1073–1083.
46. Trivers, R.L. (1971) Evolution of reciprocal altruism. *Quarterly Review of Biology* 46, 35–57.
47. Axelrod, R. (1984) *The Evolution of Cooperation*. Basic Books.
48. Poundstone, W. (1993) *Prisoner's Dilemma: John Von Neumann, Game Theory and the Puzzle of the Bomb*. Anchor Books, Doubleday.
49. Axelrod, R. and Hamilton, W.D. (1981) The evolution of cooperation. *Science* 211, 1390–1396.
50. Nowak, M.A. (2006) *Evolutionary Dynamics: Exploring the Equations of Life*. Harvard University Press.
51. Nowak, M. and Sigmund, K. (1993) A strategy of win stay, lose shift that outperforms tit-for-tat in the prisoners-dilemma game. *Nature* 364, 56–58.
52. Frean, M.R. (1994) The Prisoners Dilemma without synchrony. *Proceedings of the Royal Society B* 257, 75–79.
53. Roberts, G. and Sherratt, T.N. (1998) Development of cooperative relationships through increasing investment. *Nature* 394, 175–179.
54. Doebeli, M. and Knowlton, N. (1998) The evolution of interspecific mutualisms. *Proceedings of the National Academy of Sciences USA* 95, 8676–8680.
55. Noë, R. and Hammerstein, P. (1995) Biological markets. *Trends in Ecology and Evolution* 10, 336–339.
56. Noë, R. and Hammerstein, P. (1994) Biological markets: supply-and-demand determine the effect of partner choice in cooperation, mutualism and mating. *Behavioral Ecology and Sociobiology* 35, 1–11.

57. Roberts, G. (1998) Competitive altruism: from reciprocity to the handicap principle. *Proceedings of the Royal Society B* 265, 427–431.
58. Noë, R. (1990) A veto game played by baboons—a challenge to the use of the prisoners-dilemma as a paradigm for reciprocity and cooperation. *Animal Behaviour* 39, 78–90.
59. Connor, R.C. (1995) Altruism among non-relatives—alternatives to the Prisoners Dilemma. *Trends in Ecology and Evolution* 10, 84–86.
60. Nowak, M.A. *et al.* (1994) More spatial games. *International Journal of Bifurcation and Chaos* 4, 33–56.
61. Nowak, M.A. *et al.* (1994) Spatial games and the maintenance of cooperation. *Proceedings of the National Academy of Sciences USA* 91, 4877–4881.
62. Rainey, P.B. and Travisano, M. (1998) Adaptive radiation in a heterogeneous environment. *Nature* 394, 69–72.
63. Knight, C.G. *et al.* (2006) Unraveling adaptive evolution: how a single point mutation affects the protein coregulation network. *Nature Genetics* 38, 1015–1022.
64. Rainey, P.B. and Rainey, K. (2003) Evolution of cooperation and conflict in experimental bacterial populations. *Nature* 425, 72–74.
65. Xavier, J.B. and Foster, K.R. (2007) Cooperation and conflict in microbial biofilms. *Proceedings of the National Academy of Sciences USA* 104, 876–881.
66. Hauser, M.D. *et al.* (2003) Give unto others: genetically unrelated cotton-top tamarin monkeys preferentially give food to those who altruistically give food back. *Proceedings of the Royal Society B* 270, 2363–2370.
67. Connor, R.C. (1995) Impala allogrooming and the parceling model of reciprocity. *Animal Behaviour* 49, 528–530.
68. Olendorf, R. *et al.* (2004) Cooperative nest defence in red-winged blackbirds: reciprocal altruism, kinship or by-product mutualism? *Proceedings of the Royal Society B* 271, 177–182.
69. Hardin, G. (1968) The tragedy of the commons. *Science* 162, 1243–1248.
70. Rackham, O. (1986) *The History of the Countryside.* Dent.
71. Binmore, K.G. (1994) *Game Theory and the Social Contract.* MIT Press.
72. Nowak, M.A. and Sigmund, K. (2005) Evolution of indirect reciprocity. *Nature* 437, 1291–1298.
73. Rattigan, T. (1995 Edition) *The Winslow Boy.* Nick Hern Books.
74. Bateson, M. *et al.* (2006) Cues of being watched enhance cooperation in a real-world setting. *Biology Letters* 2, 412–414.
75. Alexander, R.D. (1987) *The Biology of Moral Systems.* Aldine de Gruyter.
76. Nowak, M.A. and Sigmund, K. (1998) Evolution of indirect reciprocity by image scoring. *Nature* 393, 573–577.
77. Leimar, O. and Hammerstein, P. (2001) Evolution of cooperation through indirect reciprocity. *Proceedings of the Royal Society B* 268, 745–753.
78. Nowak, M.A. and Sigmund, K. (2000) Shrewd investments. *Science* 288, 819–820.
79. Wedekind, C. and Braithwaite, V.A. (2002) The long-term benefits of human generosity in indirect reciprocity. *Current Biology* 12, 1012–1015.
80. Vogel, G. (2004) The evolution of the golden rule. *Science* 303, 1128–1131.
81. Milinski, M. *et al.* (2001) Cooperation through indirect reciprocity: image scoring or standing strategy? *Proceedings of the Royal Society B* 268, 2495–2501.
82. Wedekind, C. and Milinski, M. (2000) Cooperation through image scoring in humans. *Science* 288, 850–852.
83. Bshary, R. (2002) Biting cleaner fish use altruism to deceive image-scoring client reef fish. *Proceedings of the Royal Society B* 269, 2087–2093.

84. Bshary, R. and Grutter, A.S. (2006) Image scoring and cooperation in a cleaner fish mutualism. *Nature* 441, 975–978.
85. McGregor, P.K. (2005) *Animal Communication Networks*. Cambridge University Press.
86. Fehr, E. and Fischbacher, U. (2003) The nature of human altruism. *Nature* 425, 785–791.
87. de Waal, F.B.M. and Berger, M.L. (2000) Payment for labour in monkeys. *Nature* 404, 563.
88. Brosnan, S.F. and de Waal, F.B.M. (2003) Monkeys reject unequal pay. *Nature* 425, 297–299.
89. Flack, J.C. *et al.* (2006) Policing stabilizes construction of social niches in primates. *Nature* 439, 426–429.
90. Wedekind, C. *et al.* (1995) MHC-dependent mate preferences in humans. *Proceedings of the Royal Society B* 260, 245–249.
91. Sherratt, T.N. and Beatty, C.D. (2003) The evolution of warning signals as reliable indicators of prey defense. *American Naturalist* 162, 377–389.
92. Fehr, E. and Gächter, S. (2002) Altruistic punishment in humans. *Nature* 415, 137–140.
93. Gintis, H. (2000) Strong reciprocity and human sociality. *Journal of Theoretical Biology* 206, 169–179.
94. Bowles, S. and Gintis, H. (2004) The evolution of strong reciprocity: cooperation in heterogeneous populations. *Theoretical Population Biology* 65, 17–28.
95. Dreber, A. *et al.* (2008) Winners don't punish. *Nature* 452, 348–351.
96. Hammerstein, P. (1995) A two-fold tragedy unfolds. *Nature* 377, 478.
97. Wong, M.Y.L. *et al.* (2007) The threat of punishment enforces peaceful cooperation and stabilizes queues in a coral-reef fish. *Proceedings of the Royal Society B* 274, 1093–1099.
98. Fehr, E. (2004) Don't lose your reputation. *Nature* 432, 449–450.
99. Panchanathan, K. and Boyd, R. (2004) Indirect reciprocity can stabilize cooperation without the second-order free rider problem. *Nature* 432, 499–502.
100. Morell, V. (1995) Behavioral ecology—cowardly lions confound cooperation theory. *Science* 269, 1216–1217.
101. Heinsohn, R. and Packer, C. (1995) Complex cooperative strategies in group-territorial African lions. *Science* 269, 1260–1262.
102. Barnard, C.J. and Sibly, R.M. (1981) Producers and scroungers—a general-model and its application to captive flocks of house sparrows. *Animal Behaviour* 29, 543–550.
103. Koops, M.A. and Giraldeau, L.A. (1996) Producer–scrounger foraging games in starlings: a test of rate-maximizing and risk-sensitive models. *Animal Behaviour* 51, 773–783.
104. Backwell, P.R.Y. and Jennions, M.D. (2004) Coalition among male fiddler crabs. *Nature* 430, 417.
105. Getty, T. (1987) Dear enemies and the Prisoner's Dilemma: why should territorial neighbors form defensive coalitions? *American Zoologist* 27, 327–336.
106. Schuessler, R. (1989) Exit threats and cooperation under anonymity. *Journal of Conflict Resolution* 33, 728–749.
107. Sherratt, T.N. and Roberts, G. (1998) The evolution of generosity and choosiness in cooperative exchanges. *Journal of Theoretical Biology* 193, 167–177.
108. Connor, R.C. (1986) Pseudo-reciprocity: investing in mutualism. *Animal Behaviour* 34, 1562–1584.
109. Leimar, O. and Connor, R.C. (2003) By-product benefits, reciprocity, and pseudoreciprocity in mutualism. In *Genetic and Cultural Evolution of Cooperation* (Hammerstein, P. ed), MIT Press.
110. Leimar, O. and Axén, A.H. (1993) Strategic behaviour in an interspecific mutualism: interactions between lycaenid larvae and ants. *Animal Behaviour* 46, 1177–1182.

Chapter 4: Why Species?

Suggested general reading

The field of speciation is enormous, but thankfully there have been some excellent reviews over the past decade. We particularly recommend the following.

Coyne, J.A. and Orr, H.A. (2004) *Speciation*. Sinauer Associates.

Goldschmidt, T. (1997) *Darwin's Dreampond: Drama in Lake Victoria*. MIT Press.

Haffer, J. (2008) *Ornithology, Evolution and Philosophy; The Life and Science of Emst Mayr 1904–2005*. Springer.

Price, T. (2008) *Speciation in Birds*. Roberts & Company.

Schilthuizen, M. (2001) *Frogs, Flies and Dandelions: The Making of Species*. Oxford University Press.

Schluter, D. (2000) *The Ecology of Adaptive Radiation*. Oxford University Press.

Wood, T.E. and Rieseberg, L.H. (2005) Speciation: introduction. In *Encyclopedia of the Life Sciences*. John Wiley & Sons.

References

1. Darwin, C. (1859) *On the Origin of Species by Means of Natural Selection, or the Preservation of Favoured Races in the Struggle for Life*. First Edition, John Murray.
2. Mayr, E. (1949) Speciation and systematics. In *Genetics, Paleontology, and Evolution* (Jepsen, G.L. *et al.* eds), pp. 281–298, Princeton University Press.
3. Mayr, E. (1996) What is a species, and what is not? *Philosophy of Science* 63, 262–277.
4. Dobzhansky, T. (1937) *Genetics and the Origin of Species*. First Edition, Columbia University Press.
5. Diamond, J.M. (1966) Zoological classification system of a primitive people. *Science* 151, 1102–1104.
6. Schilthuizen, M. (2001) *Frogs, Flies and Dandelions: The Making of Species*. Oxford University Press.
7. Mousson, A. (1849) Über die land- und süsswasser-mollusken von Java. *Mittheilungen der Naturforschenden Gesellschaft in Zürich* 30, 264–273.
8. Mayr, E. (1942) *Systematics and the Origin of Species*. Columbia University Press.
9. Coyne, J.A. and Orr, H.A. (2004) *Speciation*. Sinauer Associates.
10. Cipriano, F. and Palumbi, S.R. (1999) Genetic tracking of a protected whale. *Nature* 397, 307–308.
11. Mattern, M.Y. and McLennan, D.A. (2000) Phylogeny and speciation of felids. *Cladistics* 16, 232–253.
12. Mallet, J. (2007) Hybrid speciation. *Nature* 446, 279–283.
13. Grant, P.R. and Grant, B.R. (1992) Hybridization of bird species. *Science* 256, 193–197.
14. Haas, F. and Brodin, A. (2005) The crow *Corvus corone* hybrid zone in southern Denmark and northern Germany. *Ibis* 147, 649–656.
15. Brodin, A. and Haas, F. (2006) Speciation by perception. *Animal Behaviour* 72, 139–146.
16. Hofman, S. and Szymura, J.M. (2007) Limited mitochondrial DNA introgression in a *Bombina* hybrid zone. *Biological Journal of the Linnean Society* 81, 295–306.
17. Vorndran, I.C. *et al.* (2002) Does differential susceptibility to predation in tadpoles stabilize the *Bombina* hybrid zone? *Ecology* 83, 1648–1659.
18. Mayr, E. and Diamond, J.M. (2001) *The Birds of Northern Melanesia*. Oxford University Press.
19. Brown, D.M. *et al.* (2007) Extensive population genetic structure in the giraffe. *BMC Biology* 5, 57.

20. Irwin, D.E. *et al.* (2001) Speciation in a ring. *Nature* 409, 333–337.
21. Irwin, D.E. *et al.* (2005) Speciation by distance in a ring species. *Science* 307, 414–416.
22. Hey, J. (2006) On the failure of modern species concepts. *Trends in Ecology and Evolution* 21, 447–450.
23. Mallet, J. (1995) A species definition for the modern synthesis. *Trends in Ecology and Evolution* 10, 294–299.
24. Roberts, M.S. and Cohan, F.M. (1995) Recombination and migration rates in natural populations of *Bacillus subtilis* and *Bacillus mojavensis*. *Evolution* 49, 1081–1094.
25. Cohan, F.M. (2002) What are bacterial species? *Annual Review of Microbiology* 56, 457–487.
26. Fraser, C. *et al.* (2007) Recombination and the nature of bacterial speciation. *Science* 315, 476–480.
27. Maynard Smith, J.M. *et al.* (1991) Localized sex in bacteria. *Nature* 349, 29–31.
28. Gevers, D. *et al.* (2005) Re-evaluating prokaryotic species. *Nature Reviews Microbiology* 3, 733–739.
29. Green, J. and Bohannan, B.J.M. (2006) Spatial scaling of microbial biodiversity. *Trends in Ecology and Evolution* 21, 501–507.
30. Vellai, T. *et al.* (1999) Genome economization and a new approach to the species concept in bacteria. *Proceedings of the Royal Society B* 266, 1953–1958.
31. Nee, S. and Colgrave, N. (2006) Paradox of the clumps. *Nature* 441, 417–418.
32. Margulis, L. and Schwartz, K.V. (1998) *Five Kingdoms*. Third Edition, W.H. Freeman.
33. Mignot, J.-P. and Raikov, I.B. (1992) Evidence for meiosis in the testate amoebae *Arcella*. *Journal of Protozoology* 39, 287–289.
34. Cash, J. and Hopkinson, J. (1905) *The British Freshwater Rhizopoda and Helozoa*. Vol. 1, The Ray Society.
35. Wanner, M. (1999) A review on the variability of testate amoebae: methodological approaches, environmental influences and taxonomic implications. *Acta Protozoologica* 38, 15–29.
36. Benton, M.J. and Pearson, P.N. (2001) Speciation in the fossil record. *Trends in Ecology and Evolution* 16, 405–411.
37. Sites, J.W. Jr. and Marshall, J.C. (2004) Operational criteria for delimiting species. *Annual Review of Ecology Evolution and Systematics* 35, 199–227.
38. Rieseberg, L.H. *et al.* (2006) The nature of plant species. *Nature* 440, 524–527.
39. Wagner, M. (1873) *The Darwinian Theory and the Law of Migration of Organisms*. Edward Stanford.
40. Wagner, M. (1889) *Die Entstehung der Arten durch raumliche Sinderung* (published posthumously). Gesammelte Aufsatze.
41. Leavitt, R.G. (1907) The geographic distribution of closely related species. *American Naturalist* 41, 207–240.
42. Wallace, A.R. (1876) *The Geographical Distribution of Animals*. Vol. 1, Macmillan.
43. Ayres, J.M. and Clutton-Brock, T.H. (1992) River boundaries and species range size in Amazonian primates. *American Naturalist* 140, 531–537.
44. Jordan, D.S. (1908) The law of geminate species. *American Naturalist* 42, 73–80.
45. Jordan, D.S. and Kellogg, V.L. (1907) *Evolution and Animal Life. An Elementary Discussion of Facts, Processes, Laws and Theories Relating to the Life and Evolution of Animals*. D. Appleton & Co.
46. Knowlton, N. *et al.* (1993) Divergence in proteins, mitochondrial DNA, and reproductive compatibility across the Isthmus of Panama. *Science* 260, 1629–1632.
47. Knowlton, N. and Weigt, L.A. (1998) New dates and new rates for divergence across the Isthmus of Panama. *Proceedings of the Royal Society* B 265, 2257–2263.

48. Lindstrom, S.C. (2001) The Bering Strait connection: dispersal and speciation in boreal macroalgae. *Journal of Biogeography* 28, 243–251.
49. Hughes, C. and Eastwood, R. (2006) Island radiation on a continental scale: exceptional rates of plant diversification after uplift of the Andes. *Proceedings of the National Academy of Sciences USA* 103, 10334–10339.
50. Barraclough, T.G. and Vogler, A.P. (2000) Detecting the geographical pattern of speciation from species-level phylogenies. *American Naturalist* 155, 419–434.
51. Wood, T.E. and Rieseberg, L.H. (2005) Speciation: introduction. In *Encyclopedia of Life Sciences*, John Wiley & Sons.
52. Rundle, H.D. *et al.* (1998) Single founder-flush events and the evolution of reproductive isolation. *Evolution* 52, 1850–1855.
53. Mooers, A.O. *et al.* (1999) The effects of selection and bottlenecks on male mating success in peripheral isolates. *American Naturalist* 153, 437–444.
54. Van Gossum, H. *et al.* (2007) Reproductive interference between *Nehalennia* damselfly species. *Ecoscience* 14, 1–7.
55. Dodd, D.M.B. (1989) Reproductive isolation as a consequence of adaptive divergence in *Drosophila pseudobscura. Evolution* 43, 1308–1311.
56. Rice, W.R. and Hostert, E.E. (1993) Laboratory experiments on speciation: what have we learned in 40 years? *Evolution* 47, 1637–1653.
57. Hoskin, C.J. *et al.* (2005) Reinforcement drives rapid allopatric speciation. *Nature* 437, 1353–1356.
58. Brown, W.L. Jr. and Wilson, E.O. (1956) Character displacement. *Systematic Zoology* 5, 49–64.
59. Dayan, T. and Simberloff, D. (2005) Ecological and community-wide character displacement: the next generation. *Ecology Letters* 8, 875–894.
60. Losos, J.B. (2000) Ecological character displacement and the study of adaptation. *Proceedings of the National Academy of Sciences USA* 97, 5693–5695.
61. Grant, P.R. and Grant, B.R. (2006) Evolution of character displacement in Darwin's finches. *Science* 313, 224–226.
62. Kondrashov, A.S. (2001) Speciation: Darwin revisited. Book review of: *Frogs, Flies, and Dandelions. The Making of Species* by Menno Schilthuizen. *Trends in Ecology and Evolution* 16, 412.
63. Dobzhansky, T. (1964) Biology, molecular and organismic. *American Zoologist* 4, 443–452.
64. Dobzhansky, T. (1973) Nothing in biology makes sense except in the light of evolution. *The American Biology Teacher* 35, 125–129.
65. Ritchie, M.G. (2001) Speciation: sympatric and parapatric. In *Encyclopedia of Life Sciences*, John Wiley & Sons.
66. May, R.M. (1988) How many species are there on earth? *Science* 241, 1441–1449.
67. Bush, G.L. (1969) Sympatric host race formation and speciation in frugivorous flies of the genus *Rhagoletis* (Diptera, Tephritidae). *Evolution* 23, 237–251.
68. Filchak, K.E. *et al.* (2000) Natural selection and sympatric divergence in the apple maggot *Rhagoletis pomonella. Nature* 407, 739–742.
69. Barron, A.B. (2001) The life and death of Hopkins' host-selection principle. *Journal of Insect Behaviour* 14, 725–737.
70. Linn, C. *et al.* (2003) Fruit odor discrimination and sympatric host race formation in *Rhagoletis. Proceedings of the National Academy of Sciences USA* 100, 11490–11493.
71. Feder, J.L. *et al.* (1994) Host fidelity is an effective premating barrier between sympatric races of the apple maggot fly. *Proceedings of the National Academy of Sciences USA* 91, 7990–7994.
72. Feder, J.L. *et al.* (2003) Allopatric genetic origins for sympatric host–plant shifts and race formation in *Rhagoletis. Proceedings of the National Academy of Sciences USA* 100, 10314–10319.

73. Kjellberg, F. *et al.* (1987) The stability of the symbiosis between dioecious figs and their pollinators: a study of *Ficus carica L.* and *Blastophaga psenes L. Evolution* 41, 693–704.
74. Barton, N.H. *et al.* (1988) No barriers to speciation. *Nature* 336, 13–14.
75. Mabberly, D.J. (2008) *Mabberly's Plant-Book.* Third Edition, Cambridge University Press.
76. Savolainen, V. *et al.* (2006) Sympatric speciation in palms on an oceanic island. *Nature* 441, 210–213.
77. Macnair, M.R. (1989) A new species of *Mimulus* endemic to copper mines in California (USA). *Botanical Journal of the Linnean Society* 100, 1–14.
78. Martens, K. (1997) Speciation in ancient lakes. *Trends in Ecology and Evolution* 12, 177–182.
79. Salzburger, W. and Meyer, A. (2004) The species flocks of East African cichlid fishes: recent advances in molecular phylogenetics and population genetics. *Naturwissenschaften* 91, 277–290.
80. Barlow, G.W. (2000) *The Cichlid Fishes: Nature's Grand Experiment in Evolution.* Perseus Publishing.
81. Turner, G.F. *et al.* (2001) How many species of cichlid fishes are there in African lakes? *Molecular Ecology* 10, 793–806.
82. Kornfield, I. and Smith, P.F. (2000) African cichlid fishes: model systems for evolutionary biology. *Annual Review of Ecology and Systematics* 31, 163–196.
83. Ritchie, M.G. (2007) Sexual selection and speciation. *Annual Review of Ecology, Evolution and Systematics* 38, 79–102.
84. Goldschmidt, T. (1997) *Darwin's Dreampond: Drama in Lake Victoria.* MIT Press.
85. Kocher, T.D. (2004) Adaptive evolution and explosive speciation: the cichlid fish model. *Nature Reviews Genetics* 5, 288–298.
86. Hulsey, C.D. (2006) Function of a key morphological innovation: fusion of the cichlid pharyngeal jaw. *Proceedings of the Royal Society* B 273, 669–675.
87. Turelli, M. *et al.* (2001) Theory and speciation. *Trends in Ecology and Evolution* 16, 330–343.
88. Gavrilets, S. (2003) Perspective: models of speciation: what have we learned in 40 years? *Evolution* 57, 2197–2215.
89. Dieckmann, U. and Doebeli, M. (1999) On the origin of species by sympatric speciation. *Nature* 400, 354–357.
90. Kondrashov, A.S. and Kondrashov, F.A. (1999) Interactions among quantitative traits in the course of sympatric speciation. *Nature* 400, 351–354.
91. Arnegard, M.E. and Kondrashov, A.S. (2004) Sympatric speciation by sexual selection alone is unlikely. *Evolution* 58, 222–237.
92. Doebeli, M. and Dieckmann, U. (2003) Speciation along environmental gradients. *Nature* 421, 259–264.
93. Seehausen, O. and van Alphen, J.J.M. (1998) The effect of male coloration on female mate choice in closely related Lake Victoria cichlids (*Haplochromis nyererei* complex). *Behavioral Ecology and Sociobiology* 42, 1–8.
94. Seehausen, O. *et al.* (1997) Cichlid fish diversity threatened by eutrophication that curbs sexual selection. *Science* 277, 1808–1811.
95. Schliewen, U.K. *et al.* (1994) Sympatric speciation suggested by monophyly of crater lake cichlids. *Nature* 368, 629–632.
96. Barluenga, M. *et al.* (2006) Sympatric speciation in Nicaraguan crater lake cichlid fish. *Nature* 439, 719–723.
97. Dawkins, R. (1995) *River Out of Eden.* Weidenfeld & Nicolson.
98. Tinbergen, N. (1952) Readings from Scientific American 5–9. In *Psychobiology, the Biological Bases of Behavior,* Freeman.

99. McKinnon, J.S. and Rundle, H.D. (2002) Speciation in nature: the threespine stickleback model systems. *Trends in Ecology and Evolution* 17, 480–488.
100. Taylor, E.B. and McPhail, J.D. (1999) Evolutionary history of an adaptive radiation in species pairs of threespine sticklebacks (*Gasterosteus*): insights from mitochondrial DNA. *Biological Journal of the Linnean Society* 66, 271–291.
101. Rundle, H.D. *et al.* (2000) Natural selection and parallel speciation in sympatric sticklebacks. *Science* 287, 306–308.
102. Nagel, L. and Schluter, D. (1998) Body size, natural selection, and speciation in sticklebacks. *Evolution* 52, 209–218.
103. Taylor, E.B. and McPhail, J.D. (2000) Historical contingency and ecological determinism interact to prime speciation in sticklebacks, *Gasterosteus. Proceedings of the Royal Society B* 267, 2375–2384.
104. Mayr, E. (2004) *What Makes Biology Unique?* Cambridge University Press.
105. Price, T. (2008) *Speciation in Birds.* Roberts & Company.
106. Phillimore, A.B. *et al.* (2008) Sympatric speciation in birds is rare: insights from range data and simulations. *American Naturalist* 171, 646–657.
107. Pennisi, E. (2006) Speciation standing in place. *Science* 311, 1372–1374.
108. Bolnick, D.I. and Fitzpatrick, B.M. (2007) Sympatric speciation: models and empirical evidence. *Annual Review of Ecology, Evolution and Systematics* 38, 459–487.
109. Owens, I.P.F. *et al.* (1999) Species richness among birds: body size, life history, sexual selection or ecology? *Proceedings of the Royal Society B* 266, 933–939.
110. Gage, M.J.G. *et al.* (2002) Sexual selection and speciation in mammals, butterflies and spiders. *Proceedings of the Royal Society B* 269, 2309–2316.
111. Harlan, J.R. and deWet, J.M.J. (1975) On Ö. Winge and a prayer: the origins of polyploidy. *The Botanical Review* 41, 361–390.
112. Haldane, J.B.S. (1959) Natural selection. In *Darwin's Biological Work: Some Aspects Reconsidered* (Bell, P.R. ed), pp. 101–149, Cambridge University Press.
113. Rieseberg, L.H. (1997) Hybrid origins of plant species. *Annual Review of Ecology and Systematics* 28, 359–389.
114. Soltis, D.E. *et al.* (2003) Advances in the study of polyploidy since Plant speciation. *New Phytologist* 161, 173–191.
115. Stearns, S.C. and Hoekstra, R.F. (2005) *Evolution: An Introduction.* Second Edition, Oxford University Press.
116. Ainouche, M.L. *et al.* (2004) *Spartina anglica* C.E. Hubbard: a natural model system for analysing early evolutionary changes that affect allopolyploid genomes. *Biological Journal of the Linnean Society* 82, 475–484.
117. Burton, T.L. and Husband, B.C. (1999) Population cytotype structure in the polyploid *Galax urceolata* (Diapensiaceae). *Heredity* 82, 381–390.
118. Otto, S.P. and Whitton, J. (2000) Polyploid incidence and evolution. *Annual Review of Genetics* 34, 401–437.
119. Soltis, D.E. and Soltis, P.S. (1999) Polyploidy: recurrent formation and genome evolution. *Trends in Ecology and Evolution* 14, 348–352.
120. Mavarez, J. *et al.* (2006) Speciation by hybridization in *Heliconius* butterflies. *Nature* 441, 868–871.
121. Delneri, D. *et al.* (2003) Engineering evolution to study speciation in yeasts. *Nature* 422, 68–72.
122. Spencer, H. (1862) *First Principles.* p. 359, Williams & Norgate, London.
123. Coyne, J.A. and Orr, H.A. (1998) The evolutionary genetics of speciation. *Philosophical Transactions of the Royal Society of London Series B-Biological Sciences* 353, 287–305.

124. Rainey, P.B. and Travisano, M. (1998) Adaptive radiation in a heterogeneous environment. *Nature* 394, 69–72.

125. Wilkinson, D.M. (2006) *Fundamental Processes in Ecology: An Earth Systems Approach*. Oxford University Press.

Chapter 5: Why are the Tropics so Diverse?

Suggested general reading

Colinvaux, P. (2007) *Amazon Expeditions: My Quest for the Ice-Age Equator*. Yale University Press.

Gaston, K.J. (2000) Global patterns in biodiversity. *Nature* 405, 220–227.

Gaston, K.J. and Spicer, J.I. (2004) *Biodiversity; An Introduction*. Second Edition, Blackwell Publishing.

Tokeshi, M. (1999) *Species Coexistence, Ecological and Evolutionary Perspectives*. Blackwell Science.

Willig, M.R. *et al.* (2003) Latitudinal gradients of biodiversity: pattern, process, scale, and synthesis. *Annual Reviews of Ecology, Evolution and Systematics* 34, 273–309.

Wilson, E.O. (1992) *The Diversity of Life*. Belknap Press.

References

1. Charles Darwin (1839) *Journal of Researches into the Geology and Natural History of the Various Countries Visited by H.M.S. Beagle*. Henry Colburn.
2. Browne, J. (1995) *Charles Darwin, Voyaging*. Jonathan Cape.
3. Hawkins, B.A. and Diniz-Filho, J.A.F. (2004) 'Latitude' and geographic patterns in species richness. *Ecography* 27, 268–272.
4. Smith, H.G. and Wilkinson, D.M. (1987) Biogeography of testate rhizopods in the southern temperate and Antarctic zones. *Colloque sur les Ecosystems Terrestes Subantarctic, CNFRA* 58, 83–96.
5. Pianka, E.R. (1966) Latitudinal gradients in species diversity: a review of concepts. *American Naturalist* 100, 33–46.
6. Hubbell, S.P. (2001) *The Unified Neutral Theory of Biodiversity and Biogeography*. Princeton University Press.
7. Chown, S.L. *et al.* (2004) Hemispherical asymmetries in biodiversity—a serious matter for ecology. *PLoS Biology* 2(11), e406.
8. Proctor, J. *et al.* (1983) Ecological studies in four contrasting lowland rain forests in Gunung Mulu National Park, Sarawak. I. Forest environment, structure and floristics. *Journal of Ecology* 71, 237–260.
9. Wilson, E.O. (1991) The high frontier. *National Geographic* 180(6), 78–107.
10. Henderson, P.A. *et al.* (1998) Evolution and diversity in Amazonian floodplain communities. In *Dynamics of Tropical Communities* (Newbery, D.M. *et al.* eds), pp. 385–419, Blackwell Science.
11. Maitland, P.S. and Campbell, R.N. (1992) *Freshwater Fish*. Harper Collins.
12. Attrill, M.J. *et al.* (2001) Latitudinal diversity patterns in estuarine tidal flats: indications of a global cline. *Ecography* 24, 318–324.
13. Gaston, K.J. *et al.* (1995) Large scale patterns of biogeography: spatial variation in family richness. *Proceedings of the Royal Society B* 260, 149–154.
14. MacArthur, R.H. and Wilson, E.O. (1967) *The Theory of Island Biogeography*. Princeton University Press.

15. Cox, C.B. and Moore, P.D. (2005) *Biogeography*. Seventh Edition, Blackwell Publishing.
16. Gaston, K.J. and Spicer, J.I. (2004) *Biodiversity; An Introduction*. Second Edition, Blackwell Publishing.
17. Rosenzweig, M.L. (1995) *Species Diversity in Space and Time*. Cambridge University Press.
18. Tokeshi, M. (1999) *Species Coexistence, Ecological and Evolutionary Perspectives*. Blackwell Science.
19. Collard, I.F. and Foley, R.A. (2002) Latitudinal patterns and environmental determinants of recent human cultural diversity: do humans follow biogeographic rules? *Evolutionary Ecology Research* 4, 371–383.
20. Matthews, L.H. (1977) *Penguin*. Peter Owen.
21. del Hoyo, J. *et al.* eds (1992) *Handbook of the Birds of the World*. Vol. 1, Lynx Edicions.
22. Chown, S.L. *et al.* (1998) Global patterns in species richness of pelagic seabirds: the Procellariiformes. *Ecography* 21, 342–350.
23. Janzen, D.H. (1981) The peak in North American Ichneumonid species richness lies between 38° and 42° N. *Ecology* 62, 532–537.
24. Procter, D.L.C. (1984) Towards a biogeography of free-living soil nematodes. I. Changing species richness, diversity and densities with changing latitude. *Journal of Biogeography* 11, 103–117.
25. Maraun, M. *et al.* (2007) Awesome or ordinary? Global diversity patterns of oribatid mites. *Ecography* 30, 209–216.
26. Brandt, A. *et al.* (2007) First insights into the biodiversity and biogeography of the Southern Ocean deep sea. *Nature* 447, 307–311.
27. Crame, J.A. (1997) An evolutionary framework for the polar regions. *Journal of Biogeography* 24, 1–9.
28. Smith, H.G. (1982) The terrestrial protozoan fauna of South Georgia. *Polar Biology* 1, 173–179.
29. Vincke, S. *et al.* (2006) *A Synopsis of the Testate Amoebae Fauna of Île de la Possession (Crozet Archipelago, sub-Antarctica)*. Antwerp University Press.
30. Vincke, S. *et al.* (2006) The moss dwelling testacean fauna of the Strømnness Bay (South Georgia). *Acta Protozoologica* 45, 65–75.
31. Hillebrand, H. and Azovsky, A.I. (2001) Body size determines the strength of the latitudinal diversity gradient. *Ecography* 24, 251–256.
32. Stoeck, T.S. *et al.* (2007) Protistan diversity in the Arctic: a case of paleoclimate shaping modern biodiversity? *PLoS One* 2(8), e728.
33. Buzas, M.A. *et al.* (2002) Latitudinal differences in biodiversity caused by higher tropical rate of increase. *Proceedings of the National Academy of Sciences USA* 99, 7841–7843.
34. Atlas, R.M. and Bartha, R. (1998) *Microbial Ecology; Fundamentals and Applications*. Fourth Edition, Benjamin Cummings.
35. Fierer, N. and Jackson, R.B. (2006) The diversity and biogeography of soil bacterial communities. *Proceedings of the National Academy of Sciences USA* 103, 626–631.
36. Hughes Martiny, J.B. *et al.* (2006) Microbial biogeography: putting microorganisms on the map. *Nature Reviews Microbiology* 4, 102–112.
37. Hillebrand, H. (2004) On the generality of the latitudinal diversity gradient. *American Naturalist* 163, 192–211.
38. Stehli, F.G. *et al.* (1969) Generation and maintenance of gradients in taxonomic diversity. *Science* 164, 947–949.
39. Crane, P.R. and Lidgard, S. (1989) Angiosperm diversification and paleolatitudinal gradients in Cretaceous floristic diversity. *Science* 246, 675–678.

40. Crame, J.A. (2001) Taxonomic diversity gradients through geological time. *Diversity and Distributions* 7, 175–189.
41. Colwell, R.K. and Lees, D.C. (2000) The mid-domain effect: geometric constraints in the geography of species richness. *Trends in Ecology and Evolution* 15, 70–76.
42. Diniz-Filho, J.A.F. *et al.* (2002) Null models and spatial patterns of species richness in South American birds of prey. *Ecology Letters* 5, 47–55.
43. Zapata, F.A. *et al.* (2003) Mid-domain models of species richness gradients: assumptions, methods and evidence. *Journal of Animal Ecology* 72, 677–690.
44. Cramp, S. and Perrins, C.M. eds (1994) The *Birds of the Western Palearctic*. Vol. VIII, Oxford University Press.
45. Dunn, R.R. *et al.* (2007) When does diversity fit null model predictions? Scale and range size mediates the mid-domain effect. *Global Ecology and Biogeography* 16, 305–312.
46. Currie, D.J. and Kerr, J.T. (2008) Tests of the mid-domain hypothesis: a review of the evidence. *Ecological Monographs* 78, 3–18.
47. Colwell, R.R. *et al.* (2005) The mid-domain effect: there's a baby in the bath water. *American Naturalist* 166, E149–E154.
48. Storch, D. *et al.* (2006) Energy, range dynamics and global species richness patterns: reconciling mid-domain effects and environmental determinants of avian diversity. *Ecology Letters* 9, 1308–1320.
49. Wilkinson, D.M. (2001) Dispersal: biogeography. In *Encyclopaedia of Life Sciences*, John Wiley & Sons.
50. Terborgh, J. (1973) On the notion of favourableness in plant ecology. *American Naturalist* 107, 481–501.
51. Mittelbach, G.G. *et al.* (2007) Evolution and the latitudinal diversity gradient: speciation, extinction and biogeography. *Ecology Letters* 10, 315–331.
52. Wiens, J.J. and Graham, C.H. (2005) Niche conservatism: integrating evolution, ecology, and conservation biology. *Annual Review of Ecology, Evolution and Systematics* 36, 519–539.
53. Kozak, K.H. and Wiens, J.J. (2006) Does niche conservatism promote speciation? A case study in North American salamanders. *Evolution* 60, 2604–2621.
54. Stopes, M. (1918) *Married love*. A.C. Fifield.
55. Chaloner, W.G. (2005) The palaeobotanical work of Marie Stopes. *Geological Society London Special Publications* 241, 127–135.
56. Stopes, M.C. (1915) *Catalogue of the Mesozoic Plants in the British Museum (Natural History). The Cretaceous flora. Part II—Lower Greensand (Aptian) Plants of Britain.* British Museum (Natural History).
57. Beerling, D. (2007) *The Emerald Planet*. Oxford University Press.
58. Hawkins, B.A. *et al.* (2007) Climate, niche conservatism and the global bird diversity gradient. *American Naturalist* 170, s16–s27.
59. Bowen, D.Q. and Gibbard, P.L. (2007) The Quaternary is here to stay. *Journal of Quaternary Science* 22, 3–8.
60. Jablonski, D. *et al.* (2006) Out of the tropics: evolutionary dynamics of the latitudinal diversity gradient. *Science* 314, 102–106.
61. Stebbins, G.L. (1974) *Flowering Plants: Evolution above the Species Level.* Harvard University Press.
62. Allen, A.P. *et al.* (2002) Global biodiversity, biochemical kinetics, and the energetic-equivalence rule. *Science* 297, 1545–1548.
63. Currie, D.J. *et al.* (2004) Predictions and tests of climate-based hypotheses of broad-scale variation in taxonomic richness. *Ecology Letters* 7, 1121–1134.

64. Kay, R.F. *et al.* (1997) Primate species richness is determined by plant productivity: implications for conservation. *Proceedings of the National Academy of Sciences USA* 94, 13023–13027.
65. Weir, J.T. and Schluter, D. (2007) The latitudinal gradient in recent speciation and extinction rates of birds and mammals. *Science* 315, 1574–1576.
66. Wilson, E.O. (1992) *The Diversity of Life*. Belknap Press.
67. Archiobold, O.W. (1995) *Ecology of World Vegetation*. Chapman & Hall.
68. Haffer, J. (1969) Speciation in Amazonian forest birds. *Science* 165, 131–137.
69. Bennett, K.D. (1997) *Evolution and Ecology; The Pace of Life*. Cambridge University Press.
70. Martin, P.R. and McKay, J.K. (2004) Latitudinal variation in genetic divergence of populations and the potential for future speciation. *Evolution* 58, 938–945.
71. Colinvaux, P.A. and De Oliveira, P.E. (2000) Paleoecology and climate of the Amazon basin during the last glacial cycle. *Journal of Quaternary Science* 15, 347–356.
72. Bush, M.B. *et al.* (2004) Amazonian paleoecological histories: one hill, three watersheads. *Palaeogeography, Palaeoclimotology, Palaeoecology* 214, 359–393.
73. Colinvaux, P.A. *et al.* (1996) A long pollen record from lowland Amazonia: forest and cooling in glacial times. *Science* 274, 85–88.
74. Clapperton, C. (1993) *Quaternary Geology and Geomorphology of South America*. Elsevier.
75. Bynum, W.F. and Porter, R. (2005) *Oxford Dictionary of Scientific Quotations*. Oxford University Press.
76. Dynesius, M. and Jansson, R. (2000) Evolutionary consequences of changes in species geographical distributions driven by Milankovitch climate oscillations. *Proceedings of the National Academy of Sciences USA* 97, 9115–9120.
77. Jansson, R. and Davies, T.J. (2008) Global variation in diversification rates of flowering plants: energy vs. climate change. *Ecology Letters* 11, 173–183.
78. MacArthur, R.H. (1972) *Geographical Ecology*. Princeton University Press.
79. Blackburn, T.M. and Gaston, K.J. (1996) A sideways look at patterns in species richness, or why there are so few species outside the tropics. *Biodiversity Letters* 3, 44–53.
80. Hutchinson, G.E. (1959) Homage to Santa Rosalia; or, why are there so many kinds of animals? *American Naturalist* 93, 145–159.
81. Møller, A.P. (1994) *Sexual Selection and the Barn Swallow*. Oxford University Press.
82. Janzen, D.H. and Pond, C.M. (1975) A comparison, by sweep sampling, of the arthropod fauna of secondary vegetation in Michigan, England and Costa Rica. *Transactions of the Royal Entomological Society of London* 127, 33–50.
83. Snow, D.W. (1976) *The Web of Adaptation*. Collins.
84. Snow, B. and Snow, D. (1988) *Birds and Berries*. T & AD Poyser.
85. H-Acevedo, D. and Currie, D.J. (2003) Does climate determine broad-scale patterns of species richness? A test of the causal link by natural experiment. *Global Ecology & Biogeography* 12, 461–473.
86. Chambers, F.M. (1995) Climate response, migrational lag, and pollen representation: the problems posed by *Rhododendron* and *Acer*. *Historical Biology* 9, 243–256.
87. Godwin, H. (1975) *The History of the British Flora*. Second Edition, Cambridge University Press.
88. Janzen, D.H. (1967) Why mountain passes are higher in the tropics. *American Naturalist* 101, 233–249.
89. Ghalambor, C.K. *et al.* (2006) Are mountain passes higher in the tropics? Janzen's hypothesis revisited. *Integrative and Comparative Biology* 46, 5–17.
90. Rapoport, E.H. (1975) *Areografía: Estrategias Geográficas de las Especies*. Fondo de Cultura Económica, Mexico City.

91. Stevens, G.C. (1989) The latitudinal gradient in geographical range: how so many species coexist in the tropics. *American Naturalist* 133, 240–256.
92. Rohde, K. (1999) Latitudinal gradients in species diversity and Rapoport's rule revisited: a review of recent work and what can parasites teach us about the cause of the gradients? *Ecography* 22, 593–613.
93. Janzen, D.H. (1970) Herbivores and the number of tree species in tropical forests. *American Naturalist* 104, 501–528.
94. Connell, J.H. (1978) Diversity of tropical rain forests and coral reefs. *Science* 199, 1302–1310.
95. Augspurger, C.K. (1983) Seed dispersal of the tropical tree, *Platypodium elegans*, and the escape of its seedlings from fungal pathogens. *Journal of Ecology* 71, 759–771.
96. Møller, A.P. (1998) Evidence of larger impact of parasites on hosts in the tropics: investment in immune function within and outside the tropics. *Oikos* 82, 265–270.
97. Lawton, J.H. (1996) Patterns in ecology. *Oikos* 75, 145–147.
98. Henderson, P. (2005) Life, evolution and development in the Amazonian floodplain. In *Narrow Roads of Geneland*, Vol. 3 (Ridley, M. ed), pp. 307–314, Oxford University Press.
99. Colwell, R.K. (2006) RangeModel: a Monte Carlo simulation tool for assessing geometric constraints on species richness, Version 5. http://viceroy.eeb.uconn.edu/rangemodel
100. Lyell, C. (1871) *The Student's Elements of Geology*. John Murray.

Chapter 6: Is Nature Chaotic?

Suggested general reading

Cushing, J.M. *et al.* (2003) *Chaos in Ecology: Experimental Nonlinear Dynamics*. Academic Press, London.

Gleick, J. (1988) *Chaos: Making a New Science*. Cardinal Publications.

Hastings, A. *et al.* (1993) Chaos in ecology—is mother nature a strange attractor? *Annual Review of Ecology and Systematics* 24, 1–33.

Logan, J.A. and Allen, J.C. (1992) Nonlinear dynamics and chaos in insect populations. *Annual Review of Entomology* 37, 455–477.

May, R.M. (2002) The best possible time to be alive. In *It Must be Beautiful* (Farmelo, G. ed), Granta Books.

Perry, J.N. *et al.* eds (2000) *Chaos in Real Data: The Analysis of Non-Linear Dynamics from Short Ecological Time Series*. Kluwer Academic Publishers.

Stewart, I. (1990) *Does God Play Dice: The New Mathematics of Chaos*. Penguin.

References

1. Smith, R.H. *et al.* (2000) Blowflies as a case study in non-linear dynamics. In *Chaos in Real Data: The Analysis of Non-Linear Dynamics from Short Ecological Time Series* (Perry, J.N. *et al.* eds), Kluwer Academic Publishers.
2. Poincaré, H. (1908) *Science and Method (translated by Francis Maitland). In: The Value of Science: Essential Writings of Henri Poincaré [2001]*. Random House.
3. Ma, S.J. (1958) Dynamis of *Locusta migratoria* manilensi (Meyen) in China. *Acta Entomologica Sinica* 8, 1–40.
4. Ma, S.-C. *et al.* (1965) Study on long-term prediction of locust population fluctuations. *Acta Entomologica Sinica* 14, 319–338.
5. Sugihara, G. (1995) From out of the blue. *Nature* 378, 559–560.

6. Dixon, P.A. *et al.* (1999) Episodic fluctuations in larval supply. *Science* 283, 1528–1530.
7. Grenfell, B.T. *et al.* (1998) Noise and determinism in synchronized sheep dynamics. *Nature* 394, 674–677.
8. May, R.M. (1974) Biological populations with nonoverlapping generations—stable points, stable cycles, and chaos. *Science* 186, 645–647.
9. May, R.M. (1976) Simple mathematical models with very complicated dynamics. *Nature* 261, 459–467.
10. Kingsland, S.E. (1995) *Modeling Nature.* Second Edition, University of Chicago Press.
11. Malthus, T. (1798) *An Essay on the Principle of Population.* J. Johnson.
12. Darwin, C. (1859) *Origin of Species by Means of Natural Selection or the Preservation of Favoured Races in the Struggle for Life.* First Edition, John Murray.
13. May, R.M. (2002) The best possible time to be alive. In *It Must be Beautiful* (Farmelo, G. ed), Granta Books.
14. Hamilton, W.D. (1991) Memes of Haldane and Jayakar in a theory of sex. *Journal of Genetics* 69, 17–32.
15. McGuffie, K. and Henderson-Sellers, A. (2005) *A Climate Modelling Primer.* Third Edition, John Wiley & Sons.
16. May, R.M. and Oster, G.F. (1976) Bifurcations and dynamic complexity in simple ecological models. *American Naturalist* 110, 573–599.
17. Li, T.-Y. and Yorke, J.A. (1975) Period three implies chaos. *American Mathematical Monthly* 82, 985–992.
18. Gould, S.J. (2002) *The Structure of Evolutionary Theory.* Bellnap Press of Harvard University Press.
19. Ruelle, D. (1990) Deterministic chaos: the science and the fiction. *Proceedings of the Royal Society Series A* 427, 241–248.
20. Hastings, A. *et al.* (1993) Chaos in ecology—is mother nature a strange attractor? *Annual Review of Ecology and Systematics* 24, 1–33.
21. Lorenz, E. (1963) Deterministic nonperiodic flow. *Journal of the Atmospheric Sciences* 20, 448–464.
22. Ellner, S. and Turchin, P. (1995) Chaos in a noisy world: new methods and evidence from time-series analysis. *American Naturalist* 145, 343–375.
23. Gilpin, M.E. (1979) Spiral chaos in a predator–prey model. *American Naturalist* 113, 306–308.
24. Hawkins, H. (1995) *Strange Attractors: Literature, Culture and Chaos Theory.* Prentice Hall.
25. Cushing, J.M. *et al.* (2003) *Chaos in Ecology: Experimental Nonlinear Dynamics.* Academic Press.
26. Gross, T. *et al.* (2005) Long food chains are in general chaotic. *Oikos* 109, 135–144.
27. Pool, R. (1989) Ecologists flirt with chaos. *Science* 243, 310–313.
28. Coulson, T.N. and Godfray, H.C.J. (2006) Single-species dynamics. In *Theoretical Ecology: Principles and Applications* (May, R.M. and McLean, A.R. eds), pp. 17–34, Oxford University Press.
29. Rand, D.A. and Wilson, H.B. (1991) Chaotic stochasticity: a ubiquitous source of unpredictability in epidemics. *Proceedings of the Royal Society B* 246, 179–184.
30. Dennis, B. *et al.* (2003) Can noise induce chaos? *Oikos* 102, 329–339.
31. Ellner, S.P. and Turchin, P. (2005) When can noise induce chaos and why does it matter: a critique. *Oikos* 111, 620–631.
32. Scheuring, I. and Domokos, G. (2007) Only noise can induce chaos in discrete populations. *Oikos* 116, 361–366.

33. Hassell, M.P. *et al.* (1976) Patterns of dynamical behaviour in single-species populations. *Journal of Animal Ecology* 45, 471–486.
34. Nicholson, A.J. (1954) An outline of the dynamics of animal populations. *Australian Journal of Zoology* 2, 9–65.
35. Morris, M.F. (1990) Problems in detecting chaotic behavior in natural populations by fitting simple discrete models. *Ecology* 71, 1849–1862.
36. Takens. F. (1981) Detecting strange attractors in turbulence. In *Dynamical Systems and Turbulence* (D.A. Rand and L.S. Young eds.) *Lecture Notes in Mathematics* 898, 366–381, Springer.
37. Schaffer, W.M. (2000) *Forward.* In *Chaos in Real Data: The Analysis of Non-Linear Dynamics from Short Ecological Time Series.* (Perry, J.N. *et al.* eds.) Kluwer Academic Publishers.
38. Thomas, W.R. *et al.* (1980) Chaos, asymmetric growth and group selection for dynamical stability. *Ecology* 61, 1312–1320.
39. Mueller, L.D. and Ayala, F.J. (1991) Dynamics of single-species population growth: stability or chaos. *Ecology* 62, 1148–1154.
40. Freckleton, R.P. and Watkinson, A.R. (2002) Are weed population dynamics chaotic? *The Journal of Applied Ecology* 39, 699–707.
41. Schaffer, W.M. and Kot, M. (1985) Nearly one dimensional dynamics in an epidemic. *Journal of Theoretical Biology* 112, 403–427.
42. Olsen, L.F. and Schaffer, W.M. (1990) Chaos versus noisy periodicity: alternative hypotheses for childhood epidemics. *Science* 249, 499–504.
43. Sugihara, G. *et al.* (1990) Distinguishing error from chaos in ecological time-series. *Philosophical Transactions of the Royal Society of London Series B* 330, 235–251.
44. Grenfell, B. and Keeling, M. (2006) Dynamics of infectious disease. In *Theoretical Ecology: Principles and Applications* (May, R.M. and McLean, A.R. eds), pp. 132–147, Oxford University Press.
45. Bjornstad, O.N. *et al.* (2002) Dynamics of measles epidemics. I. Estimating scaling of transmission rates using a time series SIR model. *Ecological Monographs* 72, 169–184.
46. Schaffer, W.M. (1984) Stretching and folding in lynx fur returns: evidence for a strange attractor in nature? *American Naturalist* 124, 798–820.
47. Elton, C. (1927) *Animal Ecology.* Sidgwick & Jackson.
48. Elton, C. and Nicholson, M. (1942) The ten-year cycle in numbers of the lynx in Canada. *Journal of Animal Ecology* 11, 215–244.
49. Gamarra, J.G.P. and Sole, R.V. (2000) Bifurcations and chaos in ecology: lynx returns revisited. *Ecology Letters* 3, 114–121.
50. Schaffer, W.M. and Kot, M. (1986) Chaos in ecological systems: the coals that Newcastle forgot. *Trends in Ecology and Evolution* 1, 58–63.
51. Gleick, J. (1988) *Chaos: Making a New Science.* Cardinal Publications.
52. Crone, E.E. and Taylor, D.R. (1996) Complex dynamics in experimental populations of an annual plant, *Cardamine pensylvanica. Ecology* 77, 289–299.
53. May, R.M. (1998) The voles of Hokkaido. *Nature* 396, 409–410.
54. Hansson, L. and Henttonen, H. (1988) Rodent dynamics as a community processes. *Trends in Ecology and Evolution* 3, 195–200.
55. Hanski, I. *et al.* (1993) Population oscillations of Boreal rodents—regulation by mustelid predators leads to chaos. *Nature* 364, 232–235.
56. Falck, W. *et al.* (1995) Bootstrap estimated uncertainty of the dominant lyapunov exponent for holarctic microtine rodents. *Proceedings of the Royal Society B* 261, 159–165.

57. Turchin, P. (1995) Chaos in microtine populations. *Proceedings of the Royal Society B* 262, 357–361.
58. Falck, W. *et al.* (1995) Voles and lemmings: chaos and uncertainty in fluctuating populations. *Proceedings of the Royal Society B* 262, 363–370.
59. Costantino, R.F. *et al.* (1995) Experimentally induced transitions in the dynamic behavior of insect populations. *Nature* 375, 227–230.
60. Costantino, R.F. *et al.* (1997) Chaotic dynamics in an insect population. *Science* 275, 389–391.
61. Becks, L. *et al.* (2005) Experimental demonstration of chaos in a microbial food web. *Nature* 435, 1226–1229.
62. Benincà, E. *et al.* (2008) Chaos in a long-term experiment with a plankton community. *Nature* 451, 822–826.
63. Milton, J.G. and Belair, J. (1990) Chaos, noise and extinction in models of population growth. *Theoretical Population Biology* 37, 273–290.
64. Grover, J.P. *et al.* (2000) Periodic dynamics in *Daphnia* populations: biological interactions and external forcing. *Ecology* 81, 2781–2798.
65. Woiwod, I.P. *et al.* (2000) Analysis of population fluctuation in the aphid *Hyperomyzus lactucae* and the moth *Perizoma alchemillata.* In *Chaos in Real Data: The Analysis of Non-Linear Dynamics from Short Ecological Time Series* (Perry, J.N. *et al.* eds), Kluwer Academic Publishers.
66. Zhou, X. *et al.* (1997) Detecting chaotic dynamics of insect populations from long-term survey data. *Ecological Entomology* 22, 231–241.
67. Perry, J.N. (2000) Overview. In *Chaos in Real Data: The Analysis of Non-Linear Dynamics from Short Ecological Time Series* (Perry, J.N. *et al.* eds), Kluwer Academic Publishers.
68. Begon, M. *et al.* (2000) One, two and three-species time series from a host–pathogen–parasitoid system. In *Chaos in Real Data: The Analysis of Non-Linear Dynamics from Short Ecological Time Series* (Perry, J.N. *et al.* eds), Kluwer Academic Publishers.
69. Turchin, P. and Ellner, S.P. (2000) Living on the edge of chaos: population dynamics of Fennoscandian voles. *Ecology* 81, 3099–3116.
70. Pascual, M. and Mazzega, P. (2003) Quasicycles revisited: apparent sensitivity to initial conditions. *Theoretical Population Biology* 64, 385–395.
71. Mueller, L.D. and Joshi, A. (2000) *Stability in Model Populations.* Princeton University Press.
72. Zimmer, C. (1999) Life after chaos. *Science* 284, 83–86.
73. Berryman, A.A. and Millstein, J.A. (1989) Are ecological systems chaotic—and if not, why not? *Trends in Ecology and Evolution* 4, 26–28.
74. Allen, J.C. *et al.* (1993) Chaos reduces species extinction by amplifying local-population noise. *Nature* 364, 229–232.
75. Yoshida, T. *et al.* (2003) Rapid evolution drives ecological dynamics in a predator–prey system. *Nature* 424, 303–306.
76. Nicholson, A.J. (1957) The self adjustment of populations to change. *Cold Spring Harbour Symposium in Quantitative Biology* 22, 153–173.
77. Oster, G. (1976) Internal variables in population dynamics. In *Mathematical Problems in the Life Sciences* (Levin, S. ed), pp. 37–68, American Mathematics Society.
78. Stokes, T.K. *et al.* (1988) Parameter evolution in a laboratory insect population. *Theoretical Population Biology* 34, 248–265.
79. Ferrière, R. and Gatto, M. (1993) Chaotic population dynamics can result from natural selection. *Proceedings of the Royal Society B* 251, 33–38.
80. Doebeli, M. and Koella, J.C. (1995) Evolution of simple population dynamics. *Proceedings of the Royal Society B* 260, 119–125.

81. Zeineddine, M. and Jansen, V.A.A. (2005) The evolution of stability in a competitive system. *Journal of Theoretical Biology* 236, 208–215.
82. Ebenman, B. *et al.* (1996) Evolution of stable population dynamics through natural selection. *Proceedings of the Royal Society B* 263, 1145–1151.
83. Mueller, L.D. *et al.* (2000) Does population stability evolve? *Ecology* 81, 1273–1285.
84. Prasad, N.G. *et al.* (2003) The evolution of population stability as a by-product of life-history evolution. *Proceedings of the Royal Society B (Supplement)* 270, S84–S86.
85. Scheuring, I. (2001) Is chaos due to over-simplification in models of population dynamics? *Selection* 2, 179–191.
86. Ruxton, G.D. (1995) Population models with sexual reproduction show a reduced propensity to exhibit chaos. *Journal of Theoretical Biology* 175, 595–601.
87. Hastings, A. (1993) Complex interactions between dispersal and dynamics: lessons from coupled logistic equations. *Ecology* 74, 1362–1372.
88. Doebeli, M. (1995) Dispersal and dynamics. *Theoretical Population Biology* 47, 82–106.
89. Rohde, K. and Rohde, P.P. (2001) Fuzzy chaos: reduced chaos in the combined dynamics of several independently chaotic populations. *American Naturalist* 158, 553–556.
90. Anderson, C.N.K. *et al.* (2008) Why fishing magnifies fluctuations in fish abundance. *Nature* 452, 835–839.
91. Yorke, J. (2005) Big ideas: chaos. *New Scientist* 17 September 2005, p. 37.

Chapter 7: Why is the Land Green?

Suggested general reading

Hartley, S.E. and Jones, C.G. (1997) Plant chemistry and herbivory, or why is the world green? In *Plant Ecology*, Second Edition (Crawley, M.J. ed), pp. 284–324, Blackwell Science.
Polis, G.A. (1999) Why are parts of the world green? Multiple factors control productivity and the distribution of biomass. *Oikos* 86, 3–15.

References

1. 'King James' translation of *The Holy Bible*.
2. Hairston, N.G. *et al.* (1960) Community structure, population control, and competition. *American Naturalist* 94, 421–425.
3. Hardin, M.R. *et al.* (1995) Arthropod pest resurgence: an overview of potential mechanisms. *Crop Protection* 14, 3–18.
4. Majerus, M.E.N. (2002) *Moths*. Harper Collins.
5. Aplin, R.T. *et al.* (1968) Poisonous alkaloids in the body tissue of the cinnabar moth (*Callimorpha jacobaeae* L.). *Nature* 219, 747–748.
6. Hill, M.O. *et al.* (1992) Long-term effects of excluding sheep from hill pastures in North Wales. *Journal of Ecology* 80, 1–13.
7. Watt, A.S. (1981) Further observations in the effects of excluding rabbits from Grassland A in East Anglian Breckland: the pattern of change and factors affecting it (1936–73). *Journal of Ecology* 69, 509–536.
8. Jacobs, S.M. and Naiman, R.J. (2008) Large African herbivores decrease herbaceous plant biomass while increasing plant species richness in a semi-arid savanna toposequence. *Journal of Arid Environments* 72, 891–903.
9. Terborgh, J. *et al.* (2006) Vegetation dynamics of predator-free land-bridge islands. *Journal of Ecology* 94, 253–263.

10. Hölldobler, B. and Wilson, E.O. (1990) *The Ants*. Springer-Verlag.
11. Post, E. and Stenseth, N. Chr. (1998) Large-scale climatic fluctuation and population dynamics of moose and white-tailed deer. *Journal of Animal Ecology* 67, 537–543.
12. Post, E. *et al.* (1999) Ecosystem consequences of wolf behavioural response to climate. *Nature* 401, 905–907.
13. Paine, R.T. (2000) Phycology for the mammologist: marine rocky shores and mammal-dominated communities—how different are the structural processes? *Journal of Mammology* 81, 637–648.
14. Hall, S.A. (1990) Pollen evidence for historic vegetational change, Hueco Bolson, Texas. *Texan Journal of Science* 42, 399–403.
15. Hartley, S.E. and Jones, C.G. (1997) Plant chemistry and herbivory, or why is the world green? In *Plant Ecology*, Second Edition (Crawley, M.J. ed), pp. 284–324, Blackwell Science.
16. Polis, G.A. (1999) Why are parts of the world green? Multiple factors control productivity and the distribution of biomass. *Oikos* 86, 3–15.
17. Lawton, J.H. and McNeill, S. (1979) Between the devil and the deep blue sea: on the problem of being a herbivore. In *Population dynamics* (Anderson, R.M., Turner, B.D., and Taylor, L.R. eds), pp. 223–244, Blackwell Scientific Publications.
18. Fraenkel, G.S. (1959) The raison d'être of secondary plant substances. *Science* 129, 1466–1470.
19. Liu, S. *et al.* (1992) Inducible phytoalexins in juvenile soybean genotypes predict soybean looper resistance in the fully developed plants. *Plant Physiology* 100, 1479–1485.
20. Wood, J.G. and Wood, T. (1886) *The Field Naturalists Handbook*. Cassell & Co.
21. Freeland, W.J. and Janzen, D.H. (1974) Strategies in herbivory by mammals: the role of plant secondary compounds. *American Naturalist* 108, 269–289.
22. Moore, B.D. *et al.* (2005) Eucalyptus foliar chemistry explains selective feeding by koalas. *Biology Letters* 1, 64–67.
23. Crawford, R.M.M. (1989) *Studies in Plant Survival*. Blackwell Scientific Publications.
24. Walling, L.L. (2004) Plant defences against insects: constitutive and induced defences. In *Encyclopaedia of Plant and Crop Science* (Goodman, R.M. ed), pp. 939–943, Marcel Dekker.
25. Janzen, D.H. (1988) On the broadening of insect–plant research. *Ecology* 69, 905.
26. Morgan, N. and Hamilton, W.D. (1980) Low nutritive quality as defence against herbivores. *Journal of Theoretical Biology* 86, 247–254.
27. McArthur, C. and Sanson, G.D. (1993) Nutritional effects and costs of a tannin in a grazing and a browsing macropodid marsupial herbivore. *Functional Ecology* 7, 690–696.
28. Vitousek, P. (2004) *Nutrient Cycling and Limitation; Hawai'i as a Model System*. Princeton University Press.
29. Eigenbrode, S.D. (2004) Plant defences against insects: physical defences. In *Encyclopaedia of Plant and Crop Science* (Goodman, R.M. ed), pp. 944–946, Marcel Dekker.
30. Chambers, F.M. *et al.* (1999) Recent rise to dominance of *Molinia caerulea* in environmentally sensitive areas: new perspectives from palaeoecological data. *Journal of Applied Ecology* 36, 719–733.
31. Brunsting, A.M.H. and Heil, G.W. (1985) The role of nutrients in the interaction between a herbivorous beetle and some competing plant species in heathlands. *Oikos* 44, 23–26.
32. Diamond, J. (2002) Evolution, consequences and future of plant and animal domestication. *Nature* 418, 700–707.
33. Mabberly, D.J. (2008) *Mabberley's Plant Book*. Third Edition, Cambridge University Press.
34. Schaller, G.B. (1963) *The Mountain Gorilla: Ecology and Behaviour*. University Chicago Press.
35. Culham, A. (2007) Mesembryanthemaceae. In *Flowering Plant Families of the World* (Heyword, V.H. *et al.* eds), pp. 213–214, Firefly.

36. Huffaker, C.B. (1958) Experimental studies on predation: dispersion factors and predator–prey oscillations. *Hilgardia* 27, 343–383.
37. Owen Smith, N. (2008) The refuge concept extends to plants as well: storage, buffers and regrowth in variable environments. *Oikos* 117, 481–487.
38. Hoopes, M.F. and Harrison, S. (1998) Metapopulation, source-sink and disturbance dynamics. In *Conservation Science and Action* (Sutherland, W.J. ed), pp. 135–151, Blackwell Science.
39. Gripenberg, S. and Roslin, T. (2007) Up or down in space? Uniting the bottom-up versus top-down paradigm and spatial ecology. *Oikos* 116, 181–188.
40. Grubb, P.J. (1992) A positive distrust in simplicity—lessons from plant defences and from competition among plants and among animals. *Journal of Ecology* 80, 585–610.
41. Castillo, U.F. *et al.* (2007) Biologically active endophytic streptomycetes from *Nothofagus* spp and other plants in Patagonia. *Microbial Ecology* 53, 12–19.
42. Spooner, B. and Roberts, P. (2005) *Fungi.* Harper Collins.
43. Midgley, J.J. (2005) Why don't leaf-eating animals prevent the formation of vegetation? Relative vs absolute dietary requirements. *New Phytologist* 168, 271–273.
44. Berner, R.A. (2004) *The Phanerozoic Carbon Cycle.* Oxford University Press.
45. Wilkinson, D.M. (2006) *Fundamental Processes in Ecology; An Earth Systems Approach.* Oxford University Press.
46. Robinson, J.M. (1990) Lignin, land plants, and fungi: biological evolution affecting Phanerozoic oxygen balance. *Geology* 15, 607–610.
47. Scott, A.C. *et al.* (2004) Evidence of plant–insect interactions in the Upper Triassic Molteno Formation of South Africa. *Journal of the Geological Society, London* 161, 401–410.
48. Jones, C.G. and Lawton, J.H. (1991) Plant chemistry and insect species richness of British umbellifers. *Journal of Animal Ecology* 60, 767–777.
49. Richards, L.A. and Coley, P.D. (2007) Seasonal and habitat differences affect the impact of food and predation on herbivores: a comparison between gaps and understory of a tropical forest. *Oikos* 116, 31–40.
50. Grime, J.P. (2001) *Plant Strategies, Vegetation Processes, and Ecosystem Properties.* Second Edition, Wiley.
51. Bond, W.J. (2005) Large parts of the world are brown and black: a different view on the 'green world' hypothesis. *Journal of Vegetation Science* 16, 261–266.
52. Bond, W.J. and Keeley, J.E. (2005) Fire as a global 'herbivore': the ecology and evolution of flammable ecosystems. *Trends in Ecology and Evolution* 20, 387–394.
53. Scott, A.C. and Glasspool, I.J. (2006) The diversification of Paleozoic fire systems and fluctuations in atmospheric oxygen concentration. *Proceedings of the National Academy of Sciences USA* 103, 10861–10865.
54. Bond, W.J., Woodward, F.I., and Midgley, G.F. (2005) The global distribution of ecosystems in a world without fire. *New Phytologist* 165, 525–538.
55. Elton, C. (1927) *Animal Ecology.* Sidgwick & Jackson.
56. Davidson, J. and Andrewartha, H.G. (1948) The influence of rainfall, evaporation and atmospheric temperature on fluctuations in the size of a natural population of *Thrips imaginis* (Thysanoptera). *Journal of Animal Ecology* 17, 200–222.
57. Majerus, M.E.N. (1994) *Ladybirds.* Harper Collins.
58. Andrewartha, H.G. (1961) *Introduction to the Study of Animal Populations.* Methuen & Co.
59. Şekercioğlu, C.H. (2006) Ecological significance of bird populations. In *Handbook of the Birds of the World,* Vol. 11 (del Hoyo, J., Elliot, A., and Christie, D. eds), pp. 15–51, Lynx.
60. Mduma, S.A.R. *et al.* (1999) Food regulates the Serengeti wildebeest: a 40-year record. *Journal of Animal Ecology* 68, 1101–1122.

61. Peterson, R.O. (1999) Wolf–moose interaction on Isle Royale: the end of natural regulation? *Ecological Applications* 9, 10–16.
62. Page, S.E. *et al.* (2002) The amount of carbon released from peat and forest fires in Indonesia during 1997. *Nature* 420, 61–65.
63. Lal, R. (2004) Soil carbon sequestration to mitigate climatic change. *Geoderma* 123, 1–22.
64. Lenton, T.M. and Huntingford, C. (2003) Global terrestrial carbon storage and uncertainties in its temperature sensitivity examined with a simple model. *Global Change Biology* 9, 1333–1352.
65. Allison, S.D. (2006) Brown ground: a soil carbon analogue for the green world hypothesis? *American Naturalist* 167, 619–627.
66. Naeem, S. (2008) Green with complexity. *Science* 319, 913–914.
67. Krebs, C.J. (2001) *Ecology*. Fifth Edition, Benjamin Cummings.
68. Cramp, S. ed (1985) *The Birds of the Western Palearctic Vol IV*. Oxford University Press.
69. Sherratt, T.N. *et al.* (2005) Explaining Discorides' "double difference": why are some mushrooms poisonous, and do they signal their unprofitability? *American Naturalist* 166, 767–775.

Chapter 8: Why is the Sea Blue?

Suggested general reading

Arrigo, K.R. (2005) Marine microorganisms and global nutrient cycles. *Nature* 437, 349–355.
Falkowski, P.G. (2002) The oceans invisible forest. *Scientific American* 287(2), 38–45.
Falkowski, P.G. and Davis, C.S. (2004) Natural proportions. *Nature* 431, 131.
Polis, G.A. (1999) Why are parts of the world green? Multiple factors control productivity and the distribution of biomass. *Oikos* 86, 3–15.

References

1. Matthiessen, P. (1971) *Blue Meridian*. Random House.
2. Cousteau, J. (1952) *The Silent World*. Hamish Hamilton.
3. Morel, A. *et al.* (2007) Optical properties of the "clearest" natural waters. *Limnology and Oceanography* 52, 217–229.
4. Colinvaux, P. (1978) *Why Big Fierce Animals are Rare*. Princeton University Press.
5. Cox, C.B. and Moore, P.D. (2005) *Biogeography*. Seventh Edition, Blackwell.
6. Margulis, L. and Schwartz, K.V. (1998) *Five Kingdoms*. Third Edition, Freeman.
7. Cavalier-Smith, T. (2004) Only six kingdoms of life. *Proceedings of the Royal Society B* 271, 1251–1262.
8. Corliss, J.O. (2002) Biodiversity and biocomplexity of the Protists and an overview of their significant roles in maintenance of our biosphere. *Acta Protozoologica* 41, 199–219.
9. Losos, J.B. *et al.* (2008) *Biology*. Eighth Edition, McGraw-Hill.
10. Hayward, P.J. (2004) *Seashore*. Harper Collins.
11. Paine, R.T. (2000) Phycology for the mammalogist: marine rocky shores and mammal-dominated communities—how different are the structuring processes? *Journal of Mammology* 81, 637–648.
12. Burkepile, D.E. and Hay, M.E. (2006) Herbivore vs. nutrient control of marine primary producers: context-dependent effects. *Ecology* 87, 3128–3139.
13. Van der Hage, J.C.H. (1996) Why are there no insects and so few higher plants in the sea? New thoughts on an old problem. *Functional Ecology* 10, 546–547.
14. Raven, J.A. (1998) Insect and angiosperm diversity in marine environments: further comments on van der Hage. *Functional Ecology* 12, 977–979.

15. Archibold, O.W. (1995) *Ecology of World Vegetation*. Chapman & Hall.
16. Colinvaux, P. (1993) *Ecology* 2. John Wiley.
17. Ebbesmeyer, C.C. (2008) Beachcombers Alert. http://beachcombersalert.org/RubberDuckies.html (last accessed 16 May 2008).
18. Bellamy, D. (1975) *The Life-Giving Sea*. Hamish Hamilton.
19. Gower, J. *et al.* (2006) Ocean color satellites show extensive lines of floating *Sargassum* in the Gulf of Mexico. *IEEE Transactions on Geoscience and Remote Sensing* 44, 3619–3625.
20. Falkowski, P.G. (2002) The ocean's invisible forest. *Scientific American* 287(2), 38–45.
21. Arrigo, K.R. (2005) Marine microorganisms and global nutrient cycles. *Nature* 437, 349–355.
22. Whitfield, M. (2004) From philosophical dirt to agents of global climate change. *Ocean Challenge* 14, 28–37.
23. Hardy, A. (1956) *The Open Sea: The World of Plankton*. Collins.
24. Raven, J.A. (1998) The twelfth Tansley lecture. Small is beautiful: the picophytoplankton. *Functional Ecology* 12, 503–513.
25. Fuhrman, J. (2003) Genome sequences from the sea. *Nature* 424, 1001–1002.
26. Sullivan, M.B. *et al.* (2003) Cyanophages infecting the oceanic cyanobacterium *Prochlorcoccus*. *Nature* 424, 1047–1051.
27. Rocap, G. *et al.* (2003) Genome divergence in two *Prochlorococcus* ecotypes reflects oceanic niche differentiation. *Nature* 424, 1042–1057.
28. Sogin, M.L. *et al.* (2006) Microbial diversity in the deep sea and the underexplored "rare biosphere". *Proceedings of the National Academy of Sciences USA* 103, 12115–12120.
29. Rusch, D.B. *et al.* (2007) The Sorcerer II Global Sampling Expedition: Northwest Atlantic through eastern tropical Pacific. *PLoS Biology* 5(3), e77.
30. Kump, L.R. *et al.* (2004) *The Earth System*. Second Edition, Prentice Hall.
31. Boyd, P.W. *et al.* (2007) Mesoscale iron enrichment experiments 1993–2005: synthesis and future directions. *Science* 315, 612–617.
32. Jickells, T.D. *et al.* (2005) Global iron connections between desert dust, ocean biogeochemistry, and climate. *Science* 308, 67–71.
33. Watson, A.J. *et al.* (2000) Effect of iron supply on southern Ocean CO_2 uptake and implications for glacial atmospheric CO_2. *Nature* 407, 730–733.
34. Falkowski, P.G. *et al.* (1998) Biogeochemical controls and feedbacks on ocean primary production. *Science* 281, 200–206.
35. Goldblatt, C. *et al.* (2006) Biostability of atmospheric oxygen and the great oxidation. *Nature* 443, 683–686.
36. Kasting, J.F. (2006) Ups and downs of ancient oxygen. *Nature* 443, 643–645.
37. Scott, C. *et al.* (2008) Tracing the stepwise oxygenation of the Protoerzoic ocean. *Nature* 452, 456–459.
38. Blain, S. *et al.* (2007) Effect of natural iron fertilization on carbon sequestration in the Southern Ocean. *Nature* 446, 1070–1074.
39. Suda, W.G. and Huntsman, S.A. (1997) Interrelated influence of iron, light and cell size on marine phytoplankton growth. *Nature* 390, 389–392.
40. Simo, R. (2001) Production of atmospheric sulfur by oceanic plankton: biogeochemical, ecological and evolutionary links. *Trends in Ecology and Evolution* 16, 287–294.
41. Buesseler, K.O. *et al.* (2008) Ocean iron fertilization—moving forward in a sea of uncertainty. *Science* 319, 162.
42. Page, S.E. *et al.* (2002) The amount of carbon released from peat and forest fires in Indonesia during 1997. *Nature* 420, 61–65.

43. Abram, N.J. *et al.* (2003) Coral reef death during the 1997 Indian Ocean dipole linked to Indonesian wildfires. *Science* 301, 952–955.
44. Madigam, M.T. *et al.* (2000) *Brock: Biology of Microorganisms*. Ninth Edition, Prentice Hall.
45. Duce, R.A. *et al.* (2008) Impacts of atmospheric nitrogen on the open ocean. *Science* 320, 893–897.
46. Redfield, A.C. (1958) The biological control of the chemical factors in the environment. *American Scientist* 46, 205–221.
47. Redfield, A.C. (1916) The coordination of chromatophores by hormones. *Science* 43, 580–581.
48. Redfield, A.C. (1965) Ontogeny of a salt marsh estuary. *Science* 147, 50–55.
49. Polis, G.A. (1999) Why are parts of the world green? Multiple factors control productivity and the distribution of biomass. *Oikos* 86, 3–15.
50. Tyrrell, T. (1999) The relative influences of nitrogen and phosphorus on oceanic primary production. *Nature* 400, 525–531.
51. Lenton, T.M. and Watson, A.J. (2000) Redfield revisited. 1. Regulation of nitrate, phosphate and oxygen in the ocean. *Global Biogeochemical Cycles* 14, 225–248.
52. Klausmeier, C.A. *et al.* (2004) Optimal nitrogen-to-phosphorus stoichiometry of phytoplankton. *Nature* 429, 171–174.
53. Kump, L.R. (2004) Self-regulation of ocean composition by the biosphere. In *Scientists Debate Gaia* (Schneider, S.H. *et al.* eds), pp. 93–100, MIT Press.
54. Beerling, D.J. *et al.* (2002) An atmospheric pCO_2 reconstruction across the Cretaceous-Tertiary boundary from leaf megafossils. *Proceedings of the National Academy of Sciences USA* 99, 7836–7840.
55. D'Hondt, S. *et al.* (1998) Organic carbon fluxes and ecological recovery from the Cretaceous-Tertiary mass extinction. *Science* 282, 276–279.
56. Polovina, J.J. *et al.* (2008) Ocean's least productive waters are expanding. *Geophysical Research Letters* 35, L03618, doi: 10.1029/2007GL031745.
57. Kump, L.R. and Pollard, D. (2008) Amplification of Cretaceous warmth by biological cloud feedback. *Science* 320, 195.
58. Hutchinson, G.E. (1965) *The Ecological Theatre and the Evolutionary Play*. Yale University Press.
59. Lovelock, J. (2003) The living Earth. *Nature* 426, 769–770.
60. Lovelock, J. (2000) *The Ages of Gaia*. Second Edition, Oxford University Press.
61. Jones, C.G. *et al.* (1994) Organisms as ecosystem engineers. *Oikos* 69, 373–386.
62. Odling-Smee, F.J. *et al.* (2003) *Niche Construction*. Princeton University Press.
63. Seberg, O. (2007) Araceae. In *Flowering Plant Families of the World* (Heywood, V.H. *et al.* eds), pp. 345–348, Firefly.

Chapter 9: When did We Start to Change Things?

Suggested general reading

Diamond, J. (1997) *Guns, Germs and Steel*. Jonathan Cape.

Diamond, J. (2008) The last giant Kangaroo. *Nature* 454, 835–836.

Donlan, C.J. (2007) Restoring America's big, wild animals. *Scientific American* 296(960), 48–55.

Flannery, T. (1994) *The Future Eaters; An Ecological History of the Australian Lands and People*. Reed Books.

Flannery, T. (2001) *The Eternal Frontier; An Ecological History of North America and its Peoples*. Text Publishing Company.

Koch, P.L. and Barnosky, A.D. (2006) Late Quaternary extinctions: state of the debate. *Annual Review of Ecology, Evolution and Systematics* 37, 215–250.

Martin, P.S. (2005) *Twilight of the Mammoths*. University of California Press.

Ruddiman, W.F. (2005) *Plows, Plagues and Petroleum*. Princeton University Press.

References

1. Lovelock, J. (2006) *The Revenge of Gaia*. Allan Lane.
2. Flannery, T. (2005) *The Weather Makers*. Allen Lane.
3. Haag, A. (2007) Al's army. *Nature* 446, 723–724.
4. Lovelock, J. *et al.* (1972) Atmospheric sulphur and the natural sulphur cycle. *Nature* 237, 452–453.
5. Simó, R. (2001) Production of atmospheric sulfur by oceanic plankton: biogeochemical, ecological and evolutionary links. *Trends in Ecology and Evolution* 16, 287–294.
6. Charlson, R.J. *et al.* (1987) Oceanic phytoplankton, atmospheric sulphur, cloud albedo and climate. *Nature* 326, 655–661.
7. Nisbet, E.G. and Fowler, C.M.R. (2003) The early history of life. *Treatise in Geochemistry* 8, 1–39.
8. Lovelock, J. (1988) *The Ages of Gaia*. Oxford University Press.
9. Martin, P.S. (2005) *Twilight of the Mammoths*. University of California Press.
10. Diamond, J.M. (1989) Quaternary megafaunal extinctions: variations on a theme by Paganini. *Journal of Archaeological Science* 16, 167–175.
11. Owen-Smith, N. (1987) Pleistocene extinctions: the pivotal role of megaherbivores. *Paleobiology* 13, 351–362.
12. Delcourt, P.A. and Delcourt, H.R. (2004) *Prehistoric Native Americans and Ecological Change*. Cambridge University Press.
13. Graham, R.W. and Grimm, E.C. (1990) Effects of global climate change on the patterns of terrestrial biological communities. *Trends in Ecology and Evolution* 5, 289–292.
14. Grayson, D.K. and Meltzer, D.J. (2003) A requiem for north American overkill. *Journal of Archaeological Science* 30, 585–593.
15. Leaky, R. and Lewin, R. (1995) *The Sixth Extinction*. Weidenfeld & Nicolson.
16. de Castro, F. and Bolker, B. (2005) Mechanisms of disease-induced extinction. *Ecology Letters* 8, 11–126.
17. Lyons, S.K. *et al.* (2004) Was a 'hyperdisease' responsible for the late Pleistocene megafaunal extinction? *Ecology Letters* 7, 859–868.
18. van Riper, C. III *et al.* (1986) The epizootiology and ecological significance of malaria in Hawaiian land birds. *Ecological Monographs* 56, 327–244.
19. Firestone, R.B. *et al.* (2007) Evidence for an extraterrestrial impact 12,900 years ago that contributed to megafaunal extinctions and the Younger Dryas cooling. *Proceedings of the National Academy of Sciences USA* 104, 16016–16021.
20. Lyons, S.K. *et al.* (2004) Of mice, mastodons and men: human-mediated extinctions on four continents. *Evolutionary Ecology Research* 6, 339–358.
21. Steadman, D.W. *et al.* (2005) Asynchronous extinction of late Quaternary sloths on continents and islands. *Proceedings of the National Academy of Sciences USA* 102, 11763–11768.
22. Stuart, A. (1986) Who (or what) killed the giant armadillo? *New Scientist* 111 (1517), 29–32.
23. Diamond, J. (1991) *The Rise and Fall of the Third Chimpanzee*. Radius.
24. Wilson, E.O. (1992) *The Diversity of Life*. Harvard University Press.
25. Stringer, C. and McKie, R. (1996) *African Exodus*. Jonathan Cape.

26. Losos, J.B. *et al.* (2008) *Biology*. Eighth Edition, McGraw-Hill.
27. Glen, W. ed (1994) *The Mass-Extinction Debate: How Science Works in a Crisis*. Stanford University Press.
28. Baillie, M. (2007) The case for significant numbers of extraterrestrial impacts through the late Holocene. *Journal of Quaternary Science* 22, 101–109.
29. Frison, G.C. (1998) Paleoindian large mammal hunters on the plains of North America. *Proceedings of the National Academy of Sciences USA* 95, 14576–14583.
30. Darwin, C. (1839) *Journal of Researches into the Geology and Natural History of the Various Countries Visited by H.M.S. Beagle, Under the Command of Captain Fitzroy, RN from 1832 to 1836*. Henry Colburn.
31. Waters, M.R. and Stafford, T.W. Jr. (2007) Redefining the age of Clovis: implications for the peopling of the Americas. *Science* 315, 1122–1125.
32. Frison, G.C. (1990) Clovis, Goshen, and Folsom: lifeways and cultural relationships. In *Megafauna and Man* (Agenbroad, L.D. *et al.* eds), pp. 100–108, Mammoth Site of Hot Springs.
33. Frison, G.C. (1989) Experimental use of Clovis weaponry and tools on African Elephants. *American Antiquity* 54, 766–784.
34. Alroy, J. (2001) A multispecies overkill simulation of the end-Pleistocene megafaunal mass extinction. *Science* 292, 1893–1896.
35. Goebel, T. *et al.* (2008) The late Pleistocene dispersal of modern humans in the Americas. *Science* 319, 1497–1502.
36. Gonzalez, S. *et al.* (2003) Earliest humans in the Americas: new evidence from México. *Journal of Human Evolution* 44, 379–387.
37. Gilbert, M.T.P. *et al.* (2008) DNA from Pre-Clovis human coprolites in Oregon, North America. *Science* 320, 786–789.
38. Cole, S. (1975) *Leakey's Luck*. Collins.
39. González, S. *et al.* (2006) Human footprints in Central Mexico older than 40,000 years. *Quaternary Science Reviews* 25, 201–222.
40. O'Regan, H.J. *et al.* (2002) European Quaternary refugia: a factor in large carnivore extinction? *Journal of Quaternary Science* 17, 789–795.
41. Berger, J. *et al.* (2001) Recolonizing carnivores and naïve prey: conservation lessons from Pleistocene extinctions. *Science* 291, 1036–1039.
42. Koch, P.L. and Barnosky, A.D. (2006) Late Quaternary extinctions: state of the debate. *Annual Review of Ecology, Evolution and Systematics* 37, 215–250.
43. Janzen, D.H. and Martin, P.S. (1981) Neotropical anachronisms: the fruits the Gomphotheres ate. *Science* 215, 19–27.
44. Jackson, S.T. and Weng, C. (1999) Late Quaternary extinction of a tree species in eastern North America. *Proceedings of the National Academy of Sciences USA* 96, 13847–13852.
45. Donlan, J. *et al.* (2005) Re-wilding North America. *Nature* 436, 913–914.
46. Louys, J. *et al.* (2007) Characteristics of Pleistocene megafauna extinctions in Southeast Asia. *Palaeogeography, Palaeoclimotology, Palaeoecology* 243, 152–173.
47. Stuart, A.J. *et al.* (2004) Pleistocene to Holocene extinction dynamics in giant deer and woolly mammoth. *Nature* 431, 684–689.
48. Guthrie, R.D. (2004) Radiocarbon evidence of mid-Holocene mammoths stranded on an Alaskan Bering Sea island. *Nature* 429, 746–749.
49. Nogués-Bravo, D. *et al.* (2008) Climate change, humans, and the extinction of the woolly mammoth. *PLoS Biology* 6(4), e79.
50. Steadman, D.W. and Martin, P.S. (2003) The late Quaternary extinction and future resurrection of birds on Pacific islands. *Earth-Science Reviews* 61, 133–147.

51. Bovy, K.M. (2007) Global human impacts or climate change?: explaining the Sooty Shearwater decline at the Minard site, Washington State, USA. *Journal of Archaeological Science* 34, 1087–1097.
52. Webb, R.E. (1997) Megamarsupial extinction: the carrying capacity argument. *Antiquity* 72, 46–55.
53. Bowler, J.M. *et al.* (2003) New ages for human occupation and climatic change at Lake Mungo, Australia. *Nature* 421, 837–840.
54. David, B. *et al.* (2007) Sediment mixing at Noda Rock: investigations of stratigraphic integrity at an early archaeological site in northern Australia and implications for the human colonisation of the continent. *Journal of Quaternary Science* 22, 449–479.
55. Miller, G.H. *et al.* (1999) Pleistocene extinction of *Genyornis newtoni*: human impact on Australian megafauna. *Science* 283, 205–208.
56. Roberts, R.G. *et al.* (2001) New ages for the last Australian megafauna: continent-wide extinction about 46,000 years ago. *Science* 292, 1888–1892.
57. Wrangham, R. and Conklin-Brittain, N.L. (2003) Cooking as a biological trait. *Comparative Biochemistry and Physiology A* 136, 35–46.
58. Preece, R.C. *et al.* (2006) Humans in the Hoxnian: habitat, context and fire use at Beeches Pit, West Stow, Suffolk, UK. *Journal of Quaternary Science* 21, 485–496.
59. Karkanas, P. *et al.* (2007) Evidence for habitual use of fire at the end of the Lower Palaeolithic: site-formation processes at Qesem Cave, Israel. *Journal of Human Evolution* 53, 197–212.
60. Shapiro, H.L. (1976) *Peking Man*. George Allen & Unwin.
61. Brain, C.K. ed (1993) The occurrence of burnt bones at Swartkrans and their implications for the controlled use of fire by early hominids. In *Swartkrans. A Caves Chronicle of Early Man*, pp. 229–242, Transvaal Museum.
62. Mayr, E. (2001) *What Evolution Is*. Weidenfeld & Nicolson.
63. Pitts, M. and Roberts, M. (1997) *Fairweather Eden*. Century.
64. Diamond, J. (2002) Evolution, consequences and future of plant and animal domestication. *Nature* 418, 700–707.
65. Lev-Yadum, S. *et al.* (2000) The cradle of agriculture. *Science* 288, 1602–1603.
66. Fuller, D.Q. (2007) Contrasting patterns in crop domestication and domestication rates: recent archaeobotanical insights from the old world. *Annals of Botany* 100, 903–924.
67. Gkiasta, M. *et al.* (2003) Neolithic transition in Europe: the radiocarbon record revisited. *Antiquity* 77, 45–62.
68. Willis, K.J. and Bennett, K.D. (1994) The Neolithic transition—fact or fiction? Palaeoecological evidence from the Balkans. *The Holocene* 4, 326–330.
69. Mannion, A.M. (1999) Domestication and the origins of agriculture: an appraisal. *Progress in Physical Geography* 23, 37–56.
70. Diamond, J. (1997) *Guns, Germs and Steel*. Jonathan Cape.
71. Moore, A.M.T. *et al.* (2000) *Village on the Euphrates*. Oxford University Press.
72. Grove, A.T. and Rackham, O. (2001) *The Nature of Mediterranean Europe; An Ecological History*. Yale University Press.
73. Roberts, N. (2002) Did prehistoric landscape management retard the post-glacial spread of woodland in Southwest Asia? *Antiquity* 76, 1002–1101.
74. Clare, T. *et al.* (2001) The Mesolithic and Neolithic landscapes of Barfield Tarn and Eskmeals in the English Lake District. *Journal of Wetland Archaeology* 1, 83–105.
75. Crutzen, P.J. (2002) Geology of mankind. *Nature* 415, 23.
76. Crutzen, P.J. and Steffen, W. (2003) How long have we been in the anthropocene era? *Climatic Change* 61, 251–257.

77. Woodward, F.I. (1987) Stomatal numbers are sensitive to increases in CO_2 from preindustrial levels. *Nature* 327, 617–618.
78. Beerling, D.J. and Chaloner, W.G. (1993) Stomatal density responses of Egyptian *Olea europaea* L. Leaves to CO_2 change since 1327 BC. *Annals of Botany* 71, 431–435.
79. de la Bédoyère, G. (2001) *Eagles over Britannia; The Roman Army in Britain*. Tempus.
80. Garfield, S. (2002) *The Last Journey of William Huskisson*. Faber & Faber.
81. Gregory, A. (2001) *Eureka! The Birth of Science*. Icon.
82. Ruddiman, W.F. (2003) The anthropogenic greenhouse era began thousands of years ago. *Climatic Change* 61, 261–293.
83. Ruddiman, W.F. (2005) *Plows, Plagues and Petroleum*. Princeton University Press.
84. Ruddiman, W.F. (2005) The early anthropogenic hypothesis a year later. *Climatic Change* 69, 427–434.
85. Bowen, D.Q. and Gibbard, P.L. (2007) The Quaternary is here to stay. *Journal of Quaternary Science* 22, 3–8.
86. Oldroyd, D. (1996) *Thinking about the Earth: A History of Geology*. Athlone.
87. Geikie, J. (1874) *The Great Ice Age*. W. Isbister & Co.
88. Hays, J.D. *et al.* (1976) Variations in the Earth's orbit: pacemaker of the ice ages. *Science* 194, 1121–1132.
89. Ruddiman, W.F. and Thomson, J.S. (2001) The case for human causes of increased atmospheric CH_4 over the last 5000 years. *Quaternary Science Reviews* 20, 1769–1777.
90. Ruddiman, W.F. *et al.* (2005) A test of the overdue-glaciation hypothesis. *Quaternary Science Reviews* 24, 1–10.
91. Claussen, M. *et al.* (2005) Did humankind prevent a Holocene glaciation? *Climatic Change* 69, 409–417.
92. Cheddadi, R. *et al.* (2005) Similarity of vegetation dynamics during interglacial periods. *Proceedings of the National Academy of Sciences USA* 102, 13939–13943.
93. Ruddiman, W.F. (2005) Cold climate during the closest stage 11 analog to recent millennia. *Quaternary Science Reviews* 24, 1111–1121.
94. Lu, T.L.-D. (2004) Crop domestication in China. In *Encyclopaedia of Plant and Crop Science*. (Goodman, R.M. ed), pp. 307–309, Marcel Dekker.
95. Fenchel, T. *et al.* (1998) *Bacterial Biogeochemistry: The Ecophysiology of Mineral Cycling*. Second Edition, Academic Press.
96. Behre, K.-E. (1988) The rôle of man in European vegetation history. In *Vegetation History* (Huntley, B. and Webb, T. III eds), pp. 633–672, Kluwer.
97. Joos, F. *et al.* (2004) Transient simulations of Holocene atmospheric carbon dioxide and terrestrial carbon since the Last Glacial Maximum. *Global Biogeochemical Cycles* 18, GB2002, doi: 10.1029/2003GB002156.
98. Olofsson, J. and Hickler, T. (2008) Effects of human land-use on the global carbon cycle during the last 6,000 years. *Vegetation History and Archaeobotany*. Advanced publication on line, doi 10.1007/s00334–007-0126–6.
99. Dearing, J.A. (2006) Climate–human–environment interactions: resolving our past. *Climates of the Past* 2, 187–203.
100. Wilkinson, D.M. (2006) *Fundamental Processes in Ecology; An Earth Systems Approach*. Oxford University Press.
101. Lal, R. (2004) Soil carbon sequestration to mitigate climate change. *Geoderma* 23, 1–22.
102. Moore, P.D. (1993) The origin of blanket mire revisited. In *Climatic Change and Human Impact on the Landscape* (Chambers, F.M. ed), pp. 217–224, Chapman & Hall.

103. Li, C. *et al.* (2005) Carbon sequestration in arable soils is likely to increase nitrous oxide emissions, offsetting reductions in climate radiative forcing. *Climatic Change* 72, 321–338.
104. Houghton, R.A. *et al.* (1983) Changes in the carbon content of terrestrial biota and soils between 1860 and 1980: a net release of CO_2 to the atmosphere. *Ecological Monographs* 53, 235–262.
105. Gibbard, S. *et al.* (2005) Climate effects of global land cover change. *Geophysical Research Letters* 32, L23705, doi: 10.1029/2005GL024550.
106. Keppler, F. *et al.* (2006) Methane emissions from terrestrial plants under aerobic conditions. *Nature* 439, 187–191.
107. Dueck, T.A. *et al.* (2007) No evidence for substantial aerobic methane emission by terrestrial plants: a ^{13}C-labelling approach. *New Phytologist* 175, 29–35.
108. Foster, D.R. (1992) Landuse history (1730–1990) and vegetation dynamics in central New England, USA. *Journal of Ecology* 80, 753–772.
109. Cohen, J.E. (1995) *How Many People Can the Earth Support?* Norton.
110. Yeloff, D. and van Geel, B. (2007) Abandonment of farmland and vegetation succession following the Eurasian plague pandemic of AD 1347–52. *Journal of Biogeography* 34, 575–582.
111. Hoskins, W.G. (1955) *The Making of the English Landscape.* Hodder & Stoughton.
112. Faust, F.X. *et al.* (2006) Evidence for the post conquest demographic collapse of the Americas in historical CO_2 levels. *Earth Interactions* (http://EarthInteractions.org), paper 11, 1–12.
113. Lamb, H.H. (1982) *Climate, History and the Modern World.* Methuen.
114. Bush, M.B. and Colinvaux, P.A. (1994) Tropical forest disturbance: paleoecological records from Darien, Panama. *Ecology* 75, 1761–1768.
115. Willis, K.J. *et al.* (2004) How "virgin" is virgin rainforest? *Science* 304, 402–403.
116. Shorrocks, B. (2007) *The Biology of African Savannahs.* Oxford University Press.
117. Fuller, E. (2002) *Dodo: From Extinction to Icon.* Harper Collins.
118. Walker, D. (1965) The post-glacial period in the Langdale fells, English Lake District. *New Phytologist* 64, 488–510.

Chapter 10: How will the Biosphere End?

Suggested general reading

Morton, O. (2007) *Eating the Sun.* Fourth Estate (see Chapters 6 and 7).
Ward, P. and Brownlee, D. (2002) *The Life and Death of Planet Earth.* Times Books.

References

1. Hoyle, F. and Hoyle, G. (1973) *The Inferno.* William Heinemann.
2. Narlikar, J.V. (1999) *Seven Wonders of the Cosmos.* Cambridge University Press.
3. Mitton, S. (2005) *Fred Hoyle, A Life in Science.* Aurum.
4. Haldane, J.B.S. (1927) *Possible Worlds and Other Essays.* Chatto & Windus.
5. Maynard Smith, J. and Szathmáry, E. (1995) *The Major Transitions in Evolution.* W.H. Freeman.
6. Huggett, R.J. (1999) Ecosphere, biosphere, or Gaia? What to call the global ecosystem. *Global Ecology and Biogeography* 8, 425–431.
7. Lekevičius, E. (2006) The Russian paradigm in ecology and evolutionary biology: pro et contra. *Acta Zoologica Lituanica* 16, 3–19.
8. Lovelock, J. (1979) *Gaia.* Oxford University Press.
9. Lovelock, J. (2003) The living Earth. *Nature* 425, 769–770.

10. Holliday, K. (1999) *Introductory Astronomy*. Wiley.
11. Crutzen, P.J. and Brühl, C. (1996) Mass extinctions and supernova explosions. *Proceedings of the National Academy of Sciences USA* 93, 1582–1584.
12. Gehrels, N. *et al.* (2003) Ozone depletion from nearby supernovae. *Astrophysical Journal* 585, 1169–1176.
13. Kump, L.R. *et al.* (2004) *The Earth System*. Second Edition, Prentice Hall.
14. Foster, C.B. and Afonin, S.A. (2005) Abnormal pollen grains: an outcome of deteriorating atmospheric conditions around the Permian-Triassic boundary. *Journal of the Geological Society, London* 162, 653–659.
15. Smith, A.B. (2007) Marine diversity through the Phanerozoic: problems and prospects. *Journal of the Geological Society, London* 164, 731–745.
16. Beerling, D. (2007) *The Emerald Planet*. Oxford University Press.
17. Lovelock, J. (1995) *The Ages of Gaia*. Second Edition, Oxford University Press.
18. Gould, S.J. (1983) *Hen's Teeth and Horse's Toes*. W.W. Norton & Co.
19. Alvarez, L.W. *et al.* (1980) Extraterrestrial cause for the Cretaceous-Tertiary extinction. *Science* 208, 1095–1108.
20. Pope, K.O. *et al.* (1998) Meteorite impact and the mass extinction of species at the Cretaceous/Tertiary boundary. *Proceedings of the National Academy of Sciences USA* 95, 11028–11029.
21. Lehto, H.J. (2007) From the Big Bang to the molecules of life. In *Complete Course in Astrobiology* (Horneck, G. and Rettberg, P. eds), pp. 23–54, Wiley.
22. Sleep, N.H. *et al.* (1989) Annihilation of ecosystems by large asteroid impacts on the early Earth. *Nature* 342, 139–142.
23. Nisbet, E.G. *et al.* (2007) Creating habitable zones, at all scales, from planets to mud microhabitats, on Earth and Mars. *Space Science Reviews* 129, 79–121.
24. Wells, L.E. *et al.* (2003) Reseeding of early Earth by impacts of returning ejecta during the late heavy bombardment. *Icarus* 162, 38–46.
25. Stevenson, D.J. (2008) A planetary perspective on the deep Earth. *Nature* 451, 261–265.
26. Watson, A.J. (1999) Coevolution of the Earth's environment and life: Goldilocks, Gaia and the anthropic principle. *Geological Society London, Special Publication* 150, 75–88.
27. Wilkinson, D.M. (2006) *Fundamental Processes in Ecology; An Earth Systems Approach*. Oxford University Press.
28. Fenchel, T. and Finlay, B.J. (1995) *Ecology and Evolution in Anoxic Worlds*. Oxford University Press.
29. Hoyle, F. (1950) *The Nature of the Universe*. Blackwell.
30. Ward, P. and Brownlee, D. (2002) *The Life and Death of Planet Earth*. Times Books.
31. Sagan, C. and Mullen, G. (1972) Earth and Mars: evolution of atmospheres and surface temperatures. *Science* 177, 52–177.
32. Sagan, C. and Chyba, C. (1997) The early faint sun paradox: organic shielding of ultraviolet-labile greenhouse gases. *Science* 276, 1217–1221.
33. Nisbet, E.G. and Sleep, N.H. (2001) The habitat and nature of early life. *Nature* 409, 1083–1091.
34. Begon, M., Townsend, C.R., and Harper, J.L. (2006) *Ecology, from Individuals to Ecosystems*. Fourth Edition, Blackwell.
35. Krebs, C.J. (2001) *Ecology*. Fifth Edition, Benjamin Cummings.
36. Walker, J.C.G. *et al.* (1981) A negative feedback mechanism for the long-term stabilization of Earth's surface temperature. *Journal of Geophysical Research* 86, 9776–9782.
37. Lovelock, J.E. and Watson, A.J. (1982) The regulation of carbon dioxide and climate: Gaia or geochemistry. *Planet Space Science* 30, 795–802.

38. Lovelock, J.E. and Whitfield, M. (1982) Life span of the biosphere. *Nature* 296, 561–563.
39. Schwartzman, D.W. and Volk, T. (1989) Biotic enhancement of weathering and the habitability of Earth. *Nature* 340, 457–460.
40. Berner, R.A. (2004) *The Phanerozoic Carbon Cycle*. Oxford University Press.
41. Schwartzman, D. (1999) *Life, Temperature, and the Earth*. Columbia University Press.
42. Aghamiri, R. and Schwartzman, D.W. (2002) Weathering rates of bedrock by lichens: a mini watershed study. *Chemical Geology* 188, 249–259.
43. Bormann, B.T. *et al.* (1998) Rapid, plant induced weathering in an aggrading experimental ecosystem. *Biogeochemistry* 43, 129–155.
44. Lovelock, J.E. Conversation with DMW, November 2007.
45. Nisbet, E.G. and Fowler, C.M.R. (2003) The early history of the Earth. *Treatise on Geochemistry* 8, 1–39.
46. Caldeira, K. and Kasting, J.F. (1992) The life span of the biosphere revisited. *Nature* 360, 721–723.
47. Lenton, T.M. and von Bloh, W. (2001) Biotic feedback extends the life span of the biosphere. *Geophysical Research Letters* 28, 1715–1718.
48. Stan-Lotter, H. (2007) Extremophiles, the physicochemical limits of life (growth and survival). In *Complete Course in Astrobiology* (Horneck, G. and Rettberg, P. eds), pp. 121–150, Wiley.
49. Franck, S. *et al.* (2006) Causes and timings of future biosphere extinctions. *Biogeosciences* 3, 85–92.
50. Volk, T. (1992) When climate and life finally devolve. *Nature* 360, 707.
51. House, K.Z. *et al.* (2007) Electrochemical acceleration of chemical weathering as an energetically feasible approach to mitigating anthropogenic climate change. *Environmental Science and Technology* 41, 8464–8470.
52. Hoffert, M.I. *et al.* (2002) Advanced technological paths to global climate stability: energy for a greenhouse planet. *Science* 298, 981–987.
53. Stone, R. (2008) Preparing for doomsday. *Science* 319, 1326–1329.
54. Gilbert, O. (2000) *Lichens*. Harper Collins.
55. Lenton, T.M. and Watson, A.J. (2004) Biotic enhancement of weathering, atmospheric oxygen and carbon dioxide in the Neoprotozoic. *Geophysical Research Letters* 31, L05202, doi: 10.1029/2003GL018802.

Index

Note: emboldened page numbers refer to terms in the glossary

Printed and bound by CPI Group (UK) Ltd, Croydon, CR0 4YY